Earth Observation by GNSS and GIS Techniques

Earth Observation by GNSS and GIS Techniques

Guest Editors

Aggeliki Kyriou
Lia Bárbara Cunha Barata Duarte
Christos Pikridas

MDPI

Basel • Beijing • Wuhan • Barcelona • Belgrade • Novi Sad • Cluj • Manchester

Guest Editors
Aggeliki Kyriou
GIS and Remote Sensing Laboratory
University of Patras
Patras
Greece

Lia Bárbara Cunha Barata Duarte
Faculty of Sciences
University of Porto
Porto
Portugal

Christos Pikridas
School of Rural and Surveying Engineering
Aristotle University of Thessaloniki
Thessaloniki
Greece

Editorial Office
MDPI AG
Grosspeteranlage 5
4052 Basel, Switzerland

This is a reprint of the Special Issue, published open access by the journal *Geosciences* (ISSN 2076-3263), freely accessible at: https://www.mdpi.com/journal/geosciences/special_issues/1N7BZGH801.

For citation purposes, cite each article independently as indicated on the article page online and as indicated below:

Lastname, A.A.; Lastname, B.B. Article Title. *Journal Name* **Year**, *Volume Number*, Page Range.

ISBN 978-3-7258-6288-7 (Hbk)
ISBN 978-3-7258-6289-4 (PDF)
https://doi.org/10.3390/books978-3-7258-6289-4

Contents

About the Editors

Aggeliki Kyriou

Aggeliki Kyriou is a postdoctoral researcher at the GIS and Remote Sensing Laboratory in the Department of Geology at the University of Patras. In 2015, she received her Bachelor's degree in Geology and later obtained a Master's in Geosciences and the Environment in 2017. She holds a Ph.D. in Remote Sensing and GIS (2021), which was funded by the Hellenic Foundation for Research and Innovation (HFRI). Her doctoral research was focused on the synergistic use of different remote sensing data (i.e., SAR, multispectral, LiDAR, GNSS) for landslide monitoring. Since 2021, she has participated in research projects (i.e., PROION, UNMASK, m4mining, Greek National Satellite project) funded by European and Greek funds. Her main research interests include Earth observation using remote sensing data obtained from different sensors (SAR, multispectral, UAVs, TLS, etc.), GNSS measurements, and GIS applications. In her career thus far, she has authored 59 scientific publications featured in scientific journals and conference proceedings.

Lia Bárbara Cunha Barata Duarte

Lia Bárbara Cunha Barata Duarte is an Assistant Professor in the Department of Geosciences, Environment and Land Planning at the Faculty of Sciences of the University of Porto and a researcher at the Institute of Earth Sciences. Her work focuses on surveying engineering, remote sensing, GIScience, and the development of open-source geospatial tools, with applications in environmental monitoring, natural hazards, biodiversity assessment, and land management. She has participated in several national and international research projects, co-supervised over ten MSc students, and is the author of more than 70 peer-reviewed scientific publications. Her contributions also include several book chapters and numerous communications at international conferences. Duarte has served as a guest editor for multiple scientific journals and currently sits on the Editorial Boards of *Sustainability* (MDPI) and *Scientific Reports* (Nature). She is a reviewer for over twenty journals and an active member of academic committees, contributing to the advancement of geospatial science and its applications.

Christos Pikridas

Christos Pikridas is a Professor at the Department of Geodesy and Surveying, Faculty of Engineering, Aristotle University of Thessaloniki, Greece, and Director of the Laboratory of Geodetic Methods and Satellite Applications. His main interests are GNSS data collection, processing and analysis, error theory and modeling in GNSS measurements, algorithm development, data quality control and specifications for installing GNSS monitoring networks. His research interests also include GNSS applications in large-scale projects as well as monitoring and managing natural disasters using GNSS systems. He teaches the courses Positioning with GNSS Satellite Systems, GNSS NRTK Applications and Geometrical Geodesy and Networks and has supervised three doctoral studies and many diploma theses. He is member of the study committee and is an academic staff member of the Interdepartmental Program of Postgraduate Studies "Aerial Autonomous Systems"; an administrator of the Hermes NTRIP Caster for the dissemination of RTCM data via the Internet; responsible for the management of three Greek permanent GNSS stations participating in the European Network (EUREF); head of the AUT1 Analysis Center in the European EUMETNET EIG GNSS-EGVAP program; and an active member of the Working Group for the creation of a European velocity field by combining existing geodetic velocities from permanent GNSS stations at the national level. He is co-manager of the permanent GNSS network in the national program entitled "Hellenic Plate

Observing System". As a partner, he was the winner of the Madrid Challenge award in the European Satellite Navigation Competition (ESNC) in 2017. He is the co-author of *GPS and Geodetic Applications* 2nd Edition; ISBN: 978-960-456-346-3, and of two chapters of the book *3D Printing: Applications in Medicine and Surgery* published by Elsevier; ISBN: 978-0-323-66164-5. He is the author of more than 200 papers in peer-reviewed journals and conference proceedings.

geosciences

MDPI

Editorial

Special Issue "Earth Observation by GNSS and GIS Techniques"

Aggeliki Kyriou [1,*], Lia Bárbara Cunha Barata Duarte [2] and Christos Pikridas [3]

[1] GIS and Remote Sensing Laboratory, Division of Applied Geology and Geophysics, University of Patras, 265 04 Patra, Greece
[2] Faculty of Sciences, University of Porto, 4169-007 Porto, Portugal; liaduarte@fc.up.pt
[3] Laboratory of Geodetic Methods and Satellite Observations, School of Rural and Surveying Engineering, Aristotle University of Thessaloniki, 541 24 Thessaloniki, Greece; cpik@topo.auth.gr
* Correspondence: a.kyriou@ac.upatras.gr

Over the past few decades, Geographic Information Systems (GIS) and Global Navigation Satellite Systems (GNSS) have undergone transformative developments that have profoundly influenced positioning/navigation, data science, and geospatial technologies. The origins of GIS can be traced to the 1960s, when advances in computer technology and the emergence of computational and quantitative geography provided the foundation for spatial data analysis. In 1963, geographer Roger Tomlinson initiated a national land-use management program for the Canadian government aimed at cataloging natural resources. During this project, he introduced the term Geographic Information System, establishing the conceptual and methodological basis for digital spatial analysis. The United States Department of Defense initiated the Global Positioning System (GPS) project in 1973 to develop a satellite-based navigation system capable of overcoming the limitations of earlier ground-based positioning technologies. The first commercial GPS receivers became available in 1989, though their large size and high cost initially restricted widespread adoption. Nowadays, four fully operational GNSS constellations, i.e., GPS, GLONASS, Galileo, and BeiDou, collectively provide precise global positioning and navigation capabilities, forming the backbone of modern geospatial science and applications.

In this framework, billions of people worldwide rely on GIS and GNSS every day to enhance convenience and improve their quality of life. These technologies are embedded in a wide range of commercial devices, including smartphones, tablets, automobiles, buses, airplanes, unmanned aerial vehicles (UAVs), and many other systems. The integration of GIS and GNSS supports a variety of services, such as car navigation, public transport optimization, logistics and goods delivery, supply chain management, daily fitness tracking, as well as travel and tourism. In fact, the global GIS market was valued at approximately $77 billion in 2021 and is expected to grow to $174 billion by 2027. Similarly, the market for Location-Based Services (LBS), which combines GIS and GNSS technologies, was valued at $37.22 billion in 2025, with projections indicating that it will increase to $125.92 billion by 2032 [1].

At the same time, GIS and GNSS technologies are increasingly utilized by scientists and researchers across a wide range of environmental and scientific fields to deepen our understanding of Earth processes. These applications include environmental management [2,3], precision agriculture [4,5], landslide mapping [6,7], crustal motion monitoring [8], archeological site reconstruction and conservation [9,10], water resources management [11], coastal evolution studies [12], and a broad array of Earth observation research. In light of this, the current Special Issue has gathered eight high-quality original research articles, focused on the application of GNSS and GIS methodologies for Earth observation purposes. It incorporates articles focusing on different scientific topics such as vegetation

https://doi.org/10.3390/geosciences15120450

monitoring [13], rockfalls [14], coastal monitoring [15], geoid computations [16], landform erosion [17], wetlands mapping [18], open ocean and inland water conditions mapping [19], and deformation monitoring [20].

Specifically, Badioui et al. [13] integrated medium-resolution multispectral data with various numerical models within a GIS environment to analyze the evolution of vegetation in the challenging and hard-to-access environment of an oasis. They processed PROBA-V and Sentinel-3 satellite data, using the Tool for Raster Data Exploration to generate two vegetation indices: the Leaf Area Index (LAI) and the Normalized Difference Vegetation Index (NDVI). To ensure the robustness of their methodology, the authors validated the results through in situ measurements, aiming to develop a reliable approach for future applications.

Another study examined the influence of the spatial resolution of various Digital Surface Models (DSMs) on the simulation of rockfall trajectories within a GIS environment [14]. The HY-STONE software, designed for 3D numerical modeling of rockfall processes, was utilized to simulate rock movements, including free fall, impact, and rolling dynamics. The study investigated two well-documented rockfall events in Western Greece as test sites. In total, seven freely available DSMs, with spatial resolutions ranging from centimeters to 90 m, were evaluated to determine their impact on the accuracy of the rockfall trajectory simulations.

Additionally, four diachronic datasets, spanning the period from 2017 to 2021, consisting of very high-resolution satellite and aerial imagery, were visually analyzed to map the morphometric features and monitor the annual mobility of boulders along the eastern coast of the Gulf of Taranto. QGIS, an open-source software, was used to detect and map the boulder movements, while GNSS measurements were conducted to verify the accuracy of the results [15].

The objective of the next study [16] was to investigate the influence of interpolation methods on the gridding of terrestrial gravity anomalies and their implications for geoid model determination. Four distinct interpolation methods were employed: geostatistical Kriging, nearest neighbor, inverse distance to a power, and artificial neural networks. The accuracy of the interpolation results was validated against independent benchmark points, measured using GNSS. Their finding highlights the critical role of interpolation strategy in accurate geoid determination, with important implications for GNSS-based height systems and geodetic reference frames.

Moreover, Unmanned Aerial Vehicle (UAV) data, GNSS measurements, and 3D spatial analysis techniques within a GIS environment were integrated to evaluate how UAV flight geometry influences landform erosion [17]. Three photogrammetric flights with varying geometries were conducted, and the acquired imagery was processed to produce orthomosaics, DSMs, and 3D point clouds. These datasets were subsequently analyzed to map small-scale erosional landforms, such as rills and gullies, using root mean square error assessment, length measurements, and the Multiscale Model-to-Model Cloud Comparison algorithm.

In terms of wetlands mapping, eleven land-use/land cover maps derived from various remote sensing data with spatial resolutions ranging from 10 m to 1000 m (i.e 10 m, 30 m, 100 m, 500 m and 1000 m) were compared and evaluated for their consistency and accuracy in capturing the spatial and temporal dynamics of wetlands [18] within the Irtysh River Basin, which covers an vast area of 17,000 km^2. The datasets were processed within a GIS environment, reprojected to a common coordinate system, and analyzed using similarity metrics, such as the Jaccard Similarity Coefficient, as well as classification accuracy measures, including the Kappa Coefficient.

A total of 173 Sentinel-1 SAR images were processed, comprising 143 images for open-ocean conditions and 40 for inland-water conditions [19]. The primary objective of the study was to evaluate the applicability of algorithms, widely used in current practice and primarily developed for scatterometers, under SAR conditions. Additionally, the study examined the influence of various factors on the accuracy of wind speed and direction calculations derived from these algorithms. To validate the wind speed magnitudes derived from SAR image processing, wind direction data were obtained from NOAA NDBC oceanographic buoys. This integration of SAR remote sensing and geophysical modeling enhances the precision of GNSS- and GIS-informed environmental monitoring.

Finally, SAR data with varying spatial resolutions from the Sentinel-1 and Cosmo-SkyMed satellite missions were validated using GNSS measurements to map subsidence phenomena across the Emilia-Romagna Region in Italy [20]. The Small Baseline Subset (SBAS) interferometric technique was applied to both datasets over a ten-year period, from 2012 to 2022. The resulting displacement maps revealed several critical sites exhibiting significant vertical displacements, which may pose risks for future infrastructure stability.

In conclusion, the value of GIS and GNSS in Earth observation has been clearly demonstrated across a wide range of studies and research topics. GNSS remains indispensable for the precise monitoring of Earth's dynamic systems, while GIS provides the analytical framework that gives these observations context and meaning. The widespread integration of GNSS sensors into daily life, combined with GIS capabilities in data management, harmonization, and spatial analysis, continues to open new opportunities for research. A unifying theme emerging from these studies is the growing importance of interdisciplinary approaches, which leverage the strengths of both GNSS and GIS to advance our understanding of Earth processes and to support innovative applications in environmental monitoring, natural hazard assessment, and sustainable resource management.

Acknowledgments: The Guest Editors of this Special Issue sincerely thank all the authors for their valuable contributions, sharing both their scientific results and expertise. We also extend our gratitude to the journal's editorial team and reviewers for their support throughout the peer-review process.

Conflicts of Interest: The authors declare no conflicts of interest.

References

1. Fortune Business Insights. Location-Based Services Market Size, Share & COVID-19 Impact Analysis, By Component (Solutions, Services), By Application (Retail & E-Commerce, Travel & Tourism, Healthcare, Transport & Logistics, Others), and Regional Forecast. 2025. Available online: https://www.fortunebusinessinsights.com/industry-reports/location-based-services-market-101060 (accessed on 26 November 2025).
2. Kochanek, A.; Generowicz, A.; Zacłona, T. The Role of Geographic Information Systems in Environmental Management and the Development of Renewable Energy Sources—A Review Approach. *Energies* **2025**, *18*, 4740. [CrossRef]
3. Faka, A.; Tserpes, K.; Chalkias, C. Environmental Sensing: A Review of Approaches Using GPS/GNSS. In *GPS and GNSS Technology in Geosciences*; Petropoulos, G.P., Srivastava, P.K., Eds.; Elsevier: Amsterdam, The Netherlands, 2021; Chapter 10. [CrossRef]
4. Radočaj, D.; Plaščak, I.; Jurišić, M. Global Navigation Satellite Systems as State-of-the-Art Solutions in Precision Agriculture: A Review of Studies Indexed in the Web of Science. *Agriculture* **2023**, *13*, 1417. [CrossRef]
5. Panda, S.S.; Siddique, A.; Terrill, T.H.; Mahapatra, A.K.; Morgan, E.; Pech-Cervantes, A.A.; van Wyk, J.A. Decision Support System for Lespedeza cuneata Production and Quality Evaluation: A WebGIS Dashboard Approach to Precision Agriculture. *Front. Plant Sci.* **2025**, *16*, 1520163. Available online: https://www.frontiersin.org/journals/plant-science/articles/10.3389/fpls.2025.1520163 (accessed on 15 November 2025). [CrossRef] [PubMed]
6. Nikolakopoulos, K.G.; Kyriou, A.; Koukouvelas, I.K. UAV, GNSS, and GIS for the Rapid Assessment of Multi-Occurrence Landslides. *Geosciences* **2024**, *14*, 160. [CrossRef]
7. Nikolakopoulos, K.G.; Kyriou, A.; Koukouvelas, I.K.; Tomaras, N.; Lyros, E. UAV, GNSS, and InSAR Data Analyses for Landslide Monitoring in a Mountainous Village in Western Greece. *Remote Sens.* **2023**, *15*, 2870. [CrossRef]
8. Yücel, M.A.; Pırtı, A.; Bayburt, S.; Bektaş, Ö.; Büyüksaraç, A. Monitoring Crustal Motions Using GNSS Techniques. *Bull. Geophys. Oceanogr.* **2024**, *65*, 309–326. [CrossRef]

9. D'Urso, M.G.; Aldrighettoni, J. Geodatabase, Metric Reconstruction and a GIS Platform of Historical-Archaeological Sites in Aquino. *Acta IMEKO* **2024**, *13*, 3. [CrossRef]
10. Treccani, D.; Adami, A.; Brunelli, V.; Fregonese, L. Mobile Mapping System for Historic Built Heritage and GIS Integration: A Challenging Case Study. *Appl. Geomat.* **2024**, *16*, 293–312. [CrossRef]
11. Coruhlu, Y.E.; Altas, S.S. Establishing a Geo-Database for Drinking Water and Its Delivery and Storage Components with an Object-Based Approach. *Water* **2024**, *16*, 1753. [CrossRef]
12. Marin Mićunović, S.; Faivre, S. Evolution of Hvar Island Pocket Beaches during the Last 200 Years (Eastern Adriatic Coast, Croatia). *Geomorphology* **2024**, *447*, 109023. [CrossRef]
13. Badioui, K.; Van Griensven, A.; Verbeiren, B. Vegetation Monitoring of Palm Trees in an Oasis Environment (Boudenib, Morocco) Using Automatic Processing of Medium-Resolution Remotely Sensed Data. *Geosciences* **2025**, *15*, 104. [CrossRef]
14. Kakavas, M.P.; Frattini, P.; Previati, A.; Nikolakopoulos, K.G. Evaluating the Impact of DEM Spatial Resolution on 3D Rockfall Simulation in GIS Environment. *Geosciences* **2024**, *14*, 200. [CrossRef]
15. Delle Rose, M. Annual Coastal Boulder Mobility Detected in 2017–2021 Remote Sensing Imagery and Its Relation to Marine Storms (Gulf of Taranto, Mediterranean Sea). *Geosciences* **2024**, *14*, 136. [CrossRef]
16. Karaca, O.; Erol, B.; Erol, S. Assessments of Gravity Data Gridding Using Various Interpolation Approaches for High-Resolution Geoid Computations. *Geosciences* **2024**, *14*, 85. [CrossRef]
17. Nikolova, V.; Gospodinova, V.; Kamburov, A. Assessment of Unmanned Aerial System Flight Plans for Data Acquisition from Erosional Terrain. *Geosciences* **2024**, *14*, 75. [CrossRef]
18. Luo, K.; Samat, A.; Abuduwaili, J.; Li, W. Evaluation of Remote Sensing Products for Wetland Mapping in the Irtysh River Basin. *Geosciences* **2024**, *14*, 14. [CrossRef]
19. Sergeev, D.; Ermakova, O.; Rusakov, N.; Poplavsky, E.; Gladskikh, D. Verification of C-Band Geophysical Model Function for Wind Speed Retrieval in the Open Ocean and Inland Water Conditions. *Geosciences* **2023**, *13*, 361. [CrossRef]
20. Beccaro, L.; Cianflone, G.; Tolomei, C. InSAR-Based Detection of Subsidence Affecting Infrastructures and Urban Areas in Emilia-Romagna Region (Italy). *Geosciences* **2023**, *13*, 138. [CrossRef]

geosciences

MDPI

Article

InSAR-Based Detection of Subsidence Affecting Infrastructures and Urban Areas in Emilia-Romagna Region (Italy)

Lisa Beccaro [1,*], Giuseppe Cianflone [2,3] and Cristiano Tolomei [1]

1 Istituto Nazionale di Geofisica e Vulcanologia, Via di Vigna Murata 605, 00143 Rome, LZ, Italy
2 Dipartimento di Biologia Ecologia e Scienze della Terra (DiBEST), Università della Calabria, Via Ponte Bucci, 87036 Rende, CS, Italy
3 E3 (Earth, Environment, Engineering) Spin-Off, Università della Calabria, Via Ponte Bucci, 87036 Rende, CS, Italy
* Correspondence: lisa.beccaro@ingv.it

Abstract: The study of deformation signals associated with seismicity in alluvial plain areas is a challenging topic that, however, is increasingly studied thanks to the great aid given by remote sensing techniques that exploit Synthetic Aperture Radar (SAR) data. This study focuses on the determination of the deformation field within the Emilia-Romagna Region (northern Italy), in the area comprising Modena, Reggio Emilia, and Parma cities. SAR data acquired along both orbits during the Sentinel-1 and Cosmo-SkyMed satellite missions were processed with the Small Baseline Subset interferometric technique from June 2012, after the serious seismic swarm of May 2012, to January 2022, just before the two earthquakes occurred in February 2022. The results, validated with Global Navigation Satellite System measurements, do not highlight displacements correlated with the seismicity but, thanks to their high spatial resolution, it was possible to discriminate areas affected by noticeable subsidence phenomena: (i) the highly industrialized areas located north of the municipalities of Reggio Emilia and Modena cities and (ii) a sector of the high-speed railway sited north of the Reggio Emilia city centre, close to the Reggio Emilia AV Mediopadana station. Here we show that, at least since 2012, the latter area is affected by subsidence which can be related to the secondary consolidation process of the fine soils loaded by the railway embankment. The piezometric level analysis also suggests that the lowering of the groundwater table could accelerate the subsidence rate, affecting the stability of infrastructures in highly populated and industrialized areas.

Keywords: InSAR; remote sensing; subsidence; alluvial plain; Emilia-Romagna; GNSS

1. Introduction

During the past few decades, synthetic aperture radar interferometry (InSAR) has been effectively applied for the study of natural phenomena such as landslides [1–4], subsidence [5–7], volcanic activity [8–11] or earthquakes [12,13]. Advanced interferometric techniques (e.g., Small Baseline Subset (SBAS) [14], Persistent Scatterers [15]) exploiting long stacks of synthetic aperture radar (SAR) data have also proven to be useful tools for studying buildings and infrastructures affected by thermal effects [16] or land subsidence [17,18], evaluating displacements that could compromise their stability [19].

Our study area comprises Modena, Reggio Emilia, and Parma municipalities, located in the NW portion of the Emilia-Romagna alluvial plain, which occupied the southern portion of the Po Plain in northern Italy (Figure 1).

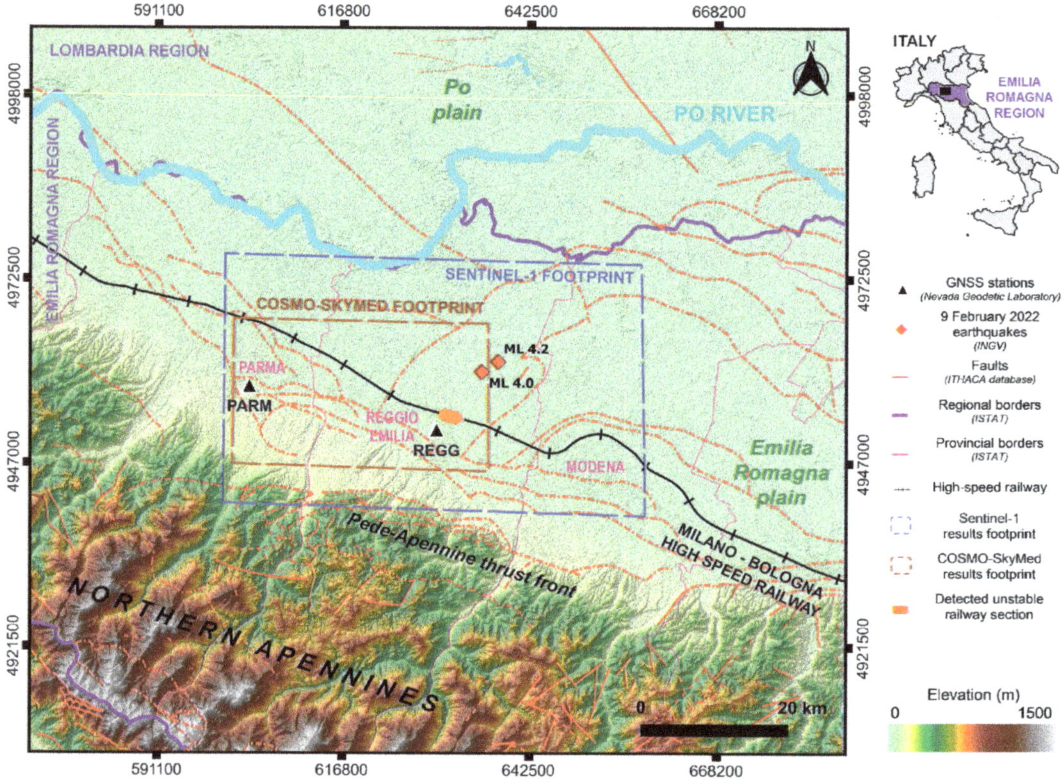

Figure 1. Location and topography of the study area inside the Emilia Romagna alluvial plain (northern Italy). The map contains the footprints of SAR images used in this study. Global Navigation Satellite System (GNSS) data are from [20], faults are from the ITHACA database [21] and 2022 earthquakes epicentres are from http://terremoti.ingv.it/ (accessed on 28 March 2023). Geographic coordinate system is UTM 32N, WGS 84.

The Emilia-Romagna plain is one of the areas in Italy where the greatest attention is paid to ground movements as it is a highly urbanized and industrialized zone, with important cultural heritages and intensive farming [19]. Primary ground movements that have been recorded in this area over the years using different techniques (e.g., levelling [22] and Global Positioning System (GPS) measurements [23], InSAR analyses [24–26]), are mainly directed in an up-down direction, therefore related to subsidence or uplift phenomena. Previous work [24] correlated the occurrence of uplifts in the plain with the presence of buried tectonic structures considered active or potentially active: the Emilia folds, the Ferrara folds, and the Pede–Apennine thrust front (Figure 1), as well as the reduction of water withdrawal from the subsoil for various purposes (drinking, industrial and irrigation use). Instead, subsidence phenomena recorded inside the Emilia-Romagna plain over the years have been attributed to both natural and anthropogenic causes, operating on different time scales [27,28]. Natural subsidence, acting in millions to thousands of years, is quantified in a few millimetres per year [23] and could be related to tectonic/sediment loading, sediment compaction, and post-glacial rebound. Instead, anthropogenic subsidence takes place over hundreds to tens of years and is the result of human activities. Anthropogenic subsidence can exceed natural subsidence rates by more than one order of magnitude [27]. The main cause of this type of movement has been attributed to groundwater extraction [23,28], but also hydrocarbons extraction [24,29], underground excavation for mining activities, construction of underground facilities, and carbonate rock solution can significantly create soil subsidence [30].

In this contribution, we show ground displacement maps over the NW part of the Emilia-Romagna plain obtained using the SBAS multi-temporal InSAR method over a long-time-interval, from June 2012 to February 2022. The period was specifically chosen after the infamous earthquake occurred at the end of May 2012 [31,32] and before the recent shocks of Ml 4.0 and Ml 4.3 happened on 9 February 2022 (Figure 1), in order to exclude the coseismic deformation signal from the long time series. It was thus possible to correctly study the displacements that occurred in the area during the last decade, focusing on the subsiding movements identified both in the results obtained with Cosmo-SkyMed and Sentinel-1 data, at many industrial areas and along a portion of the high-speed railway on the Milano–Bologna line as well (Figure 1).

In the past, alongside the nearest section of the Milano—Bologna regional railway, running parallel and located about 1.5 km south of the high-speed line, railway embankment segments affected by instability phenomena were already identified. These phenomena were characterized by deformations and lowering of the ballast, causing dangerous level defects in the tracks [33].

Therefore, the evaluation of displacements, which have been measured in this work through InSAR methods, is considered interesting and useful for the purposes of future in-depth assessment by the railway network managers on possible actions to be undertaken, given the great importance that the stability of the infrastructures has in guaranteeing people safety.

2. Study Area

The Po Plain is the biggest alluvial plain in Italy [24] and is bounded by the Alps to the north and by the Apennines to the south [34]. The floodplain, representing the foreland sedimentary basin system of the two fold-and-thrust belts [35], covers an area extended about 40.000 km^2, spreading EW for approximately 650 km, from the western Alps to the Adriatic Sea [30]. The Emilia-Romagna plain, in particular, lies on a series of fold-faulted Apenninic terrains dipping northward in a tectonic regime characterized by NE compression [36]. The upper part of the stratigraphic succession of the Emilia-Romagna plain is made of recent (Middle Pleistocene–Holocene) alluvial fine and unconsolidated sediments, i.e., sands, silts, and clays, that are dropped off by the Po River and its tributaries [24]. This sequence, which locally exceeds 500 m of thickness, hosts a complex multi-aquifer freshwater system [30]. In particular, the subsoil of the investigated area is characterized by the Emiliano-Romagnolo super-synthema [37,38]. The latter includes the Lower Emiliano-Romagnolo synthema, with middle Pleistocene age, made by sediments deposited in an alluvial environment (alluvial plain, interfluvial basin, distal alluvial fan). Upward, the previous synthema passes to the Upper Emiliano-Romagnolo synthema, over 100 m thick with Upper Pleistocene–Holocene age [37,38]. This upper synthema includes coarse and fine-grained sediments in the upper valley and fine-grained deposits in the middle valley. In the investigated area the Upper Emiliano-Romagnolo is divided into the following (from the bottom) sub-synthemas: (i) Agazzano sub-synthema mainly made by gravelly alluvial fan deposits; (ii) Villa Verucchio sub-synthema consisting of a cyclic succession of gravel, sand, silt, and clay of alluvial environment; (iii) the Ravenna sub-synthema mainly includes silt, clayey silt, and sandy silt.

3. Materials and Methods

In order to detect potential deformation signals, SAR data acquired by the Cosmo-SkyMed (CSK) and Sentinel-1 (S1) satellite missions were collected. CSK images were requested from the Italian Space Agency (ASI) and downloaded from the ASI portal (https://www.asi.it/en/earth-science/cosmo-skymed/ (accessed on 28 March 2023)). S1 data are freely distributed by the European Space Agency and were collected from the Alaska Satellite Facility catalogue (https://asf.alaska.edu (accessed on 28 March 2023)).

All the exploited SAR datasets were processed adopting the SBAS advanced InSAR technique [14] implemented into the SARscape software (5.6 version, sarmap SA). The SBAS technique adopted in this work combines a large amount of SAR differential interferograms formed according to user-defined spatio-temporal constraints, aiming at preserving SAR phase coherence and spatial coverage, and at the same time limiting the bias effects of the interferometric phase [39]. Detailed information on the SBAS algorithm can also be found in [40,41]. This multi-temporal method permits the retrieval of ground displacement maps and deformation time series along the satellite's Line Of Sight (LOS) over the study area in a very long temporal range, i.e., between 2012 and 2022.

Specifically, two stacks of X-band (wavelength equal to 3.1 cm) CSK data, acquired along both the descending and ascending tracks, were analysed from June 2012 to December 2016. In addition, two datasets of C-band (wavelength equal to 5.6 cm) S1 data, acquired along both acquisition geometries, were analysed during the January 2016–February 2022 temporal span. Table 1 summarizes the main parameters of the considered dataset: name of the satellite mission, data acquisition mode, number of images used for each processing, number of interferograms evaluated as reliable for the results generation, processing time interval, ground resolution, and angle of incidence at the scene centre.

Table 1. Main features of the datasets used in the multi-temporal interferometric processing.

Satellite	Orbit Type	N. of Images	N. of Pairs	Temporal Span	Processing Resolution (m)	Incidence Angle (°)
Cosmo-SkyMed	Descending	82	560	15 June 2012–19 December 2016	15	26.7
	Ascending	79	482	17 July 2012–11 December 2016		
Sentinel-1A	Descending	181	777	4 January 2016–1 February 2022	30	36.7
	Ascending	179	766	12 January 2016–28 January 2022		33.4

The CSK images were processed according to the following constraints: (i) maximum orbital separation of 830 m for reducing spatial decorrelation; (ii) maximum temporal distance between the two passes of 180 days to minimize temporal decorrelation. Therefore, 560 descending and 482 ascending interferograms have been generated. In order to improve the signal-to-noise ratio, a multilooking operation was applied equal to 7 and 6 for the range and azimuth direction respectively, to obtain a final ground resolution of 15 m.

Regarding S1 data processing, we found out that the best results, i.e., not affected by phase bias errors (according to [39]), are obtained considering the minimum and maximum time interval between acquisitions equal to 36 and 180 days. It was considered a maximum orbital separation not exceeding 98 m for the ascending track and 113 m for the descending one. Such settings allowed us to obtain 766 and 777 interferograms for ascending and descending orbits, respectively. To get the 30 metres ground resolution the applied multilooking factors were equal to 7 and 2 for the range and azimuth direction respectively.

The ALOS 30 m DEM was adopted as the reference elevation data in all processing to remove the topographic phase component from the interferograms. Moreover, in order to minimize the interferometric noise the Goldstein method [42] was used. The Delaunay MCF unwrapping method [43] was applied and the atmosphere phase contribution was estimated and removed, applying double filtering in space and time, to correctly reconstruct the surface movement at each considered date. Pixels having coherence thresholds lower than 0.25 and 0.35, for the CSK and S1 processing respectively, were ignored during the final result generation.

In order to estimate the interferometric results errors, the precision associated with the ascending and descending displacement maps obtained during both CSK and S1 processing was calculated [44]. The maximum precision values, which are derived from parameters such as coherence and wavelength, are equal to 1.3 mm/yr and 1.0 mm/yr, respectively for the CSK and S1 LOS results. Considering that the higher the precision value, the lower the measurement precision, these values testify to the high accuracy of the SBAS results.

Finally, the availability of ascending and descending tracks for both CSK and S1 data allowed the calculation also of the horizontal (East–West) and vertical components of the displacement during the common time period [45].

We validated SBAS results using geodetic measurements [20] from two Global Navigation Satellite System (GNSS) stations located inside Parma (PARM) and Reggio Emilia (REGG) municipalities (Figure 1). Due to the different reference systems of the two measuring systems, it was necessary to calculate the measurements difference recorded at the two GNSS stations (e.g., GPS_{PARM}–GPS_{REGG}), and, likewise, at about 10 SAR measurement points located next to the two geodetic stations (e.g., SAR_{PARM}–SAR_{REGG}). Therefore, the comparison between the vertical components of SAR and GNSS displacement time series was correctly performed.

4. Results and Discussion

Firstly, we show the overall ground displacement maps performed using the SBAS InSAR technique within the Emilia-Romagna plain. Cosmo-SkyMed data allowed the signal recording over an area of about 40 km × 40 km comprising the Parma and Reggio Emilia municipalities and surroundings. Sentinel-1 data, instead, permit us to evaluate displacements also at Modena city centre and its nearby areas. Then, we focused the analysis on four distinct areas, characterized by peculiar displacements in the vertical direction. We computed the combined study between vertical displacement time series and piezometric level measurements. Finally, we focused on a specific Milano–Bologna high-speed railway section and we calculated the vertical consolidation related to the railway embankment load.

4.1. InSAR Displacement Maps

Following the convention, negative LOS velocities indicate ground movements away from the satellite sensor, while positive LOS velocities imply ground movements approaching the satellite sensor. Moreover, vertical displacement maps show negative values if the ground motion is downward, while positive values if surface uplift occurs. Moreover, maps showing displacements in the horizontal direction display negative values if the movement is oriented to the west, while positive values if the movement is oriented towards the east.

Cosmo-SkyMed displacement maps, computed for the 2012–2016 temporal span, are visible in Figure 2. Both maps showing displacements measured along descending (Figure 2a) and ascending (Figure 2b) LOS highlighted negative average displacement rates in the area north of the Reggio Emilia Municipality and positive values south of its city centre. The negative and positive displacements recorded by both tracks indicate predominantly downward and upward movements, respectively north and south of Reggio Emilia city. This is confirmed by the displacement pattern exhibited by the vertical displacement map (Figure 2c). The horizontal displacement map (Figure 2d) does not show particular trends in the EW direction.

Uplift identified in the southern part of Reggio Emilia is well documented also by previous works [24] and it will be analysed in detail through the displacement time series analysis (Section 4.3.1). Instead, continuous subsidence movements are identified in the northern and eastern part of the city of Reggio Emilia and are located in correspondence with: (i) highly industrialized areas (e.g., Mancasale industrial area; from 6 to 10 mm/yr) and (ii) a part of the high-speed railway network approximately 2.5 km long near the train station known as Reggio Emilia Mediopadana (from 10 to 15 mm/yr). This last area will be treated specifically inside Sections 4.3.3 and 4.4.

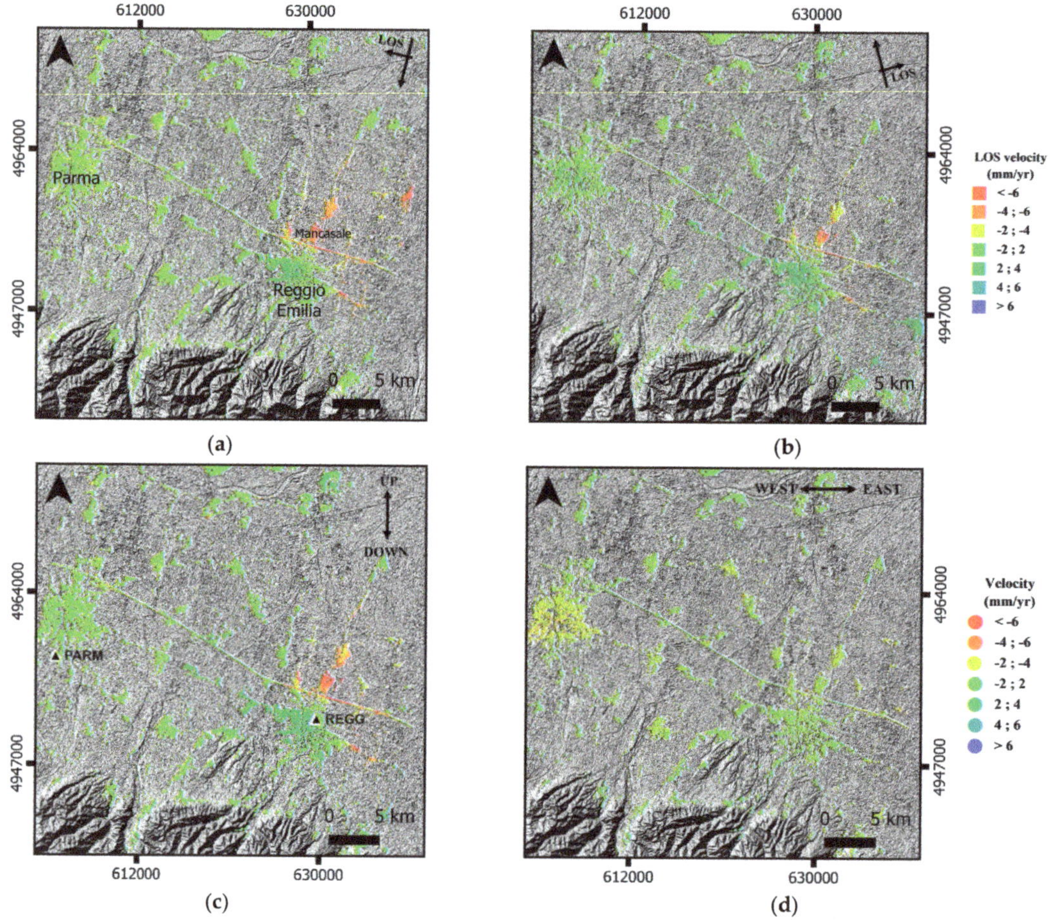

Figure 2. Descending (**a**) and ascending (**b**) LOS displacement maps computed by using Cosmo-SkyMed data. Vertical (**c**) and horizontal (**d**) Cosmo-SkyMed displacement maps.

Sentinel-1 displacement maps, retrieved for the 2016–2022 period, are presented in Figure 3. Descending (Figure 3a) and ascending (Figure 3b) displacements maps show negative values mainly north of Reggio Emilia city and in the highly industrialized area located between the municipalities of Reggio Emilia and Modena. These areas are characterized by high subsidence in the vertical displacement map (Figure 3c). The horizontal displacement map (Figure 3d) shows slight westward rates (from 2 to 4 mm/yr) throughout the area north of the main regional country seats. This signal is characterized by strong oscillations, probably deriving from atmospheric residuals in the descending results affecting coherent agricultural areas south of the Po River.

It is interesting to note that, unlike the results obtained with Cosmo-SkyMed data, no uplifts in the entire analysed area were recorded. Subsidence is clearly visible in the industrial area north of Reggio Emilia (Mancasale industrial area), in the Correggio district, and in the industrial area east of it, where subsidence rates are in the order of 6 and 15 mm/yr. These values are well comparable with the subsidence rates detected in the same areas until 2016 [26], suggesting that the subsidence trend in these areas has continued steadily from 2016 until the present. Regarding the province of Modena, high subsidence movements are recorded in correspondence with the industrial area west of Carpi (about

20 mm/yr) and south of the Soliera district (about 10 mm/yr). The values confirm what was identified up to 2016 [26], i.e., the continuous reduction of subsidence within the aforementioned areas. Finally, subsidence values greater than 10 mm/yr were recorded along a ~10 km long portion of high-speed railway located north of the town of Reggio Emilia, where also the Cosmo-SkyMed outcomes showed the same displacement rates. Spatially, the displacement evolution over time in this area can be observed in Video S1.

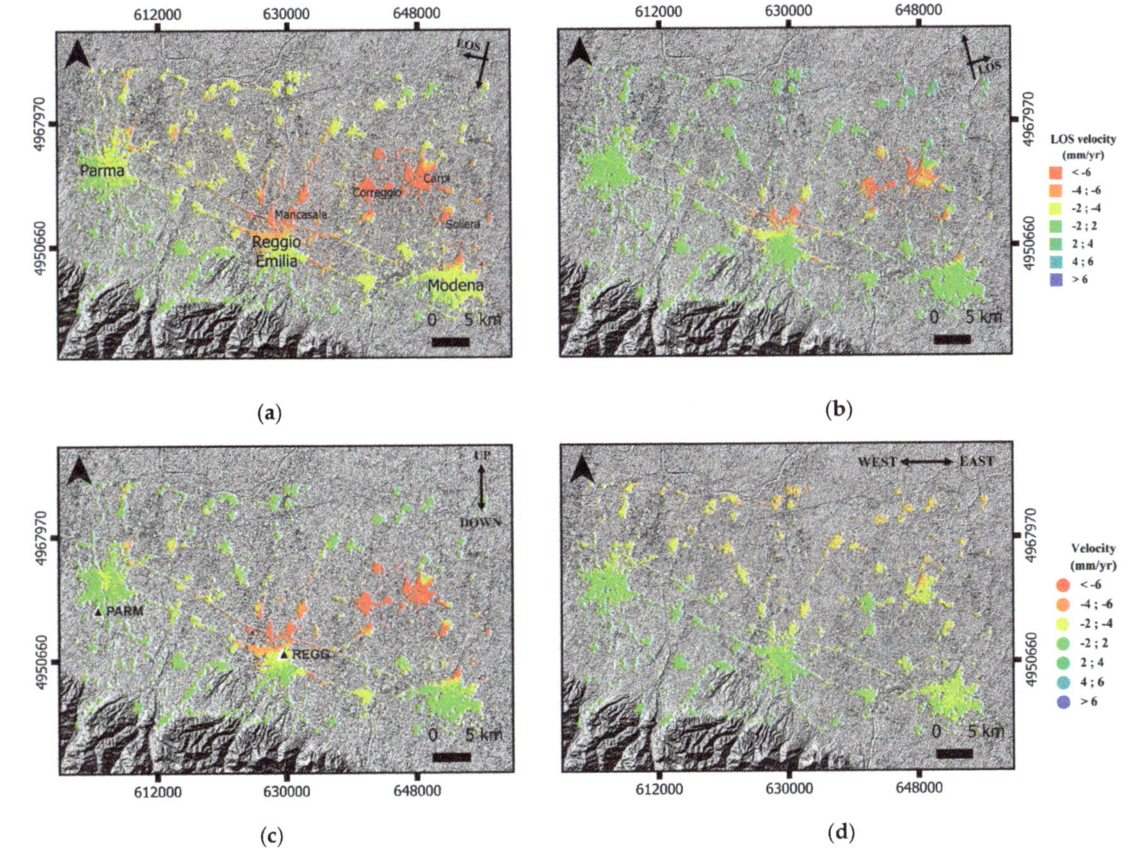

Figure 3. Descending (**a**) ascending (**b**) LOS displacement maps computed by using Sentinel-1 data. Vertical (**c**) and horizontal (**d**) Sentinel-1 displacement maps.

4.2. InSAR Results Validation

Figure 4a shows the graph that compares the vertical displacement time series obtained by processing CSK data (red line) and those obtained from the analysis of GNSS measurements (light blue crosses). Likewise, Figure 4b reveals the comparison between the Sentinel-1 vertical displacements time series (purple line) and those obtained from the GNSS processing (light blue crosses). Note that the comparisons shown in Figure 4 represent the differences between GPS measurements acquired at the PARM and REGG stations and, similarly, between SAR measurements at the two previously mentioned stations. Therefore, the behaviours obtained from the analysis do not depict a real phenomenon but permit us to appreciate that trends are very similar, testifying to the high accuracy of the InSAR results. The errors associated with SAR and GNSS measurements are estimated through the standard deviation (Σ) of satellite and ground measurements, respectively. Regarding CSK and GNSS data comparison, Σ is equal to 3.4 and 6.3 mm/yr, correspondingly.

Instead, S1 and GNSS measurements comparisons have respectively an error of 4.4 and 5.7 mm/yr.

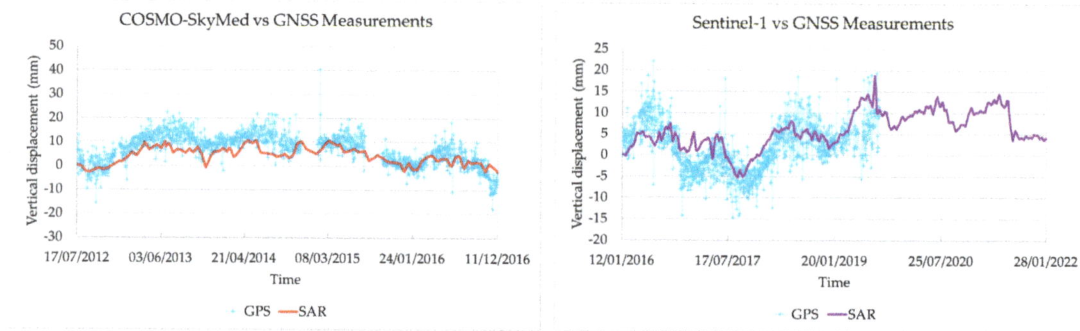

(**a**) (**b**)

Figure 4. Comparison between GNSS vertical displacement rates (light blue crosses) and those derived through SBAS processing, marked by the red and purple lines for Cosmo-SkyMed (**a**) and Sentinel-1 (**b**) data, respectively. The time interval analysed coincides with the period common to the ascending and descending LOS processing, therefore 17 July 2012–11 December 2016 for the comparison with CSK data (**a**) and 12 January 2016–28 January 2022 for the one with S1 data (**b**). The latter shows GPS measurements until September 2019 due to the limited data available at the REGG station.

4.3. *InSAR Displacement Time Series and Piezometric Level Analysis*

For those areas affected by specific deformations for the vertical direction, i.e., (i) the southern part of the city centre of Reggio Emilia, (ii) the industrial areas of Mancasale (north Reggio Emilia) and Carpi (northwest Modena) and (iii) the section of the Milano–Bologna high-speed railway (Figure 5), it was considered important to deepen the study by generating the vertical displacement time series (DTS). DTS permitted to follow the evolution of the vertical displacements over time within the aforementioned areas, integrating also piezometric measurements from ARPAE (https://dati.arpae.it (accessed on 28 March 2023)). For the sake of completeness, we compared the deformation rates detected in our elaborations with those published within the European Ground Motion Service (EGMS) (https://land.copernicus.eu/pan-european/european-ground-motion-service (accessed on 28 March 2023)) at the aforementioned areas. Overall, we can state that average vertical velocities recorded by EGMS and retrieved by our results are comparable. The only difference is the less oscillatory behaviour of EGMS results which could be due to processing strategies different from those used by us. In particular, we think that different filtering approaches could strongly influence those oscillations. The use of slight signal filtering, as performed in our case, allows a more detailed analysis with external data (e.g., piezometric level).

4.3.1. Southern Reggio Emilia City

At the south-western portion of the centre of Reggio Emilia Municipality, where the CSK results showed uplift (Figure 2c) and the Sentinel-1 outcomes recorded subsidence (Figure 3c), the displacement time series were generated by averaging all the points falling within a 100-metre neighbourhood of the RE 55-00 piezometric station (Figure 5).

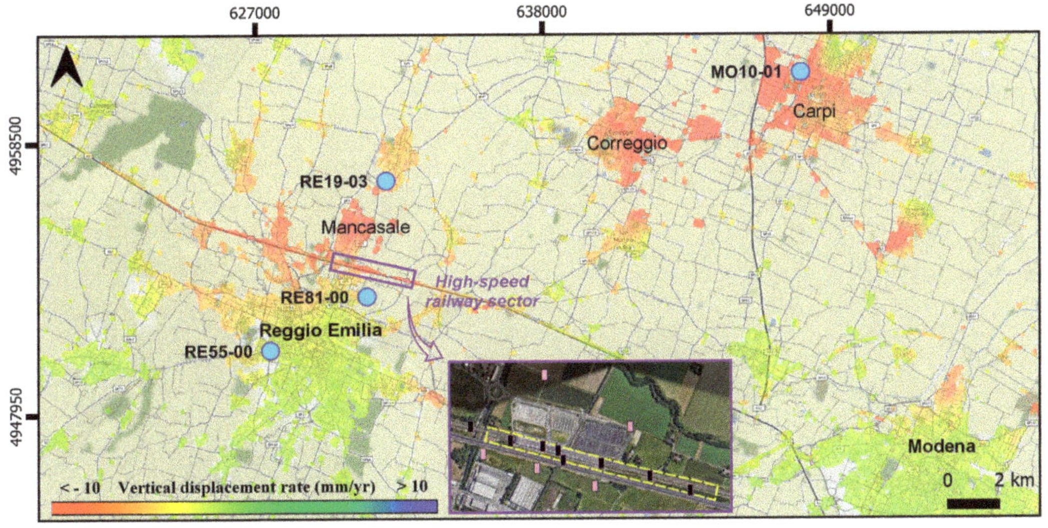

Figure 5. Sentinel-1 vertical displacement map over Stamen Terrain basemap. The map contains the zones investigated in detail through DTS and piezometric analysis. Piezometric measurement stations are represented by light blue circles on the map. In the inset, black and pink rectangles represent respectively the geotechnical core drilling and the static penetration tests used for reconstructing the geotechnical model (see Section 4.4). Geotechnical data are from https://geoportale.regione.emilia-romagna.it (accessed on 28 March 2023).

CSK vertical DTS visible in Figure 6 shows that the uplift reaches about 2 cm from the beginning of the considered time period (July 2012) up to mid-2015, then, gradual subsidence starts. This suggests that according to movements recorded NW of Bologna city by other authors [24], uplifts observed south of the Reggio Emilia Municipality are man-induced and not driven by tectonics. Here the behaviour of the uplift is intermittent and reflects the recovery of piezometric levels (blue line in Figure 6) due to a decrease in groundwater withdrawals. Sentinel-1 vertical displacement time series visible in the same graph (red diamonds in Figure 6) show that the subsidence identified by CSK data starting from mid-2015 continues here and is still ongoing at the end of the analysed period (February 2022). In addition, in this case, Sentinel-1 DTS oscillations seem to correspond to water table variation. Moreover, the reversal trend (from uplift to subsidence) identified in mid-2015 appears to be driven by the sharp decrease in the groundwater level, equal to 5 m in just 12 months, from October 2015 to the same month of 2016. Overall, this result suggests that since mid-2015 groundwater withdrawal has exceeded aquifer recharge at most of the urban area south of the municipality of Reggio Emilia.

4.3.2. Industrial Areas

Figure 7 shows vertical displacement time series plotted at two of the industrial zones sited in our study area: Mancasale (Figure 7a) and Carpi (Figure 7b), the latter located in the middle valley where the thickness of the Ravenna sub-synthema (made up by fine compressible sediments) is greater than the upper valley hosted Reggio Emilia centre. For the first case, DTS were generated by averaging all the points falling within a 200-metre neighbourhood of the RE 19-03 piezometric station, instead for the second area DTS were plotted considering all the points inside a 500 m neighbourhood of the MO 10-01 piezometer (Figure 5).

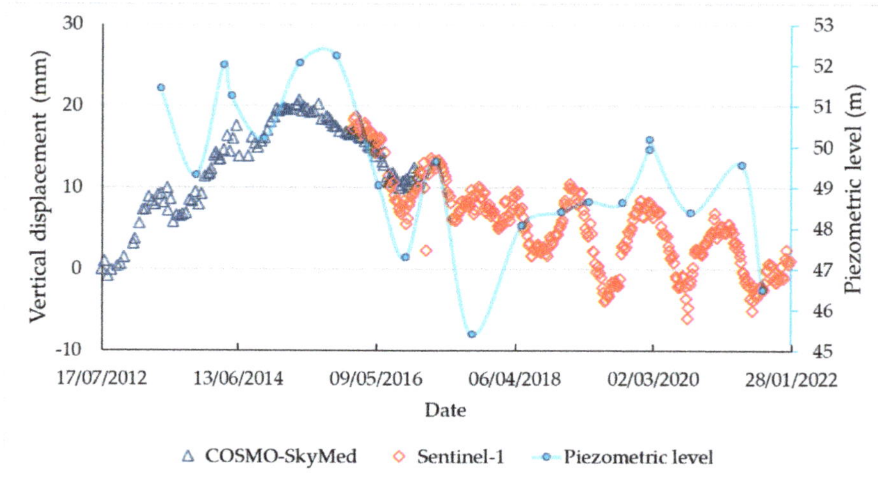

Figure 6. Cosmo-SkyMed (blue triangles) and Sentinel-1 (red diamonds) vertical displacement time series plotted at the southern part of the Reggio Emilia Municipality, around the location of RE 55-00 piezometric station. In the graph is evident the uplift that reaches about 2 cm in July 2015 and the following subsidence trend related to the decrease of the piezometric level (light blue line) due to higher groundwater withdrawals.

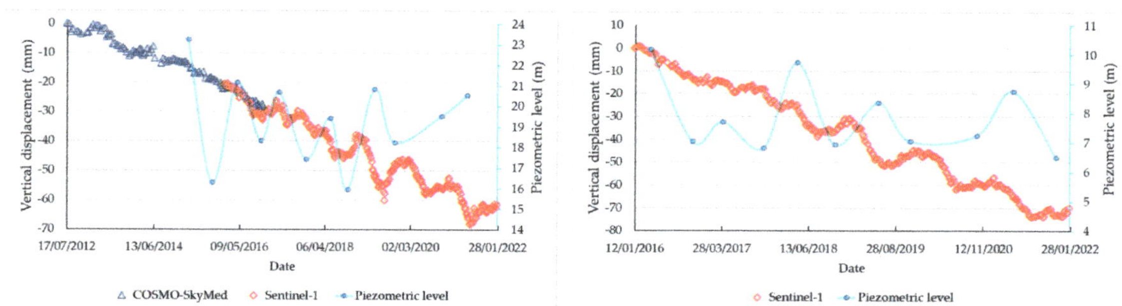

(**a**) (**b**)

Figure 7. Cosmo-SkyMed and Sentinel-1 vertical displacement time series plotted north of Mancasale industrial zone, around RE 19−03 piezometric station (**a**), and Sentinel-1 displacement time series retrieved at Carpi district, around MO 10−01 piezometric station (**b**).

DTS at Mancasale show a constant subsidence trend throughout the entire period (2012–2022). CSK and Sentinel-1 subsidence rates are equal to 6.73 mm/yr and 6.77 mm/yr, respectively. The constant subsidence rate throughout the analysed period can be explained by considering both the decrease in groundwater withdrawals in recent years which translates into a piezometric level rise, and the industrial expansion of the area which certainly took place before August 2016 (from historical Google Earth images) which creates an overload on the subsoil.

Sentinel-1 DTS at Carpi industrial area (Figure 7b) highlighted 11.5 mm/yr of subsidence. In this zone, the recorded movements can be mainly related to a recent increase in groundwater withdrawal used for normal industrial consumption. Moreover, the groundwater level since 2016 is around 6 m a.s.l., while before 2016 it was around 10 m a.s.l. Furthermore, in this area, the subsoil is characterized by the presence of highly compressible peat and organic matter, as inferred by geotechnical core drilling available in the literature [38,46].

4.3.3. High-Speed Railway Sector

At the section of the Milano–Bologna high-speed railway, where both CSK and Sentinel-1 vertical displacement maps showed remarkable ground motions (Figures 2c and 3c, respectively), DTS were generated by averaging all the points falling within a 1300 × 80 m area, next to AV Mediopadana train station (yellow dashed area within the inset in Figure 5). This area is occupied by a railway embankment above which the rails rest (the effects of embankment overload are debated in Section 4.4).

The time series plotted in this area show continuous subsidence with a mostly constant trend. The average subsidence velocity in the entire period analysed is equal to 11.3 mm/yr. The vertical displacement rate considering only the CSK data is equal to 9.7 mm/yr, while the one obtained with Sentinel-1 data is equal to 12.2 mm/yr. The acceleration of the subsidence rate in the railway embankment area may be conditioned by the rapid lowering of the piezometric level in 2016, measured at the RE-81 piezometer and visible by the light blue line in Figure 8. Moreover, in just 6 months, from March to September 2016, the water table dropped by about 2 m.

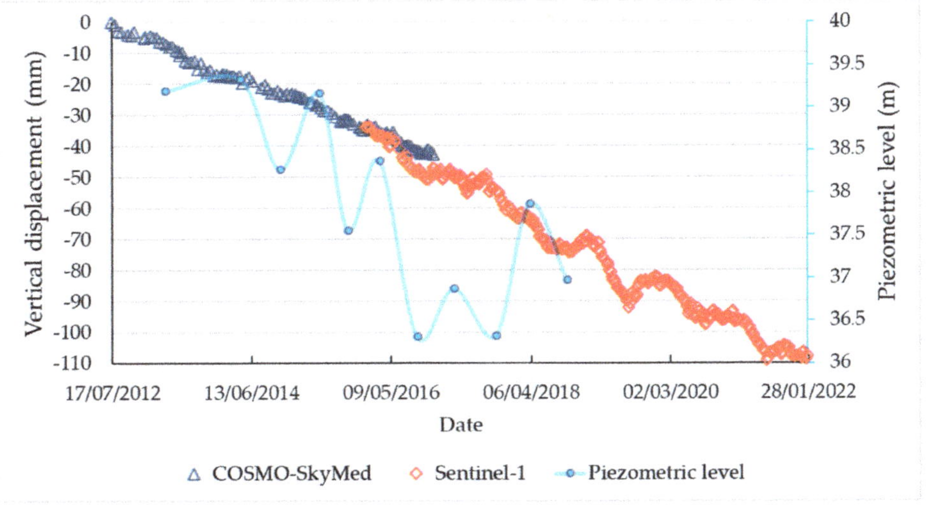

Figure 8. Cosmo-SkyMed and Sentinel-1 vertical displacement time series plotted at the high-speed railway sector. The trend over the entire period is almost linear, with an acceleration of the subsidence trend starting from the early months of 2016, reflecting the rapid decrease of the piezometric level (light blue line).

4.4. Vertical Consolidation along Railway

Along the high-speed railway, we computed the subsidence rate induced by the railway embankment load following the classical one-dimensional vertical consolidation theory by Terzaghi [47] and using the Geostru Loadcap software. For the railway embankment close to the Reggio Emilia AV Mediopadana station, we reconstructed a simplified geotechnical model based on an official geological map (sheet 200 "Reggio nell'Emilia" [37]), eight geotechnical core drilling and five static penetration tests (CPT) both available from Emilia Romagna Region (https://geoportale.regione.emilia-romagna.it (accessed on 28 March 2023)). The CTPs were re-elaborated to investigate geotechnical parameters. The location of the geotechnical investigations used for the reconstruction of the model is visible in the inset of Figure 5.

The geotechnical model (Figure 9a) consists of 20 m-thick deposits of Ravenna sub-synthema lying on the sands and gravels (unit 3) of the Villa Verucchio sub-synthema. The Ravenna sub-synthema was divided into two 10 m-thick litho-geotechnical units made, starting from the ground level, by silty clay (unit 1) and clayey-sandy silt (unit 2).

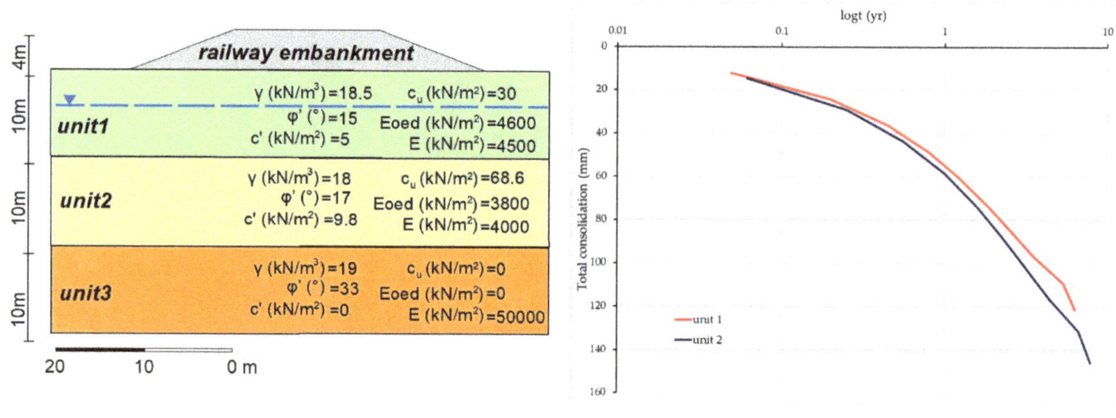

(a) (b)

Figure 9. (**a**) Simplified geotechnical model (γ = specific weight; φ' = friction angle; c' = cohesion; c_u = undrained cohesion; Eoed = oedometric modulus; E = elastic modulus). (**b**) Consolidation curve, for units 1 and 2, derived from the simulation.

For the railway embankment, we considered an isosceles trapezoid section characterized by parallel sides of 40 and 16 m, a height of 4 m, and a specific weight of 20 kN/m^3. The embankment produces an overload of about 50 kN/m^2. Moreover, we took into account a water table depth of 4 m referring to the last measurement available from the ARPAE piezometer RE-81.

The analysis of the time-dependent consolidation (Figure 9b), evaluated for a period of 10 years, shows a total consolidation of ~27 cm which involves units 1 and 2. The total consolidation includes primary (occurring in a short period) and secondary consolidation processes [48]. The primary consolidation occurred during the first ~2.5 and ~3.2 yr (with a rate higher than 30 mm/yr) in units 1 and 2, respectively. The secondary consolidation is instead characterized by rates of ~10 mm/yr which agrees with the vertical displacement rates inferred by S1 and CSK data. Therefore, SAR data allows monitoring the secondary consolidation due to the embankment overload which, unlike the primary consolidation, is realized over a longer period.

5. Conclusions

We analysed ground displacements which occurred from June 2012 to February 2022 by exploiting the SBAS InSAR technique applied to Cosmo-SkyMed and Sentinel-1 data on a heavily urbanized and populated portion of the Emilia Romagna Region. Displacement maps highlighted some sensitive sites characterized by non-negligible displacements in the vertical direction, as they could result in future infrastructure instabilities: (i) the south-western part of the Reggio Emilia city, (ii) the industrial areas of Mancasale (north Reggio Emilia) and Carpi (northwest Modena) and (iii) the section of the Milano–Bologna high-speed railway.

Overall, the displacement time series analysis in such areas, together with the examination of the water table heights, suggests that the displacements trend over time is mainly conditioned by the oscillations of the piezometric level due to variations in groundwater withdrawals and by the different compressibility of sediments constituting the subsoil.

Specifically, the geotechnical analysis carried out considering the embankment load in the railway area north of the city of Reggio Emilia allowed us to attest that the subsidence rate detected by the InSAR analysis (~11 mm/yr) matches the modelled secondary vertical consolidation (~10 mm/yr). Therefore, it is believed that in the 10 km long (at the beginning of 2022) sector of the railway line the subsidence identified derives mainly from secondary consolidation processes of the fine-grained soils loaded by the railway embankment and that rapid changes in the water table level can accelerate the displacement rate.

Since 2001, in the Po Plain, the drought episodes have become stronger in terms of frequency and length [49]. An exceptional drought episode is assessed in 2022 and the possible future increase of frequency for events of this intensity is supposed [50]. In this climate-change scenario, the ever-increasing drought and water stress negatively affect many human activities, producing ever-growing exploitation of groundwater resources. Considering the shown relationship between piezometric level oscillation and vertical ground motion, a possible increase in groundwater withdrawal could exacerbate the subsidence. The knowledge of these phenomena is therefore essential in order to prevent negative impacts on the population and socio-economic activities.

Supplementary Materials: The following supporting information can be downloaded at: https://www.mdpi.com/article/10.3390/geosciences13050138/s1, Video S1: Cumulated displacements from 2016 to 2022 retrieved with S1 data in the area North of the Reggio Emilia city including the railway section analysed and the Mancasale industrial area. The basemap is a high-resolution CSK amplitude image acquired on 19 December 2016.

Author Contributions: Conceptualization, L.B. and C.T.; methodology, L.B., C.T. and G.C.; software, L.B. and G.C.; validation, L.B.; writing—original draft preparation, L.B.; writing—review and editing, C.T. and G.C; funding acquisition, C.T. All authors have read and agreed to the published version of the manuscript.

Funding: The present work is supported by the ATTEMPT—Integrated System for Multi-Hazard from Space over Mediterranean—INGV research project. The Sentinel-1 data are distributed by the European Space Agency free of charge. COSMO-SkyMed data were provided by Italian Space Agency (Projects ID: 641 and 682).

Data Availability Statement: The generated products are available on request from the corresponding author. Sentinel-1 original data can be found in the Alaska Satellite Facility catalog (https://search.asf.alaska.edu (accessed on 28 March 2023)).

Conflicts of Interest: The authors declare no conflict of interest.

References

1. Tessari, G.; Floris, M.; Pasquali, P. Phase and amplitude analyses of SAR data for landslide detection and monitoring in non-urban areas located in the North-Eastern Italian pre-Alps. *Environ. Earth Sci.* **2017**, *76*, 85. [CrossRef]
2. Cianflone, G.; Tolomei, C.; Brunori, C.A.; Monna, S.; Dominici, R. Landslides and subsidence assessment in the Crati Valley (Southern Italy) using InSAR data. *Geosciences* **2018**, *8*, 67. [CrossRef]
3. Ietto, F.; Conforti, M.; Tolomei, C.; Cianflone, G. Village relocation as solution of the landslide risk, is it always the right choice? The case study of Cavallerizzo ghost village (Calabria, southern Italy). *Int. J. Disaster Risk Reduct.* **2022**, *81*, 103267. [CrossRef]
4. Bekaert, D.P.; Handwerger, A.L.; Agram, P.; Kirschbaum, D.B. InSAR-based detection method for mapping and monitoring slow-moving landslides in remote regions with steep and mountainous terrain: An application to Nepal. *Remote Sens. Environ.* **2020**, *249*, 111983. [CrossRef]
5. Cianflone, G.; Tolomei, C.; Brunori, C.A.; Dominici, R. InSAR time series analysis of natural and anthropogenic coastal plain subsidence: The case of Sibari (southern Italy). *Remote Sens.* **2015**, *7*, 16004–16023. [CrossRef]
6. Cianflone, G.; Vespasiano, G.; Tolomei, C.; De Rosa, R.; Dominici, R.; Apollaro, C.; Walraevens, K.; Polemio, M. Different Ground Subsidence Contributions Revealed by Integrated Discussion of Sentinel-1 Datasets, Well Discharge, Stratigraphical and Geomorphological Data: The Case of the Gioia Tauro Coastal Plain (Southern Italy). *Sustainability* **2022**, *14*, 2926. [CrossRef]
7. Cigna, F.; Tapete, D. Satellite InSAR survey of structurally-controlled land subsidence due to groundwater exploitation in the Aguascalientes Valley, Mexico. *Remote Sens. Environ.* **2021**, *254*, 112254. [CrossRef]
8. Beccaro, L.; Tolomei, C.; Gianardi, R.; Sepe, V.; Bisson, M.; Colini, L.; De Ritis, R.; Spinetti, C. Multitemporal and Multisensor InSAR Analysis for Ground Displacement Field Assessment at Ischia Volcanic Island (Italy). *Remote Sens.* **2021**, *13*, 4253. [CrossRef]

9. Pezzo, G.; Palano, M.; Beccaro, L.; Tolomei, C.; Albano, M.; Atzori, S.; Chiarabba, C. Coupling Flank Collapse and Magma Dynamics on Stratovolcanoes: The Mt. Etna Example from InSAR and GNSS Observations. *Remote Sens.* **2023**, *15*, 847. [CrossRef]
10. Pino, N.A.; Carlino, S.; Beccaro, L.; De Martino, P. Aseismic Creep, Coseismic Slip, and Postseismic Relaxation on Faults in Volcanic Areas: The Case of Ischia Island. *Remote Sens.* **2023**, *15*, 1791. [CrossRef]
11. Poland, M.P.; Zebker, H.A. Volcano geodesy using InSAR in 2020: The past and next decades. *Bull. Volcanol.* **2022**, *84*, 27. [CrossRef]
12. Atzori, S.; Hunstad, I.; Chini, M.; Salvi, S.; Tolomei, C.; Bignami, C.; Stramondo, S.; Trasatti, E.; Antonioli, A.; Boschi, E. Finite fault inversion of DInSAR coseismic displacement of the 2009 L'Aquila earthquake (central Italy). *Geophys. Res. Lett.* **2009**, *36*, 15. [CrossRef]
13. Şentürk, S.; Çakır, Z.; Ergintav, S.; Karabulut, H. Reactivation of the Adıyaman Fault (Turkey) through the Mw 5.7 2007 Sivrice earthquake: An oblique listric normal faulting within the Arabian-Anatolian plate boundary observed by InSAR. *J. Geodyn.* **2019**, *131*, 101654. [CrossRef]
14. Berardino, P.; Fornaro, G.; Lanari, R.; Sansosti, E. A new algorithm for surface deformation monitoring based on small baseline differential interferograms. *IEEE Trans. Geosci. Remote Sens.* **2002**, *40*, 2375–2383. [CrossRef]
15. Ferretti, A.; Prati, C.; Rocca, F. Permanent scatterers in SAR interferometry. *IEEE Trans. Geosci. Remote Sens.* **2001**, *39*, 8–20. [CrossRef]
16. Milillo, P.; Perissin, D.; Salzer, J.T.; Lundgren, P.; Lacava, G.; Milillo, G.; Serio, C. Monitoring dam structural health from space: Insights from novel InSAR techniques and multi-parametric modeling applied to the Pertusillo dam Basilicata, Italy. *Int. J. Appl. Earth Obs. Geoinf.* **2016**, *52*, 221–229. [CrossRef]
17. Polcari, M.; Moro, M.; Romaniello, V.; Stramondo, S. Anthropogenic subsidence along railway and road infrastructures in Northern Italy highlighted by Cosmo-SkyMed satellite data. *J. Appl. Remote Sens.* **2019**, *13*, 024515. [CrossRef]
18. Ciampalini, A.; Solari, L.; Giannecchini, R.; Galanti, Y.; Moretti, S. Evaluation of subsidence induced by long-lasting buildings load using InSAR technique and geotechnical data: The case study of a Freight Terminal (Tuscany, Italy). *Int. J. Appl. Earth Obs. Geoinf.* **2019**, *82*, 101925. [CrossRef]
19. Solari, L.; Del Soldato, M.; Bianchini, S.; Ciampalini, A.; Ezquerro, P.; Montalti, R.; Raspini, F.; Moretti, S. From ERS 1/2 to Sentinel-1: Subsidence monitoring in Italy in the last two decades. *Front. Earth Sci.* **2018**, *6*, 149. [CrossRef]
20. Blewitt, G.; Hammond, W.C.; Kreemer, C. Harnessing the GPS data explosion for interdisciplinary science. *Eos* **2018**, *99*. [CrossRef]
21. ITHACA Working Group (2019). ITHACA (ITaly HAzard from CApable Faulting), A Database of Active Capable Faults of the Italian Territory. Version December 2019. ISPRA Geological Survey of Italy. Web Portal. Available online: http://sgi2.isprambiente.it/ithacaweb/Mappatura.aspx (accessed on 28 March 2023).
22. D'Anastasio, E.; De Martini, P.M.; Selvaggi, G.; Pantosti, D.; Marchioni, A.; Maseroli, R. Short-term vertical velocity field in the Apennines (Italy) revealed by geodetic levelling data. *Tectonophysics* **2006**, *418*, 219–234. [CrossRef]
23. Baldi, P.; Casula, G.; Cenni, N.; Loddo, F.; Pesci, A. GPS-based monitoring of land subsidence in the Po Plain (Northern Italy). *Earth Planet. Sci. Lett.* **2009**, *288*, 204–212. [CrossRef]
24. Severi, P. Soil uplift in the Emilia-Romagna plain (Italy) by satellite radar interferometry. *Boll. Geof. Teor. Appl.* **2021**, *62*, 527–542.
25. Stramondo, S.; Saroli, M.; Tolomei, C.; Moro, M.; Doumaz, F.; Pesci, A.; Loddo, F.; Baldi, P.; Boschi, E. Surface movements in Bologna (Po plain—Italy) detected by multitemporal DInSAR. *Remote Sens. Environ.* **2007**, *110*, 304–316. [CrossRef]
26. ARPAE (Agenzia Prevenzione Ambiente Energia Emilia-Romagna). *Rilievo Della Subsidenza Nella Pianura Emiliano-Romagnola—Seconda Fase*; ARPAE: Bologna, Italy, 2018.
27. Carminati, E.; Di Donato, G. Separating natural and anthropogenic vertical movements in fast-subsiding areas: The Po Plain (N. Italy) case. *Geophys. Res. Lett.* **1999**, *26*, 2291–2294. [CrossRef]
28. Carminati, E.; Martinelli, G. Subsidence rates in the Po Plain, northern Italy: The relative impact of natural and anthropogenic causation. *Eng. Geol.* **2002**, *66*, 241–255. [CrossRef]
29. Modoni, G.; Darini, G.; Spacagna, R.L.; Saroli, M.; Russo, G.; Croce, P. Spatial analysis of land subsidence induced by groundwater withdrawal. *Eng. Geol.* **2013**, *167*, 59–71. [CrossRef]
30. Eid, C.; Benetatos, C.; Rocca, V. Fluid Production Dataset for the Assessment of the Anthropogenic Subsidence in the Po Plain Area (Northern Italy). *Resources.* **2022**, *11*, 53. [CrossRef]
31. Bignami, C.; Burrato, P.; Cannelli, V.; Chini, M.; Falcucci, E.; Ferretti, A.; Gori, S.; Kyriakopoulos, C.; Melini, D.; Moro, M.; et al. Coseismic deformation pattern of the Emilia 2012 seismic sequence imaged by Radarsat-1 interferometry. *Ann. Geophys.* **2012**, *55*, 789–795. [CrossRef]
32. Pezzo, G.; Merryman Boncori, J.P.; Tolomei, C.; Salvi, S.; Atzori, S.; Antonioli, A.; Trasatti, E.; Novali, F.; Serpelloni, E.; Candela, L.; et al. Coseismic deformation and source modeling of the May 2012 Emilia (northern Italy). *Seismol. Res. Lett.* **2013**, *84*, 645–655. [CrossRef]
33. Consolidation of the Railroad Embankment by Construction of Jet Grouting Columns, in the Bologna—Piacenza Section, between km 69 + 250 and km 69 + 875. Drill Pac, S.r.l. Available online: https://www.drillpac.com/en/projects/reggio-emilia (accessed on 28 March 2023).
34. Bruno, L.; Campo, B.; Costagli, B.; Stouthamer, E.; Teatini, P.; Zoccarato, C.; Amorosi, A. Factors controlling natural subsidence in the Po Plain. *Proc. Int. Assoc. Hydrol. Sci.* **2020**, *382*, 285–290. [CrossRef]
35. Doglioni, C. Some remarks on the origin of foredeeps. *Tectonophysics* **1993**, *228*, 1–20. [CrossRef]

36. Martinelli, G.; Minissale, A.; Verrucchi, C. Geochemistry of heavily exploited aquifers in the Emilia-Romagna region (Po Valley, northern Italy). *Environ. Geol.* **1998**, *36*, 195–206. [CrossRef]
37. ISPRA. Foglio 200-Reggio nell'Emilia. In *Carta Geologica d'Italia Alla Scala 1:50.000*; Istituto Superiore per la Protezione e la Ricerca Ambientale: Roma, Italy, 2010.
38. ISPRA. Foglio 201-Modena. In *Carta Geologica d'Italia Alla Scala 1:50.000*; Istituto Superiore per la Protezione e la Ricerca Ambientale: Roma, Italy, 2009.
39. Ansari, H.; De Zan, F.; Parizzi, A. Study of systematic bias in measuring surface deformation with SAR interferometry. *IEEE Trans. Geosci. Remote Sens.* **2020**, *59*, 1285–1301. [CrossRef]
40. Casu, F.; Manzo, M.; Lanari, R. A quantitative assessment of the SBAS algorithm performance for surface deformation retrieval from DInSAR data. *Remote Sens. Environ.* **2006**, *102*, 195–210. [CrossRef]
41. Lanari, R.; Casu, F.; Manzo, M.; Zeni, G.; Berardino, P.; Manunta, M.; Pepe, A. An overview of the small baseline subset algorithm: A DInSAR technique for surface deformation analysis. In *Deformation and Gravity Change: Indicators of Isostasy, Tectonics, Volcanism, and Climate Change*; Birkhäuser: Basel, Switzerland, 2007; pp. 637–661. [CrossRef]
42. Goldstein, R.M.; Werner, C.L. Radar interferogram filtering for geophysical applications. *Geophys. Res. Lett.* **1998**, *25*, 4035–4038. [CrossRef]
43. Costantini, M. A novel phase unwrapping method based on network programming. *IEEE Trans. Geosci. Remote Sens.* **1998**, *36*, 813–821. [CrossRef]
44. Just, D.; Bamler, R. Phase statistics of interferograms with applications to synthetic aperture radar. *Appl. Opt.* **1994**, 33, 4361–4368. [CrossRef]
45. Dalla Via, G.; Crosetto, M.; Crippa, B. Resolving vertical and east-west horizontal motion from differential interferometric synthetic aperture radar: The L'Aquila earthquake. *J. Geophys. Res. Solid Earth.* **2012**, 117. [CrossRef]
46. ISPRA. *Note Illustrative Della Carta Geologica d'Italia Alla Scala 1:50.000 Foglio 201-Modena*; Istituto Superiore per la Protezione e la Ricerca Ambientale: Roma, Italy, 2009.
47. Terzaghi, K. Evaluation of coefficient of subgrade reaction. *Geotechnique* **1955**, *5*, 41–50. [CrossRef]
48. Terzaghi, K.; Peck, R.B.; Mesri, G. *Soil Mechanics in Engineering Practice*; John Wiley & Sons: Berlin/Heidelberg, Germany, 1996.
49. Baronetti, A.; González-Hidalgo, J.C.; Vicente-Serrano, S.M.; Acquaotta, F.; Fratianni, S. A weekly spatio-temporal distribution of drought events over the Po Plain (North Italy) in the last five decades. *Int. J. Clim.* **2020**, *40*, 4463–4476. [CrossRef]
50. Bonaldo, D.; Bellafiore, D.; Ferrarin, C.; Ferretti, R.; Ricchi, A.; Sangelantoni, L.; Vitelletti, M.L. The summer 2022 drought: A taste of future climate for the Po valley (Italy)? *Reg. Environ. Chang.* **2023**, *23*, 1. [CrossRef]

geosciences

MDPI

Article

Verification of C-Band Geophysical Model Function for Wind Speed Retrieval in the Open Ocean and Inland Water Conditions

Daniil Sergeev *, Olga Ermakova, Nikita Rusakov, Evgeny Poplavsky and Daria Gladskikh

Federal Research Center Institute of Applied Physics of the Russian Academy of Sciences (IAP RAS), 603950 Nizhny Novgorod, Russia; ermakova@ipfran.ru (O.E.); nikitarusakov@ipfran.ru (N.R.); poplavsky@ipfran.ru (E.P.); daria.gladskikh@ipfran.ru (D.G.)

* Correspondence: daniil@ipfran.ru

Abstract: Verification of the C-band geophysical model functions (GMF) for the open ocean and the Gorky reservoir was carried out using Sentinel-1 IW-mode Synthetic aperture radar (SAR) images. CMOD5.N, CMOD7, GMF for the Caspian Sea, and CSARMOD2 were considered. The motivation for this study is concerned with the clarification of applying C-band GMFs for SAR images including for the conditions of inland water bodies, as well as with the study of the influence of various wind speed direction sources on the results of wind speed magnitude retrieval for ocean conditions. Comparison of wind speed from the CMOD5.N algorithm using wind direction data from NOAA NDBC oceanographic buoys together with the data provided by NCEP reanalysis data showed that regardless of the geographic location, the result does not depend significantly on the choice of the wind direction source. Novel results of CMOD5.N, CMOD7, GMF for the Caspian Sea, and CSARMOD2 applications to the conditions of the Gorky reservoir are presented. The comparison of these results with the meteorological station measurements showed the best agreement for CMOD5.N. The preliminary results on the construction of new C-band GMF adjusted to the Gorky Reservoir have shown statistical parameters better than for Caspian Sea GMF and CSARMOD2.

Keywords: inland waters; SAR image; Sentinel-1 satellite; geophysical model function; oceanographic buoy

1. Introduction

SAR images of the Earth's surface from spacecraft are one of the most important sources of information about the state of the subsurface layer of hydrosphere objects and the atmospheric boundary layer above them [1]. Unlike images in the optical range, microwave remote sensing can be performed regardless of the daytime and weather conditions. It should be noted that earlier SAR images were not publicly available (see, for example, data from Radarsat-1/2 or Terrasar-X), but at the present time, the new Sentinel-1 A/B mission makes it possible to obtain images free of charge with relatively small time delays.

One of the most important characteristics of the meteorological state is the surface wind speed (usually at a height of 10 m). Numerous attempts to retrieve the wind speed over the sea surface in a wide range of geophysical and geographical conditions have been carried out for the last 30 years and are still an urgent problem [2–5]. One of the main tools involved in solving this problem is active remote sensing using satellite-based microwave scatterometers and SAR. Since these instruments measure the backscatter on the sea surface covered with wind waves, and not the wind speed directly, so-called geophysical model functions have been proposed relating the characteristics of the microwave backscatter to the wind speed. The construction of these functions becomes possible since the amount of backscatter depends on the roughness of the water surface and as a result the wind surface stress (or the wind friction velocity) [6,7], which is related to the wind speed at a height of

10 m in the framework of the Monin–Obukhov theory [8]. Currently, for low and moderate wind speeds, geophysical model functions developed for the C-band are widely used: CMOD4 [9,10], CMOD5 [11], CMOD5.N [12], CMOD_IFR2 [13], CMOD7 [14]. CMOD4 was developed on the basis of ERS-1 data [9] and relates the NRCS of the sea surface to the wind speed, incidence angle and the angle between azimuth angle and wind direction. However, CMOD4 demonstrated the existence of biases in the range of low to moderate winds [15] and a significant underestimation of wind speed values due to saturation of backscatter for extreme wind speeds [10]. The above problems served as an impetus for the improvement of this model, resulting in CMOD5 developed on the basis of adding up more than 22,000,000 collocations from ERS-2 scatterometer backscatter and ECMWF wind speeds for the period from 1 August 1998 till 31 December 1998 which nevertheless also had noticeable biases. The next step was the development of CMOD5.N model on the base of CMOD5 and the refit of 28 coefficients which, unlike its predecessors, made it possible to obtain neutral wind speeds [12]. As a result of verification on the basis of ERS-2 and ASCAT data the CMOD5.N function was constructed from the condition that the model function for neutral winds should give the wind speeds 0.7 m/s stronger than those obtained from CMOD5 for all wind speeds and incident angles. The proposed bias accounts for the stability correction (0.2 m/s) and the bias 0.5 m/s from CMOD5 itself [12]. At present CMOD5.N is successfully used for the wind speed retrieval from the backscatter measured by the scatterometers using corrections provided by the Numerical Weather Prediction model (NWP) [16–18]. The CMOD7 was developed on the basis of ERS and ASCAT scatterometer data to avoid the mismatch for CMOD5.N in the range of low winds which helped to improve its performance for the ASCAT data [14].

All of the above models were developed mainly using data from scatterometers, nevertheless, there were numerous attempts to use these models to retrieve the wind speed from SAR data [19–31], some of which were made for the Radarsat-1,2 satellites [28–31]. At the same time, geophysical model functions C-SARMOD and C-SARMOD2, directly based on the accumulation of SAR data from Radarsat-2 and Sentinel-1 satellites, collocated with data from buoys were developed [32,33]. The biases given by the C-SARMOD geophysical model function were compared with the results of calculations in the framework of CMOD5.N and showed that in the range of low winds the errors according to the C-SARMOD were smaller, while for the entire range of incidence angles for the moderate winds they had values higher than the errors for CMOD5.N. In this regard, the question of which of the existing models is the most suitable for wind speed retrieval from SAR imagery remains open. In addition, in order to correctly retrieve the value of the wind speed based on the family of CMOD models, it is necessary to obtain information about the wind speed direction, which is usually obtained from external sources (for example, buoys or reanalysis data). At the same time, it is necessary to have an understanding of which of the sources is more adequate in terms of correct wind speed retrieval. In this regard, in the first part of our study, we made an attempt to analyze the wind speed values obtained using the CMOD5.N model applied to the SAR data from the Sentinel-1 satellite for various geographic locations. In this case, the wind direction was taken from the data from the NDBC buoys and from NCEP data to assess the influence of one or another measurement source on the result of wind speed retrieval.

It should also be noted that all existing models of wind speed retrieval based on remote sensing data were created for open ocean conditions, while some of the buoys used for the model calibration were located in shallow water conditions. Thus, this may suggest that these models might be relevant for inland water bodies. However, the issue of their applicability to inland water bodies remains unexplored. At the same time, this algorithm, as applied to wind speed retrieval from SAR data, in the case of inland seas, lakes or reservoirs could be very useful due to a significantly higher resolution than that of scatterometers. These wind speed data for inland water bodies are very important for monitoring and predicting high-resolution weather conditions, which have recently become an increasingly urgent task [34]. It should also be noted that remote sensing from the Sentinel 1 A/B satellite is carried out more often for the land surface than for the open ocean conditions, which causes a sufficient amount of data for inland waters. However, there is still a very small number of studies containing attempts to retrieve the magnitude of the wind speed over lakes, inland seas and reservoirs from SAR data. In particular, attempts to retrieve the wind speed based on the use of SAR images for a number of lakes from Hungary and Austria (Salzkammergut, Neusiedl and Balaton lakes), obtained from the Sentinel-1 satellite mission, were madein [35]. The authors proposed the empirical relationships between NRCS and the wind speed for VV and VH polarizations represented by exponential functions. SAR data were calibrated using the measurements from the meteorological station. However, it has been shown that there are some differences between the results for the ascending and descending orbits.

The study presented in [36] included an application of geophysical model functions CMOD7 and C_SARMOD2 for the southern Caspian Sea to the Sentinel-1A VV SAR images based on comparison with 25 km standard resolution reanalysis data. It was demonstrated that CMOD7 showed the smallest RMSEs compared to C_SARMOD2 which gave errors in the area of low wind speeds. However, it should be noted that on the one hand the vast majority of inland water bodies, including large ones, have scales much smaller than the resolution of reanalysis data used in [36] for the verification, on the other hand, the advantage of SAR images over high-resolution scatterometer measurements was not used. So in general inhomogeneities in the surface conditions typical for inland water bodies might adversely influence the result of wind speed retrieval which is a noticeable problem for the lake cases.

Despite rare attempts to apply geophysical model functions developed primarily for oceanic conditions to inland water conditions, the question of which model is more adequate in this case remains open. So the cross-verification of CMOD algorithms for Sentinel 1 A/B SAR images based on comparison with the results of numerical modeling of the meteorological situation at these objects is still an actual problem.

In the present study, we have validated and verified the C-band GMFs (CMOD5.N, CMOD7, CSARMOD2 and the GMF from [36]) model both for the conditions of coastal ocean zones (and open ocean), including a wide range of geographic locations, and for the conditions of an inland water body using the example of the Gorky Reservoir. The paper is organized as follows. In Section 2 the main instruments and datasets for the analysis were described. Section 3 contains the verification results of C-band GMFs (CMOD5.N, CMOD7, CSARMOD2 and the GMF from [36]) using buoy measurements for ocean conditions. When applying the CMOD5.N model to ocean conditions, estimates of the sensitivity of the results to the choice of a wind speed direction data source based on measurements from oceanographic buoys or reanalysis data were made. The results of verification of the CMOD5.N algorithm for inland water conditions using the example of the Gorky Reservoir and its comparison with the calculations using other C-band models (CMOD7, CSARMOD2 and the GMF from [36]) are reported in Section 4. A procedure of new C-band GMF construction applicable to inland waters (in particular to Gorky Reservoir) is reported in Section 5. The model is based on calculations using a two-scale model of the scattering of electromagnetic radiation on a rough sea surface with the Elfouhaily spectrum.

2. Instruments and Datasets

2.1. Satellite Data

In the present study, we analyze images from C-band SAR based on the board of the Sentinel-1 satellites (S1) downloaded from https://scihub.copernicus.eu/dhus/#/home accessed on 15 August 2023. The Interferometric Wide swath mode (IW) was selected as it provides a good resolution together with a large area (250 km in the range and 200 km in the azimuth direction) and regular ocean surface survey.

For the ocean wind speed retrieval, we used S1 IW GRDH (Ground Range Detected High-resolution) level-1 products for HH or VV polarization that has been calibrated and thermal noise removed. The geographic location of the SAR images was chosen in such a way that, first, they could be collocated in time and space with oceanographic buoys, and, second, only those areas were considered where images could be obtained regularly during the year—on average, once every two weeks.

For the case of inland waters wind speed retrieval, we considered an S1 IW GRDH level-1 product for VV and VH polarization for a fragment of the Gorky reservoir taken for a period from September 2017 to June 2020. 52 SAR images were selected for this period (excluding the freeze-up period), which contained the southern part of the reservoir, where the lighthouse was located. The average value of the NRCS for the VV polarized SAR images was calculated for a square area 1 × 1 km in size, the center of which was located 1.5 km from the coast (see Figure 1) and in close proximity (at the distance of about 1 km) from the lighthouse. The azimuth angle was constant at 256 degrees, and the incident angle for one group of images was 34.27 and for the other 41.75 degrees. The direction of the wind was taken according to the data of the meteorological station.

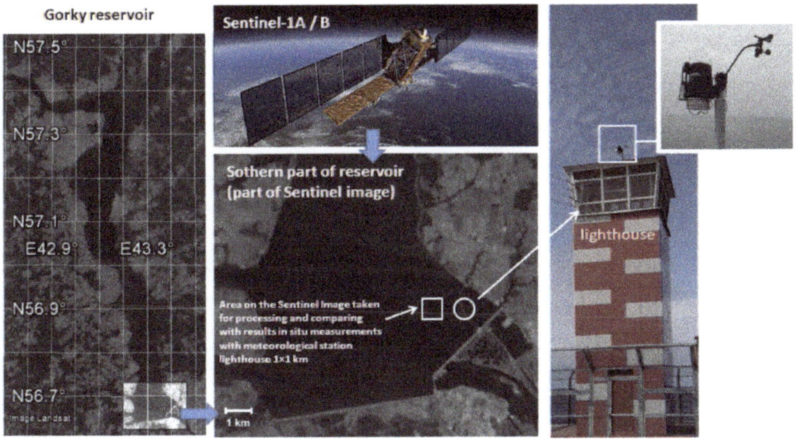

Figure 1. A map of the Gorky reservoir (see the **left** panel of the illustration); central panel—the southern (enlarged) part indicating the location of the lighthouse where the station is installed and the area where the NRCS is calculated, the lighthouse with the station installed on the roof is illustrated on the **right** panel.

For the main analysis, a VV polarization SAR image was used, while a cross-polarization VH image was utilized to define the boundaries of the reservoir (see Figure 2), as in contrast with VV, the NRCS for cross-polarized VH mode does not demonstrate dependence on the wind (wave) conditions.

Figure 2. SAR image in VH mode used for retrieval of the Gorky reservoir boundary.

2.2. CMOD5.N Algorithm

In the present study, we used the CMOD5.N algorithm to retrieve the wind speed from SAR images. According to this algorithm, the value of the Normalized Radar Cross Section (NRCS) of the sea surface σ_0 is represented as a function of neutral wind speed v, wind direction χ and incidence angle θ [12]:

$$\sigma_0 = B_0(c_0, v, \theta)(1 + B_1(c_1, v, \theta)\cos\varphi + B_2(c_2, v, \theta)\cos(2\varphi))^{1.6} \quad (1)$$

where $\varphi = \chi - \alpha$ is the relative direction between the wind direction χ and the radar azimuth look angle α [12]. The other parameters, B_i (i = 0, 1, 2) depend on the radar incidence angle, wind speed and coefficients c_i. The value of the B_0 parameter determines the order of the wind speed, while the value of the B_2 parameter reflects the asymmetry along the wind and crosswind, and B_1 allows refining the value of the wind speed for the wind direction in the range of 180 degrees. In [12] for the CMOD5.N model, 28 coefficients c_i were recalculated from the condition that this model should provide winds by 0.7 m/s more than for the calculations within CMOD5 for all wind speeds and incidence angles, as a result, the following values for c_i were used (see Table 1) [12]. In order to obtain the value of NRCS from SAR images using CMOD5.N the wind direction should be acquired from some external sources, for example, in situ measurements or reanalysis calculations may be used as input to the CMOD5.N algorithm.

Table 1. Coefficients c_i for CMOD5.N used for the calculation of wind speed value.

No	Value	No	Value
c_1	−0.6878	c_{15}	0.0066
c_2	−0.7957	c_{16}	0.3222
c_3	0.3380	c_{17}	0.0120
c_4	−0.1728	c_{18}	22.700
c_5	0.0000	c_{19}	2.0813
c_6	0.0040	c_{20}	3.0000
c_7	0.1103	c_{21}	8.3659
c_8	0.0159	c_{22}	−3.3428
c_9	6.7329	c_{23}	1.3236
c_{10}	2.7713	c_{24}	6.2437
c_{11}	−2.2885	c_{25}	2.3893
c_{12}	0.4971	c_{26}	0.3249
c_{13}	−0.7250	c_{27}	4.1590
c_{14}	0.0450	c_{28}	1.6930

2.3. Buoy Data

To verify the magnitudes of wind speed from SAR images using the CMOD5.N algorithm, the data on the wind speed direction were collected using NOAA NDBC oceanographic buoys. The choice of buoy location was associated with the possibility of collocation them with SAR data in order to subsequently synchronize the acquisition data on wind velocity and its direction with NRCS of SAR images. Due to the fact that regular (monthly) remote sensing data can only be obtained in certain geographic regions, the following oceanographic buoys were selected according to the location of these regions: Station 41,043—NE Puerto Rico 21.030 N 64.790 W; Station 44,137—East Scotia Slope 42.260 N 62.030 W; Station 51,003—Western Hawaii, 19.196 N 160.639 W; Station 46,184—North Nomad, 53.920 N 138.850 W; and Station 46,072—Central Aleutians, 51.672 N 172.088 W.

3. Verification of CMOD5.N Using Buoy Measurements for Ocean Conditions

For complex verification of the CMOD5.N algorithm, we formed a representative data array containing both SAR images of sea surface areas and collocated data on wind speed and wind direction with the maximum sampling rate in time, in the maximally wide range of geographic locations. Also, the array was supplemented with NCEP reanalysis data (from the website https://earth.nullschool.net, accessed on 4 July 2023) for wind speed as close as possible to the time the image was taken. A feature of the Sentinel 1 A/B for IW mode is that on a regular basis, the Earth's surface areas occupied by land are surveyed at least once every two weeks, and the images for the open ocean are made from time to time in different modes depending on the conditions. For example, data collection becomes more frequent during adverse weather events such as tropical hurricanes, typhoons, storms, etc. However, together with regular surveys of coastal areas, significant zones of the sea surface are also recorded. At the same time, we often chose observation points near small islands in order to minimize the influence of the coastal zone. On the other hand, an oceanographic buoy should be present in the selected area of regular surveys in order to perform the collocation of data. Based on these requirements, we selected 5 observation areas, see Table 2. It should be noted that these areas corresponded to different climatic regions, including subpolar, temperate, and tropical regions, which were located in the Atlantic as well as in the Pacific Ocean. The location of the selected areas is shown on the map (see Figure 3).

For each area, Sentinel 1A/B imagery was taken with a duty cycle of 2 weeks, during the time period throughout 2020, with the exception of area No. 4, where, due to insufficient data, the dataset was supplemented with images from 2019. The procedure for determining the NRCS in the area of the buoy was as follows: an area of size 2 × 2 km was taken, which necessarily captured the location of the buoy, and on the other hand was completely inside the image, and then the NRCS was averaged inside this area. The CMOD5.N algorithm was used to retrieve the velocity magnitude. It should be noted that due to the selection of different areas, the data were obtained at different values of the incidence angle and azimuth angle (see Table 2).

Table 2. Analyzed geographic areas and corresponding oceanographic buoys.

Region No	1	2	3	4	5
General characteristics	Pacific Ocean, Tropic of the North, Western Hawaii	Atlantic Ocean, temperate, West coast of Canada	Atlantic Ocean, Tropic of the North, Puerto Rico	Pacific Ocean, northern subpolar, Aleutian Islands	Pacific Ocean, north temperate, East coast of Canada
Station type and NDBC identification number, coordinates, anemometer installation height z_a	Station 51,003—Western Hawaii, 19.196 N 160.639 W z_a = 3.8 m	Station 44,137—East Scotia Slope, 42.260 N 62.030 W z_a = 5 m	Station 41,043—NE Puerto Rico, 21.026 N 64.793 W z_a = 4.1 m	Station 46,072—Central Aleutians, 51.666 N 172.114 W z_a = 4.1 m	Station 46,184—North Nomad 53.920 N 138.850 W z_a = 5 m
Number of Sentinel 1 A/B images in the selected region over a period of time	24 (2020)	24 (2020)	24 (2020)	39 (2019–2020)	22 (2020)
Incidence angle	43.38	37.26	38.32	31.14	35.72
Azimuth angle	282	77	282	285	286

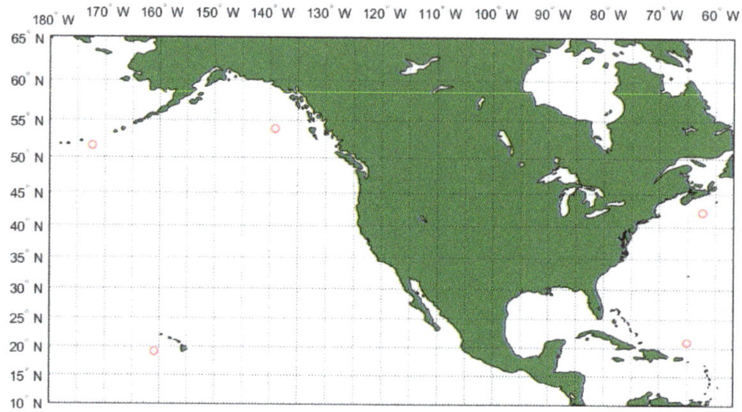

Figure 3. The location of 5 selected observation areas (red circles).

Direct comparison of wind speed measurements U_a by anemometers from buoys with the results of U_{10} calculations using CMOD5.N is not correct, because it should be taken into account that the anemometers from different buoys are installed at different heights from the surface z_a, which, generally speaking, is different from 10 m. Therefore, U_a was recalculated from the height z_a (see Table 2) to a height of 10 m, according to the logarithmic law:

$$U_{10} = U_a \ln(z_a/z_0) / \ln(H_{10}/z_0) \quad (2)$$

where z_0 is the roughness length. Here we used a constant value of z_0 equal to 1.52×10^{-4} m, as it was proposed in [27]). Despite the fact that the dependency of roughness length on the wind speed for the selected wind range is strong (it has values from 10^{-5} up to 10^{-3} m (see [27] and the references in it)), the calculated wind speed magnitude vary slightly with the roughness length (less than 1%).

Figure 5a shows the results of comparing the wind speed U_{10} obtained from the measurements from the buoy anemometers with the results of calculations using the CMOD5.N algorithm. A linear fit gives a straight-line slope of 0.99.

It can be seen from Figure 5a that the results are in good agreement with each other, regardless of the survey area and the buoy where they were obtained. Thus, the universality of using the algorithm is demonstrated regardless of the conditions (incidence angle, azimuth angle, wind speed).

A comparison of the buoy data were made also with the calculations within a number of widely used geophysical model functions–CMOD7, GMF from [36] and CSARMOD2. It can be seen that for the selected values the discrepancies between the results calculated with CMOD5.N and with CMOD7 are negligible (see Figure 4a,b). As for the GMF from [36], it describes ocean conditions quite well (see Figure 4c,d), and slightly better results are obtained in the case of calculations for the Elfouhaily spectrum (see the details in [36]). At the same time, the CSARMOD2 showed itself to be somewhat worse (see Figure 4e); for its values of bias and RMSE are the largest. Since the CMOD5.N model for our dataset performed better than the others, we will analyze below the sensitivity of the calculations to the choice of the wind direction source.

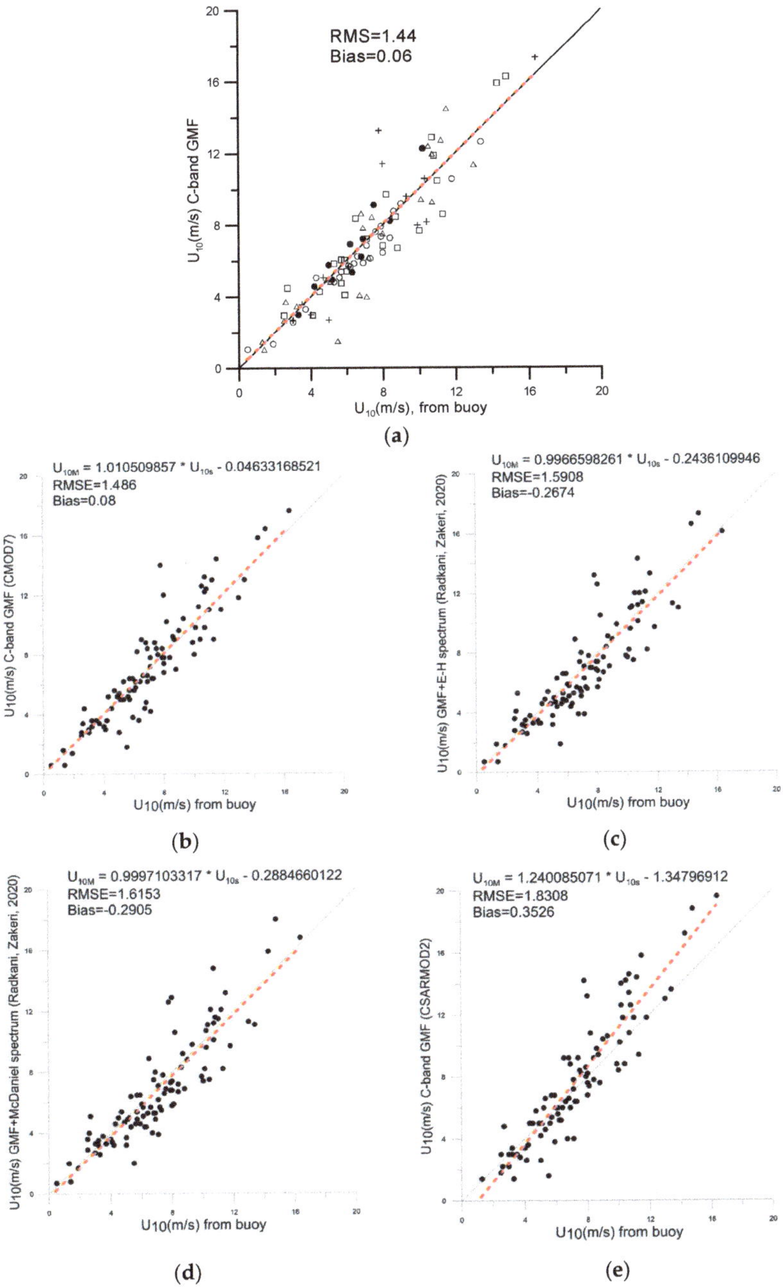

Figure 4. (**a**) Comparison of the results of measurements by anemometers from buoys and results of calculation by CMOD5.N. Symbols correspond to areas: black circles—No. 1, triangles—No. 2, open circles—No. 3, squares—No. 4, pluses—No. 5. Black line y = x, red dashed red line corresponds to linear best-fit y = 1.0099x (applied for the full dataset). The results of comparison of the buoy data with wind speeds from CMOD7 (**b**); GMF from [36] for Elfouhaily directional spectrum (**c**) McDaniel spectrum (**d**); CSARMOD2 (**e**).

In order to retrieve wind speed from the CMOD5.N algorithm the direction of wind speed should be obtained. Instead of data from buoys, which are few and extremely unevenly distributed over the ocean, it is more reliable to use data on the global meteorological situation. In this regard, a comparison was made for the results obtained by CMOD5.N, both using wind direction data from the buoy anemometer measurements and according to the data of NCEP CFSv2 data reanalysis (0.50). The comparison demonstrated in Figure 5 shows that the results of U_{10} calculations by CMOD5.N are weakly dependent on the source of data on the wind speed direction. The good agreement is likely due to the selected range of wind speeds, containing predominantly low wind speeds, which are characterized by a weak level of variability in the magnitude of the wind speed and its direction. This causes a slight difference between the results of modeling and measurements. Therefore, it is possible to use the reanalysis data available in the public domain without using formulas for recalculating wind speed.

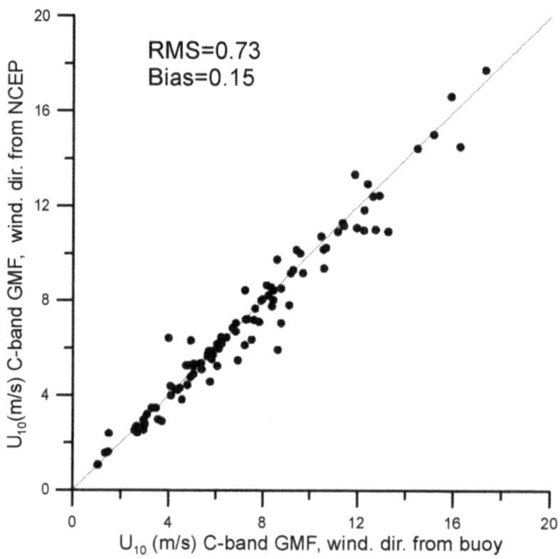

Figure 5. Comparison of U_{10} calculation from the CMOD5.N algorithm using wind direction data from buoy measurements and data provided by NCEP.

4. Verification of the CMOD5.N for Inland Water Conditions Using the Example of the Gorky Reservoir and the Comparison with the Calculations Using Other C-Band Models

To verify the CMOD5.N algorithm under the conditions of an inland water body, we used data from observations of the meteorological state from the station installed on the roof of a lighthouse (height 9.5 m from the water surface) in the southern part of the Gorky reservoir. The lighthouse itself is located at the end of a long narrow jetty (see Figure 1), which limits the outport in front of the locks, reducing the effects of airflow sheltering due to the topography of the coastline.

A comparison of the results (see Figure 6) demonstrates a satisfactory agreement.

It can be seen from Figure 6a that the bias and RMSE turned out to be smaller compared to the results obtained for the open ocean despite the data array being 3 times smaller. The slight inclination of the best linear fit reveals that moderate wind speeds in situ measurements slightly exceed SAR processing data. Obviously, it may be due to the effect of short-wave fetches.

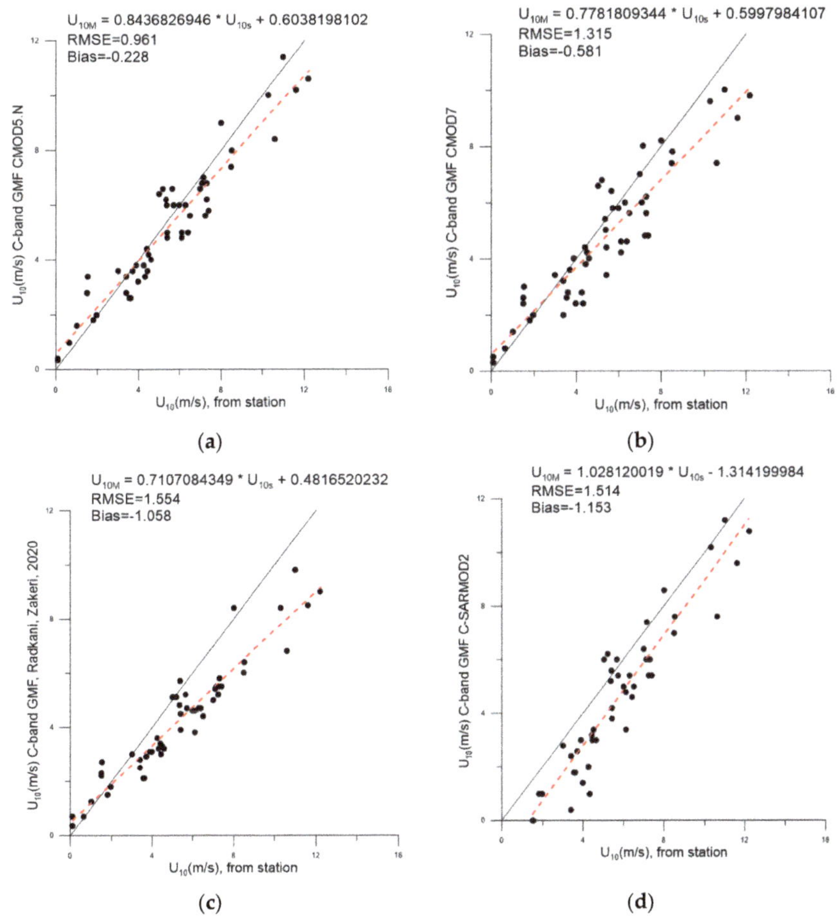

Figure 6. The results of comparison of U_{10} anemometer measurements from the meteorological station at the Gorky reservoir and the results of calculations using CMOD5.N (**a**); CMOD7 (**b**); GMF with the Elfouhaily directional spectrum (see [36]) (**c**); CSARMOD2 (**d**). Black line corresponds to y = x, red dashed line represents linear best fit.

At the same time, calculations using the CMOD7 (see Figure 6b) model give a somewhat less accurate result, since the value of RMSE, in this case, exceeds the similar value for CMOD5.N and the absolute value of bias turns out to be greater. Both models show a negative bias value, which indicates, on average, an underestimation of the calculated velocities compared to the measured ones. The GMF from [36] and CSARMOD2 models perform significantly worse (the latter demonstrates the largest absolute values of RMSE and bias) (see Figure 6c,d). For both models, bias is a negative value, while CSARMOD2 demonstrates an underestimation of speeds in the entire considered range, in contrast to the other three models. Thus, we can make a preliminary conclusion that the CMOD5.N model is best suited for calculations for the reservoir under consideration.

5. Towards a New Geophysical Model Function Based on the Data for Gorky Reservoir

Due to the fact that calculations according to the GMF proposed in [36] do not describe the measurements for the Gorky Reservoir accurately enough (see Figure 7c), we made an attempt to create a model that would be more adequate and could be used for inland water bodies alternatively to [36]. In the present study, we use a two-scale composite model (see [37,38]). In general, this model, similar to the Small-Slope Approximation (SSA2) model proposed in [39], describes electromagnetic scattering from a wavy water surface but its main feature is that it takes into account the tilting modulation due to long waves. In general, both models adequately describe scattering parameters for direct polarization, but the two-scale model is somewhat simpler for calculations, so it was chosen in the present study.

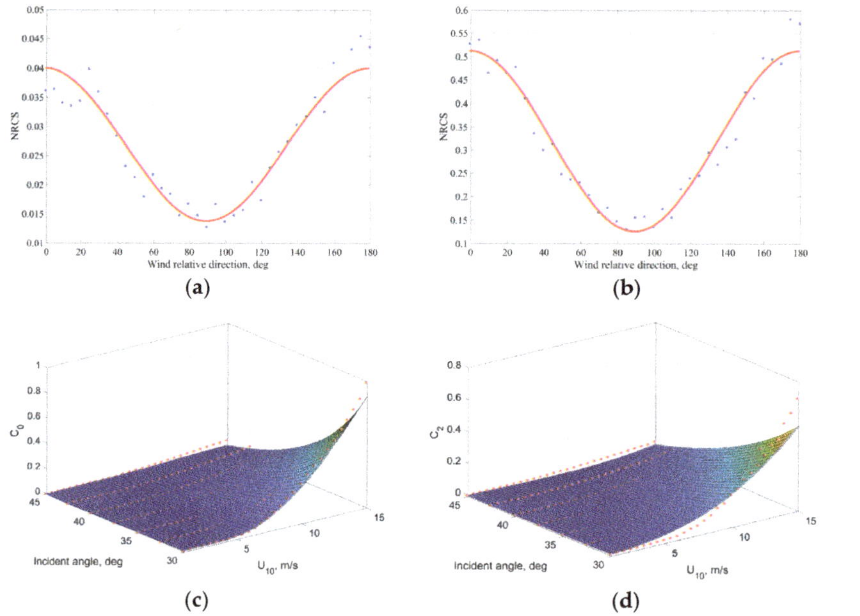

Figure 7. The dependency of modeled NRCS vs. the wind relative direction for wind speed 5 m/s (**a**) and 10 m/s (**b**), for both illustrations θ_i = 30° (points represent the modeled values, red curve corresponds to approximation (6)). The 3-D illustration of C_0 (**c**) and C_2 (**d**) coefficients depends on the incident angle and wind speed. Red points correspond to the modeled values and the surfaces represent to the approximation with the expressions (6)–(7).

We use the following expression for NRCS (see [37,38]):

$$\sigma_{VV} = \frac{\pi}{\tan^4\theta_i}\,|G_{VV}|^2(1 + g_{VV}\varsigma^2)B(k_b, \varphi) \qquad (3)$$

where θ_i is the incidence angle, $g_{VV} = \frac{\tan^4\theta_i}{2|G_{VV}|^2}\frac{\partial^2}{\partial\theta_i^2}\left(\frac{|G_{VV}|^2}{\tan^4\theta_i}\right)$ is the polarization coefficient which describes the tilting waves impact, $B(k_b, \varphi)$ is the directional curvature spectrum, $k_b = 2k_i \sin\theta_i$ is the Bragg wavenumber, k_r is the radar wavenumber, ς^2 is the mean square slope of tilting waves in the direction of incidence plane and geometrical coefficient G_{VV} has the following form:

$$G_{VV} = \frac{(\varepsilon_r - 1)(\varepsilon_r(1+\sin^2\theta_i) - \sin^2\theta_i)}{\left(\varepsilon_r \cos\theta_i + \sqrt{\varepsilon_r - \sin^2\theta_i}\right)^2} \quad (4)$$

where ε_r represents the relative permittivity which has the value of 81.5736 for the mean annual temperature equal to 16 °C [40].

The calculations are made for the frequency of 5.405 GHz and VV polarization corresponding to the Sentinel-1 SAR equipment. For the Gorky Reservoir, weak or moderate winds are the most typical—in the spring-summer period, the average wind speeds observed at the Reservoir are 3.4–5.2 m/s, and in the autumn-winter period 3.9–7.4 m/s. At the same time storm wind speeds (more than 15 m/s) are observed extremely rarely, winds of more than 28 m/s have never been recorded [41]. Thus, the considered wind speed range will include only weak or moderate winds (less than 15 m/s). We used the same as was reported in [36] ranges for incidence angles (30°–45°) and the wind relative direction (0°–180°).

In contrast to the study reported in [36] where the Hwang spectrum was used for the omnidirectional part, we used the Elfouhaily spectrum [42] for the entire range of wave numbers, since for waves in the meter range it has a k^{-3} asymptotics, which was observed in field data for the Gorky Reservoir [43]. We also used a Gaussian random process for the variations of power spectral density for the Bragg wavenumber. The calculations were performed with the step of 2.5° for the incidence angle, 5° for the wind relative direction and 0.5 m/s for wind velocity analogous to the study reported in [36].

To construct the new GMF we will use an empirical functional relation proposed for all CMOD models which contains the dependency of NRCS σ on the wind speed U_{10}, incidence angle θ_i and wind relative direction φ:

$$\sigma = C_0(U_{10}, \theta_i) + C_1(U_{10}, \theta_i)\cos(\varphi) + C_2(U_{10}, \theta_i)\cos(2\varphi) \quad (5)$$

We will assume that the C_1 is equal to zero, because we use the spectrum symmetrical from the point of view of upwind and downwind directions. To determine the values of C_0 and C_2 we use the same procedure and approximations as were used in [36]. As a result, we obtain the dependencies of C_0 and C_2 on the wind relative direction for each wind speed and incident angle (see Figure 7a,b).

In the next step, we approximated the obtained dependencies of C_0 and C_2 using the polynomial approximations for U_{10} and power functions for θ_i dependencies (see [36]):

$$C_0(U_{10}, \theta_i) = (c_{11}\theta_i{}^{\alpha} + c_{12})U_{10} + (c_{21}\theta_i{}^{\alpha} + c_{22})U_{10}^2 + (c_{31}\theta_i{}^{\alpha} + c_{32})U_{10}^3 + (c_{41}\theta_i{}^{\alpha} + c_{42})U_{10}^4 \quad (6)$$

$$C_2(U_{10}, \theta_i) = (c'_{11}\theta_i{}^{\beta} + c'_{12})U_{10} + (c'_{21}\theta_i{}^{\beta} + c'_{22})U_{10}^2 + (c'_{31}\theta_i{}^{\beta} + c'_{32})U_{10}^3 \quad (7)$$

The values of obtained coefficients c from expressions (6) and (7) are presented in Table 3.

The proposed model was validated using the dataset from the meteorological station located at the Gorky Reservoir and collocated NRCS data from Sentinel-1 SAR (see Figure 8).

Table 3. The coefficients used in (6), (7).

Coefficients from Equation (7)	Value	Coefficients from Equation (8)	Value
c_{11}	9.006×10^4	c_{11}	11
c_{12}	-0.004009	c_{12}	-0.004368
c_{21}	-4.102×10^4	c_{21}	2.892
c_{22}	0.001642	c_{22}	-0.001052
c_{31}	6432	c_{31}	0.1384
c_{32}	-0.0002178	c_{32}	8.037×10^{-7}
c_{41}	-192.3	β	-2.163
c_{42}	5.855×10^{-6}		
α	-4.47		

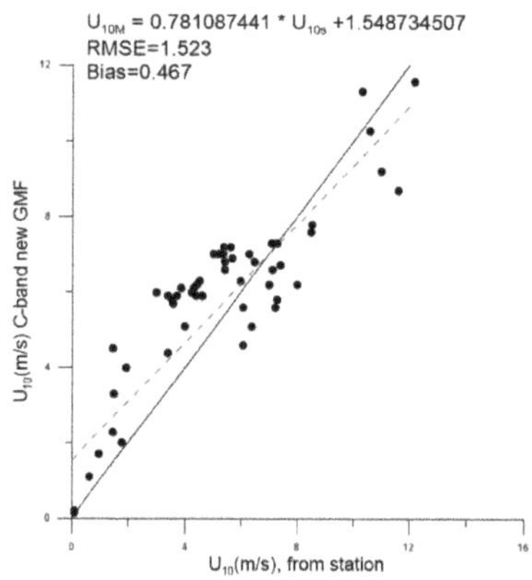

Figure 8. The results of comparison of U_{10} anemometer measurements from the meteorological station at the Gorky reservoir and the results of calculations using proposed C-band GMF. Black line corresponds to y = x, red dashed line represents linear best fit.

A comparison of the simulated velocity values with the measured ones demonstrates slightly lower RMSE and bias values than in the case of CSARMOD2 and the model reported in [36]. In addition, calculations within the proposed model demonstrate some overestimation of the calculated wind speed values, in contrast to these two models and similar to the CMOD models. However, the statistical characteristics of the data calculated using the proposed GMF turn out to be somewhat worse than in the case of CMOD5.N and CMOD7. Obviously, the model reported in [36] works worse for the case of the Gorky Reservoir, since the spectra that are used in this model are probably not entirely typical for this reservoir As measurements have shown, the Gorky Reservoir is characterized by a Phillips spectrum in the wavelength range of the order of a meter and significantly shorter (up to 12 cm) [43]. As for the long-wave part of the spectrum, it is more correct to use age-dependent wave spectra, since the reservoir is characterized by a large difference in longitudinal and transverse dimensions, and therefore the surface waves fetch can vary widely (from a few up to hundreds of kilometers). The problem of choosing the optimal spectrum to describe adequately the statistics of surface waves will be a subject of future studies.

6. Conclusions

A comprehensive verification of the C-band geophysical model functions for the open ocean as well as for the conditions of an inland water body (Gorky reservoir) was carried out on the basis of SAR images obtained from the Sentinel-1 mission. The main motivation that served as the impetus for this study is the need to clarify the possibility of applying widely used in current practice algorithms developed mainly for scatterometers for SAR conditions, as well as to study the influence of various sources on the results of calculations to obtain the magnitude of the wind speed direction. A special place in this case is occupied by the possibility of using the CMOD5.N algorithm for the conditions of inland waters.

A total of 133 SAR images for IW mode were considered in the period from 2019–2020 for open ocean conditions and 40 IW SAR images for the conditions of the Gorky reservoir. When applying the wind speed retrieval algorithms and determining the magnitude of the wind speed, it is required to obtain information about the wind direction from external sources. In our study, we used measurements of oceanographic buoys NOAA NDBC and NCEP reanalysis data for open ocean and data from a meteorological station for conditions of the Gorky Reservoir. One of the main tasks in the selection of satellite images was to expand the geography of the regions under consideration and to apply a number of widely used wind speed retrieval algorithms. 5 different geographical areas were considered within which it was possible to obtain regular images of the Sentinel-1 mission with a duty cycle of 14 days for which collocation with buoy data is possible. These areas corresponded to different climatic regions, including subpolar, temperate, and tropical regions, which were located in the Atlantic as well as in the Pacific Ocean.

A comparison of the buoy data was made with the wind speed retrievals using a number of widely used geophysical model functions–CMOD7, GMF from [36] and CSARMOD2. The discrepancies between the results calculated with CMOD5.N and with CMOD7 are negligible. As for the GMF from [36], it describes ocean conditions quite well, slightly better results are obtained in the case of calculations for the Elfouhaily spectrum. Thus, we can conclude that the model, originally intended to retrieve wind speed above enclosed seas, describes the results satisfactorily for the ocean as well. This is quite understandable since the model uses the Hwang spectrum and the Elfouhaily spreading function, which are quite typical for the ocean, and the calculations are carried out for the dielectric permittivity of salt water. As for the CSARMOD2, it showed itself to be somewhat worse than other GMFs, demonstrating larger values of bias and RMSE.

Comparison of wind speed calculation obtained from the CMOD5.N algorithm using wind direction data from buoys measurements together with the data provided by NCEP reanalysis simulation showed a good agreement demonstrating a value of RMSE = 0.73 and bias = 0.15. This result says, firstly, that regardless of the geographic location, the algorithm performs quite well, while the result does not depend significantly on the choice of the source of the wind speed angle value. As for the conditions of the inland water body for the considered Gorky reservoir, a comparison of the results of calculations carried out within the framework of CMOD5.N with the data obtained from the meteorological station showed, in general, a satisfactory agreement, at the same time, they demonstrated lower values of bias and RMSE than for the conditions of the open ocean, despite a significantly smaller amount of data. This effect may be concerned with the effect of short wave fetches typical for inland water bodies or with the impacts of water depth on wave characteristics. The data obtained from the meteorological station were also compared with the results of calculations carried out within the framework of CMOD7, GMF from [36] and CSARMOD2. It was shown that for GMF from [36] and CSARMOD2 the largest RMSE and biases were observed. At the same time, among the four models considered, the CMOD5.N showed itself best and demonstrated the smallest RMSE and bias (RMSE = 0.961, bias = −0.228).

Finally, we made an attempt to develop a new C-band geophysical model function, which is based on a two-scale model that describes the scattering of electromagnetic radiation on a rough sea surface. Initially, the Elfouhaily spectrum was chosen to describe surface waves since it takes into account the Phillips asymptotics, typical for the Gorky Reservoir, up to the wavelengths on the order of tens of centimeters (12 cm) and is widely used to describe short waves. It turned out that the proposed model fits the field data much better than the CSARMOD2 and the model reported in [36], but worse than the CMOD models. In general, we can conclude that there is a need to find the optimal spectrum (or maybe a composite one) to describe adequately the mean square slopes of tilting waves and to take into account the wave age. As the question regarding the choice of wind wave spectrum remains open it may serve as a basis for future research.

Author Contributions: Conceptualization, D.S.; methodology, D.S.; software, N.R. and E.P.; validation, D.S. and N.R.; formal analysis, D.S.; investigation, D.S., O.E., D.G., N.R. and E.P.; data curation, D.S., O.E., N.R. and E.P.; writing—original draft preparation, D.S. and O.E.; writing—review and editing, D.S. and O.E.; visualization, D.S. and N.R.; supervision, D.S.; funding acquisition, D.S., O.E., D.G., N.R. and E.P. All authors have read and agreed to the published version of the manuscript.

Funding: This work was carried out under the financial support of the Russian Science Foundation Project No 21-17-00214.

Data Availability Statement: The data presented in this study are available on request from the corresponding author.

Conflicts of Interest: The authors declare that they have no conflict of interest.

References

1. Offiler, D. The Calibration of ERS-1 Satellite Scatterometer Winds. *J. Atmos. Ocean. Technol.* **1994**, *11*, 1002–1017. [CrossRef]
2. Verhoef, A.; Portabella, M.; Stoffelen, A. High-resolution ASCAT scatterometer winds near the coast. *IEEE Trans. Geosci. Remote Sens.* **2012**, *50*, 2481–2487. [CrossRef]
3. Vachon, P.W.; Dobson, F.W. Validation of wind vector retrieval from ERS-1 SAR images over the ocean. *Glob. Atmos. Ocean Syst.* **1996**, *5*, 177–187.
4. Wackerman, C.; Rufenach, C.L.; Shuchman, R.A.; Johannessen, J.A.; Davidson, K.L. Wind vector retrieval using ERS-1 synthetic aperture radar imagery. *IEEE Trans. Geosci. Remote Sens.* **1996**, *34*, 1343–1352. [CrossRef]
5. Lin, H.; Xu, Q.; Zheng, Q. An Overview on SAR Measurements of Sea Surface Wind. *Prog. Nat. Sci.* **2008**, *18*, 913–919. [CrossRef]
6. Liu, W.T.; Xie, X. Sea surface wind/stress vector. In *Encyclopedia of Remote Sensing*; Springer: New York, NY, USA, 2014; pp. 759–767. [CrossRef]
7. Liu, W.T.; Tang, W. Relating wind and stress under tropical cyclones with scatterometer. *J. Atm. Ocean Tech.* **2016**, *33*, 1151–1158. [CrossRef]
8. Monin, A.S.; Obukhov, A.M. Basic laws of turbulent mixing in the surface layer of the atmosphere. *Tr. Akad. Nauk. SSSR Geophiz. Inst.* **1954**, *24*, 164–187.
9. Stoffelen, A.; Anderson, D. Scatterometer data interpretation: Estimation and validation of the transfer function CMOD-4. *J. Geophys. Res.* **1997**, *102*, 5767–5780. [CrossRef]
10. Donnelly, W.J.; Carswell, J.R.; McIntosh, R.E.; Chang, P.S.; Wilkerson, J.; Marks, F.; Black, P.G. Revised Ocean Backscatter Models at C and Ku-band under High-wind Conditions. *J. Geophys. Res.* **1999**, *104*, 11485–11497. [CrossRef]
11. Hersbach, H.; Stoffelen, A.; de Haan, S. An improved C-band scatterometer ocean geophysical model function: CMOD5. *J. Geophys. Res.* **2007**, *112*, C03006. [CrossRef]
12. Hersbach, H. CMOD5.N: A C-Band Geophysical Model Function for Equivalent Neutral Wind. Available online: http://www.ecmwf.int/sites/default/files/elibrary/2008/9873-cmod5n-c-band-geophysical-model-function-equivalent-neutral-wind.pdf (accessed on 17 August 2023).
13. Quilfen, Y.; Chapron, B.; Elfouhaily, T.; Katsaros, K.; Tournadre, J. Observation of tropical cyclones by high-resolution scatterometry. *J. Geophys. Res.* **1998**, *103*, 7767–7786. [CrossRef]
14. Stoffelen, A.; Verspeek, J.A.; Vogelzang, J.; Verhoef, A. The CMOD7 Geophysical Model Function for ASCAT and ERS Wind Retrievals. *IEEE J. Sel. Top. Appl. Earth Obs. Remote Sens.* **2017**, *10*, 2123–2134. [CrossRef]
15. Stoeffelen, A. Toward the true near-surface wind speed: Error modeling and calibration using triple collocation. *J. Geophys. Res.* **1998**, *103*, 7755–7766. [CrossRef]
16. Verspeek, J.; Stoffelen, A.; Verhoef, A.; Portabella, M. Improved ASCAT wind retrieval using NWP ocean calibration. *IEEE Trans. Geosci. Remote Sens.* **2012**, *50*, 2488–2494. [CrossRef]
17. Yun, R.; Stofflen, A.; Verspeek, J.; Verhoef, A. NWP ocean calibration of Ku-band scatterometers. In Proceedings of the IEEE International Geoscience and Remote Sensing Symposium, Munich, Germany, 22–27 July 2012. [CrossRef]
18. Li, Z.; Stoffelen, A.; Verhoef, A.; Verspeek, J. Numerical Weather Prediction Ocean Calibration for the Chinese-French Oceanography Satellite Wind Scatterometer and Wind Retrieval Evaluation. *Earth Space Sci.* **2021**, *8*, e2020EA001606. [CrossRef]
19. Li, Z.; Verhoef, A.; Stoffelen, A.; Shang, J.; Dou, F. First Results from the WindRAD Scatterometer on Board FY-3E: Data Analysis, Calibration and Wind Retrieval Evaluation. *Remote Sens.* **2023**, *15*, 2087. [CrossRef]
20. Lehner, S.; Horstmann, J.; Koch, W.; Rosenthal, W. Mesoscale wind measurements using recalibrated ERS SAR images. *J. Geophys. Res.* **1998**, *103*, 7847–7856. [CrossRef]
21. Lehner, S.; Schulz-Stellenfleth, J.; Schattler, B.; Breit, H.; Horstmann, J. Wind and wave measurements using complex ERS SAR wave mode data. *IEEE Trans. Geosci. Remote Sens.* **2000**, *38*, 2246–2257. [CrossRef]

22. Horstmann, J.; Schiller, H.; Schulz-Stellenfleth, J.; Lehner, S. Global wind speed retrieval from SAR. *IEEE Trans. Geosci. Remote Sens.* **2003**, *41*, 2277–2286. [CrossRef]
23. Horstmann, J.; Koch, W. Measurement of ocean surface winds using synthetic aperture radars. *IEEE J. Ocean. Eng.* **2005**, *30*, 508–515. [CrossRef]
24. Vachon, P.W.; Dobson, F.W. Wind retrieval from radarsat sar images: Selection of a suitable C-band HH polarization wind retrieval model. *Can. J. Rem. Sens.* **2000**, *26*, 306–313. [CrossRef]
25. Monaldo, F.; Jackson, C.; Li, X.; Pichel, W.G. Preliminary evaluation of Sentinel-1A wind speed retrievals. *IEEE J. Sel. Top. Appl. Earth Obs. Remote Sens.* **2016**, *9*, 2638–2642. [CrossRef]
26. Wang, H.; Yang, J.; Mouche, A.; Shao, W.; Zhu, J.; Ren, L.; Xie, C. GF-3 SAR ocean wind retrieval: The first view and preliminary assessment. *Remote Sens.* **2017**, *9*, 694. [CrossRef]
27. Wei, S.; Yang, S.; Xu, D. On accuracy of SAR wind speed retrieval in coastal area. *Appl. Ocean Res.* **2020**, *95*, 102012. [CrossRef]
28. Yang, X.; Li, X.; Pichel, W.G.; Li, Z. Comparison of ocean-surface winds retrieved from QuikSCAT scatterometer and RADARSAT-1 SAR in offshore waters of the US west coast. *IEEE Geosci. Remote Sens. Lett.* **2011**, *8*, 163–167. [CrossRef]
29. Horstmann, J.; Koch, W.; Lehner, S.; Tonboe, R. Ocean winds from RADARSAT-1 ScanSAR. *Can. J. Remote Sens.* **2002**, *28*, 524–533. [CrossRef]
30. Zhang, B.; Perrie, W.; He, Y. Wind speed retrieval from RADARSAT2 quad-polarization images using a new polarization ratio model. *J. Geophys. Res.* **2011**, *116*, C08008. [CrossRef]
31. Takeyama, Y.; Ohsawa, T.; Kozai, K.; Hasager, C.B.; Badger, M. Comparison of Geophysical Model Functions for SAR Wind Speed Retrieval in Japanese Coastal Waters. *Remote Sens.* **2013**, *5*, 1956–1973. [CrossRef]
32. Mouche, A.; Chapron, B. Global C-Band Envisat, RADARSAT-2 and Sentinel-1 SAR Measurements in Copolarization and Cross-polarization. *J. Geophys. Res.* **2015**, *120*, 7195–7207. [CrossRef]
33. Lu, Y.; Zhang, B.; Perrie, W.; Mouche, A.A.; Li, X.; Wang, H. A C-Band Geophysical Model Function for Determining Coastal Wind Speed Using Synthetic Aperture Radar. *IEEE J. Sel. Top. Appl. Earth Obs. Remote Sens.* **2018**, *11*, 2417–2428. [CrossRef]
34. Mallard, M.; Nolte, C.; Spero, T.; Bullock, R.; Alapaty, K.; Herwehe, J.; Gula, J.; Bowden, J. Technical challenges and solutions in representing lakes when using WRF in downscaling applications. *Geosci. Model Dev.* **2015**, *8*, 1085–1096. [CrossRef]
35. Katona, T.; Bartsch, A. Estimation of wind speed over lakes in Central Europe using spaceborne C-band SAR. *Eur. J. Remote Sens.* **2018**, *51*, 921–931. [CrossRef]
36. Radkani, N.; Zakeri, B.G. Southern Caspian Sea wind speed retrieval from C-band Sentinel-1A SAR images. *Int. J. Remote Sens.* **2020**, *41*, 3511–3534. [CrossRef]
37. Kudryavtsev, V.; Hauser, D.; Caudal, G.; Chapron, B. A semiempirical model of the normalized radar cross-section of the sea surface 1. Background model. *J. Geophys. Res.* **2003**, *108*, FET 2-1–FET 2-24. [CrossRef]
38. Zheng, H.; Zhang, J.; Khenchaf, A.; Li, X.-M. Study on Non-Bragg Microwave Backscattering from Sea Surface Covered with and without Oil Film at Moderate Incidence Angles. *Remote Sens.* **2021**, *13*, 2443. [CrossRef]
39. Voronovich, A.; Zavorotny, V. Theoretical model for scattering of radar signals in K u-and C-bands from a rough sea surface with breaking waves. *Waves Random Media* **2001**, *11*, 247–269. [CrossRef]
40. Sadovsky, I.N.; Kuzmin, A.V.; Sharkov, E.A.; Sazonov, D.S.; Pashinov, E.V.; Asheko, A.A.; Batulin, S.A. *The Analysis of Dielectric Permittivity's Models of Water Environment, Used in Tasks of Remote Sensing of Water Areas;* Space Research Institute of RAS: Moscow, Russia, 2002; Volume 2172, p. 60.
41. Girin, S.N. Analysis of the validity of restrictions on the operation of passenger vessels in the Gorky reservoir. *Russ. J. Water Transp.* **2022**, *72*, 167–179. [CrossRef]
42. Elfouhaily, T.; Chapron, B.; Katsaros, K.; Vandemark, D. A unified directional spectrum for long and short wind-driven waves. *J. Geophys. Res. Oceans.* **1997**, *102*, 15781–15796. [CrossRef]
43. Baydakov, G.A.; Kandaurov, A.A.; Kuznetsova, A.M.; Sergeev, D.A.; Troitskaya, Y.I. Field studies of features of wind waves at short fetches. *Izv. Ross. Akad. Nauk. Seriya Fiz.* **2018**, *82*, 1569–1572. [CrossRef]

geosciences

MDPI

Article

Evaluation of Remote Sensing Products for Wetland Mapping in the Irtysh River Basin

Kaiyue Luo [1,2,3], Alim Samat [2,3,4], Jilili Abuduwaili [2,3,4,*] and Wenbo Li [2,3,4]

1 College of Geography and Remote Sensing Science, Xinjiang University, Urumqi 830046, China; 107552203698@stu.xju.edu.cn
2 State Key Laboratory of Desert and Oasis Ecology, Key Laboratory of Ecological Safety and Sustainable Development in Arid Lands, Xinjiang Institute of Ecology and Geography, Chinese Academy of Sciences, Urumqi 830011, China; alim_smt@ms.xjb.ac.cn (A.S.); liwenbo212@mails.ucas.ac.cn (W.L.)
3 Research Center for Ecology and Environment of Central Asia, Chinese Academy of Sciences, Urumqi 830011, China
4 University of Chinese Academy of Sciences, Beijing 100049, China
* Correspondence: jilil@ms.xjb.ac.cn

Abstract: As a transboundary river with rich and unique wetland types, the Irtysh River faces various challenges and threats from human activities and climate change, which affect area, type, and function of wetland. To accurately obtain information on the spatial and temporal distribution of wetlands in this basin, this study compares and evaluates the consistency and accuracy of a total of eleven remote sensing (RS) based land use/land cover (LULC), and wetland products. The information extraction effect of each RS product was examined through methods such as wetland area and type description, thematic map comparison, and similarity coefficient and Kappa coefficient calculations, which can reflect the wetland distribution characteristics and differences among the RS products in the Irtysh River Basin. The results show that although there is a consensus among the products in the major wetland distribution areas, there are still obvious deviations in detail depiction due to differences in factors such as data sources and methods. The products of Global 30 m Wetland Fine Classification Data (GWL_FCS30) and Global 30 m Land Cover Data (GLC_FCS30-2020) released by the Institute of Space and Astronautical Information Innovation (ISAI) of the Chinese Academy of Sciences (CAS) have a clear advantage in extracting spatial morphology features of wetlands due to the use of multi-source data, while the Esri Global 10 m Land Cover Data (ESRI_Global-LULC_10m) and products such as the global 10 m land cover data (FROM_GLC10_2017) from Tsinghua University have higher classification consistency. Moreover, data resolution, classification scheme design, and validation methods are key factors affecting the quality of wetland information extraction in the Irtysh River Basin. In practical terms, the findings of this study hold significant implications for informed decision-making in wetland conservation and management within the Irtysh River Basin. By advancing wetland monitoring technologies and addressing critical considerations in information extraction, this research effectively bridges the gap between remote sensing technology and practical applications, offering valuable insights for regional wetland protection efforts.

Keywords: wetland; remote sensing; spatial and temporal distribution; consistency analysis; Irtysh River Basin

1. Introduction

Wetlands are crucial ecosystems providing sustenance, water resources, and ecological services. They face imminent threats from human activities and climate change, making them the fastest shrinking ecosystems globally [1–6]. According to the findings by Nick C. Davidson, as documented in a comprehensive study on global wetland area changes, the long-term loss of natural wetlands is estimated to be within the range of 54–57% [7]. This meta-analysis, encompassing an examination of 189 reports detailing wetland alterations, further discerns a notably accelerated rate of wetland loss during the 20th and early 21st centuries, accounting for a discernible loss of 64–71% of wetlands since 1900 AD. These outcomes underscore significant challenges prevalent across diverse geographic regions, encompassing urban, coastal, inland, and plateau areas. These challenges materialize as observable reductions in wetland area, functional decline, habitat loss, biodiversity degradation, and heightened vulnerability of ecosystems [8–11]. Notably, wetland conservation and management exhibit pronounced disparities between developed and developing countries, influenced by policy, regulatory, and technical variations [12,13]. Additionally, diverse climate zones (e.g., temperate, tropical, boreal, etc.) contribute to significant differences in wetland response and adaptive capacity to climate change [14–16]. Wetlands are dynamic and complex ecosystems that vary significantly across different regions and climates. To effectively assess and monitor their status and trends, it is essential to obtain timely and accurate information on their spatial and temporal characteristics. RS technology has emerged as a powerful tool for this purpose, as it can provide large-scale and high-resolution data on wetland distribution and types. However, existing wetland RS products, predominantly global or continental scale, struggle to capture local variations and dynamics with high spatial and temporal resolutions. To address this limitation, accurate analysis of wetland characteristics in specific regions necessitates the application of RS methods for spatial and temporal distribution mapping.

In recent years, RS for wetland research has become a focal point, with numerous scholars and institutions worldwide conducting studies across various regions and scales [17–19]. These investigations, to varying degrees, reflect the diversity and complexity of wetlands. Notable examples include Julie Betbeder et al., who meticulously mapped the land cover of forested wetlands in the expansive "central depression" of Central Africa, constituting 32% of the region's total land area. Their study unraveled four distinct forested wetland types, shedding light on the intricate interplay between wetland types, flood extent, and phenology [20]. Turning attention to the globally significant Ramsar wetland along the coast of Turkey, Kuleli et al., conducted a comprehensive examination of shoreline change rates, revealing significant retreat and erosion in specific areas. Their findings underscore the need for robust conservation and management strategies for this internationally important wetland [21]. Following international wetland studies such as the coastal wetlands of Turkey, Wang et al., shifted their focus to the arid regions of China to investigate changes in wetland area in the Heihe River Basin. They found a subtle pattern of a decrease in wetland area from 2000 to 2014, followed by an increase. Nonetheless, the wetlands are still in a degraded state, which is the result of complex interactions between climate change and human activities [22]. Exploring the transformation of natural and artificial wetlands at the Yangtze River estuary, a vital hub for biodiversity and ecosystem services, Chen et al., [17] detected substantial changes. Over the period from 1960 to 2015, they noted a remarkable reduction of natural wetland area of 574.3 km^2, counterbalanced by a corresponding expansion of artificial wetland area by 553.6 km^2. The study highlighted the primary drivers behind this shift, primarily attributable to reclamation activities and sea level rise [23]. Collectively, these studies underscore the global importance of wetlands and demonstrate the utility of RS techniques in unraveling their dynamics and complexities.

The Irtysh River, a vital transboundary river integral to the Belt and Road Initiative, spans diverse countries and regions, significantly influencing both regional and global development [24,25]. Given its complex and variable climate, the Irtysh River Basin plays a crucial role in shaping hydrological cycles, ecosystem functions, and land use patterns [26]. The basin hosts various wetland types that provide essential ecological services, including food, water resources, and biodiversity. Meanwhile, these wetlands contribute to hydrological services such as water regulation, flood mitigation, and soil erosion control [27]. Despite their pivotal roles, the wetlands in this basin face an array of threats, including agricultural expansion, industrialization, urbanization, droughts, water resource development, and glacier melting [28]. The scarcity of RS studies on these wetlands underscores the pressing need for comprehensive RS research in the Irtysh River Basin.

Technological advancements of RS have yielded a plethora of global land cover and wetland products, each with varying spatial resolutions ranging from 30 m to 1° [29–31]. However, the existence of disparities in classification criteria and accuracy among these products necessitates a meticulous evaluation of their consistency. Focusing on the Irtysh River Basin, our study leverages eleven RS-based LULC and wetland products, each designed to reflect the spatiotemporal distribution characteristics of wetlands. The primary goal is to meticulously compare and assess the relative consistency and accuracy, focusing on relative metrics rather than absolute accuracy, of these products in describing wetlands. In doing so, we aim to explore their potential and limitations in the context of wetland conservation and management. To accommodate differences in classification systems, we implemented conversions and correspondences to ensure an accurate analysis of wetland characteristics.

This research not only provides a valuable reference for mapping wetland distribution in the Irtysh River Basin, but also aims to scrutinize the strengths and weaknesses of each RS product. Through a comprehensive assessment, we seek to understand their potential and limitations in wetland conservation and management from both accuracy and spatiotemporal perspectives. Additionally, we delve into the sources of errors and factors influencing RS product quality, proposing improvement measures and outlining future research directions. These findings not only offer crucial scientific support but also establish a solid decision-making basis for wetland conservation and management, aligning with the pursuit of sustainable development goals. It is worth mentioning that the list of all acronyms used in this paper is shown in Table 1.

Table 1. List of acronyms.

Acronyms	Definition
RS	Remote sensing
LULC	Land use/land cover
ISAI	Institute of Space and Astronautical Information Innovation
CAS	Chinese Academy of Sciences
GWL_FCS30	Global 30 m Wetland Fine Classification Data released by the ISAI of the CAS
GLC_FCS30-2020	Global 30 m Land Cover Data released by the ISAI of the CAS
ESRI_Global-LULC_10m	Esri Global 10 m Land Cover Data
FROM_GLC10_2017	The global 10 m land cover data from Tsinghua University
LCCS	Land Cover Classification System
UN-FAO	United Nations Food and Agriculture Organization
ESA	European Space Agency
JRC	European Union Joint Research Centre
OA	Overall Accuracy
EA	Expected Accuracy
GEE	Google Earth Engine

2. Materials and Methods

2.1. Study Area

Situated in the Northern Hemisphere and spanning approximately 165×104 km^2, the Irtysh River Basin (Figure 1) accommodates a population of approximately 15 million people, distributed across Mongolia (<1%), China (2.9%), Kazakhstan (53.1%), and Russia (44%) [32]. Originating from the southwest slope of the Altai Mountains in China, the Irtysh River is a transboundary watercourse that traverses Kazakhstan and Russia before discharging into the Arctic Ocean, covering a distance of 4248 km. The basin's geographical features significantly influence its wetlands and ecosystems, particularly in the upper reaches located in Xinjiang, China. The headwater part of the basin is characterized by high mountainous areas with permafrost and glaciers, as well as forest and grassland ecosystems. The long-term precipitation trend in the upper reach is relatively stable, ranging from 300–320 mm/year, which is considered the primary runoff generating area for the middle reach located in Kazakhstan [33]. The annual precipitation in the Irtysh River Basin is unevenly distributed, ranging from 400–650 mm/year in the mountainous and foothill areas of the northeastern part of the basin to less than 200 mm/year in the low-lying areas of the intermountainous terrain. The climate in the Irtysh River Basin varies from mountainous areas in the southeast to flat plains in the northwest, with average temperatures ranging from −12 °C to −19 °C in January and 20–22 °C in July [32].

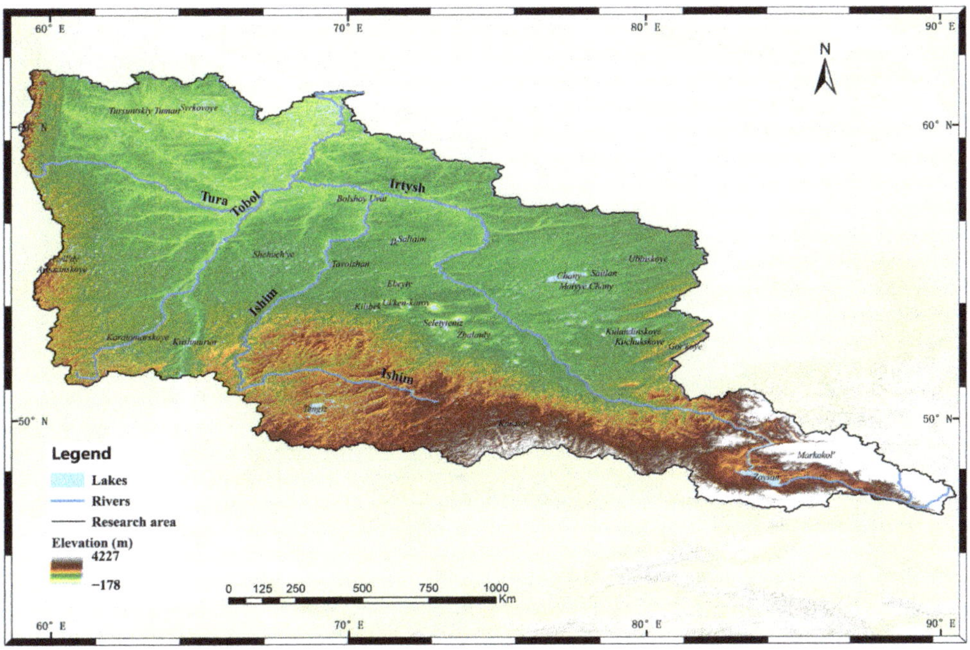

Figure 1. Map of the Irtysh River Basin. The elevations are represented relative to mean sea level, where negative values indicate heights below sea level. The legend's −178 m denotes areas with elevations lower than the reference sea level.

Amidst this diverse landscape, the Irtysh River Basin unveils its geological foundations, shaping its distinct features. Originating from the Altai mountain range, the upper reaches showcase a geological tapestry of crystalline Paleozoic complex meta-morphic rocks with magmatic intrusions. These formations not only shape the scenic terrain but serve as the primary source of both surface and groundwater for the entire catchment. As the river courses downstream, the terrain transforms into the flat expanse of the southern tip of the Siberian sedimentary basin. In the lower basin section, geological features shift, forming terraces closely hydraulically connected with the river. The Altai mountain range's Paleozoic volcanic and metamorphic rocks gradually yield to the immense sedimentary complex of the Siberian Basin, creating a smooth relief in the lower flat section—a stark contrast to the rugged upper reaches. Understanding these geological nuances is crucial as they delineate two distinctive sections within the Irtysh catchment. The upper mountainous part, characterized by crystalline rocks, serves as the wellspring of both surface and groundwater, shaping the hydrological dynamics of the entire basin. Conversely, the lower flat section, composed of sediments from the Siberian Basin, lacks significant tributaries, rendering it a relatively deficient environment from a water management perspective [34].

Wetlands within the Irtysh River Basin play crucial roles as integral components of the ecosystem, where they function as natural filters, providing habitats for biodiversity, sustaining ecological balance, and safeguarding water resources. Also, wetlands offer essential ecological services, including replenishing groundwater, regulating water quantity, mitigating floods, and controlling soil erosion. Their contributions extend to hydrological regulation and disaster prevention at the basin scale. Despite their ecological significance, these wetlands encounter formidable challenges and threats arising from human activities such as agriculture, industry, and urbanization, as well as the impacts of climate change. These pressures contribute to wetland degradation and destruction, leading to reductions in both wetland area and type, with adverse effects on wetland functions and biodiversity. The resultant degradation poses a significant threat to ecosystem health and may also precipitate economic and social challenges.

The imperative to comprehend wetland distribution and change within the Irtysh River Basin underscores the significance and value of this study. Evaluating the consistency and accuracy of wetland products is essential in this context. Precise acquisition of wetland information and analysis of wetland change trends provide insights into the current state and evolutionary processes of wetlands. Furthermore, this study establishes a scientific foundation for wetland conservation and management. Additionally, by promoting the application of RS technology in wetland monitoring and management, it contributes to enhancing the accuracy and reliability of wetland RS products. Ultimately, this research aligns with the pursuit of sustainable development goals.

2.2. Data Sources

2.2.1. GlobeLand30 V2020 Dataset

The GlobeLand30 V2020 dataset, developed by the National Geomatics Center of China, is a global land cover dataset with a 30 m resolution for the year 2020. This dataset incorporates multi-source RS images, including Landsat, HJ-1, and GF-1. Through image interpretation and classification algorithms, it encompasses 10 primary land cover types, including wetlands. The dataset boasts an overall accuracy of 85.72%, with a Kappa coefficient of 0.82 [35].

2.2.2. CGLS-LC100 Dataset

The CGLS-LC100 dataset, a global land cover change map with a 100 m resolution for 2019, is the product of the Copernicus Land Monitoring Service Global Team. Utilizing mainly PROBA-V satellite data and innovative algorithms, this dataset provides accurate and detailed information on land cover and use. Following the Land Cover Classification System (LCCS) of the United Nations Food and Agriculture Organization (UN-FAO), it encompasses 23 classification types, including wetland categories. The overall accuracy of the dataset is reported at 75% [36].

2.2.3. ESRI_Global-LULC_10m Dataset

Developed by Esri, the ESRI_Global-LULC_10m dataset is a global land use/land cover dataset with a 10 m resolution for 2020. Leveraging Sentinel-2 satellite data and deep learning models, the dataset comprises 10 primary classification types, including wetland categories. It offers detailed information on land use and cover for various applications such as conservation planning, food security, and hydrological models. The overall accuracy of the dataset is reported at 85% [37].

2.2.4. ESA WorldCover 10 m v100 Dataset

The ESA WorldCover 10 m v100 dataset, crafted by the European Space Agency (ESA), is a global land surface cover dataset with a 10 m resolution for 2020. Integrating Sentinel-1 and Sentinel-2 satellite data along with machine learning algorithms, the dataset includes

11 land surface cover categories, incorporating wetland categories. Consistent with the land surface cover classification system of the UN-FAO, it facilitates land use/land cover analysis on a global scale The overall accuracy of the dataset is reported at 75% [38].

2.2.5. GWL_FCS30 Dataset

The GWL_FCS30 dataset, developed by the Aerospace Information Innovation Institute of the Chinese Academy of Sciences, is a global wetland fine classification dataset with a 30 m resolution for 2020. Utilizing multi-source data and hierarchical adaptive models, this dataset covers nine wetland types in the Irtysh River Basin, providing detailed information on their spatial patterns and sub-class diversity. Notably, it excels in capturing the complexity and variability of wetlands The overall accuracy of the dataset is reported at 86% [39].

2.2.6. FROM_GLC10_2017 Dataset

The FROM_GLC10_2017 dataset, created by Tsinghua University, is a global land cover dataset with a 10-m resolution for 2017. Developed using Sentinel-2 satellite data and a random forest classifier, the dataset introduces the concept of stable classification, ensuring consistency in land cover types over time and space. With 10 land cover categories, it demonstrates an overall accuracy of 72.76%, offering increased spatial details and class separability compared to the 30 m Landsat-8 land cover map from the same team [40].

2.2.7. GLC_FCS30-2020 Dataset

Crafted by the Aerospace Information Innovation Institute of the Chinese Academy of Sciences, the GLC_FCS30-2020 dataset is a global land cover dataset with a 30 m resolution for 2020. Developed using Landsat time series images and an adaptive random forest model, this dataset encompasses 30 land cover types. Validated with 44,043 multi-source samples, it achieves an overall accuracy ranging from 68.7% to 82.5%, providing rich spatial and thematic details for regional or global analysis [41].

2.2.8. ESA CCI-LC Dataset

Developed by the European Space Agency, the ESA CCI-LC dataset is a global land cover dataset with a 300 m resolution for 2020. Based on multi-sensor satellite data, it offers consistent and comparable global land cover products from 1992 to 2020. Following the standardized classification system of the United Nations Food and Agriculture Organization, it encompasses 22 land cover types, with an average mapping accuracy of 71%. This dataset is valuable for analyzing regional-scale land use changes and supporting climate and ecological modeling for global change studies [42].

2.2.9. MCD12Q1 v061 Dataset

The MCD12Q1 dataset, developed by NASA, is a global land cover dataset with a 500 m resolution for 2021. Utilizing MODIS satellite data and employing data mining and machine learning algorithms, it distinguishes 17 land cover types. With an average accuracy of over 75%, the dataset is suitable for large-scale land use/cover change monitoring. Notably, it differentiates non-forest wetlands, such as swamps and river floodplains, from other land cover types, facilitating the analysis of spatiotemporal distribution and dynamics of wetlands and their interactions with other land uses [43].

2.2.10. GLC2000 Dataset

The GLC2000 dataset, developed by the European Union Joint Research Centre, is a global land cover dataset with a 1000 m resolution for the year 2000. Utilizing SPOT-4 satellite VEGETATION sensor images, the dataset adheres to the land cover classification system of the United Nations Food and Agriculture Organization, comprising 22 land cover types. Integrating other regional-scale land cover datasets, it applies standardized image classification methods to produce global land cover products, facilitating analysis of environmental issues such as land use change, biodiversity loss, and the carbon cycle [44].

2.2.11. AGLC-2015 Dataset

The AGLC-2015 dataset, developed by Sun Yat-sen University, is a land cover dataset with a 30 m resolution for 2015. Based on three global land cover products and data fusion methods, the dataset encompasses 10 types, including wetlands, with an overall accuracy of 76.10% and a Kappa coefficient of 0.72. Building upon AGLC-2015, the team also produced the global annual land cover dataset AGLC-2000-2015, providing data support for global change research and applications [45].

2.3. Data Pre-Processing

In order to facilitate a comprehensive comparison and analysis of wetland distribution across the eleven RS products and assess their accuracy, we executed the following pre-processing steps. Firstly, we extracted raster data corresponding to each RS product within the Irtysh River Basin, aligning with the vector boundary of the basin. Subsequently, we projected these raster datasets onto the WGS-84 coordinate system while preserving their original spatial resolution. Finally, thematic maps were generated for each RS product, retaining their specific wetland classification categories. Table 2 provides an overview of the key parameters associated with the eleven RS products.

The pre-processing steps were crucial to establish a standardized and consistent basis for comparison, enabling an accurate assessment of wetland distribution and RS product performance. Ensuring alignment to a common coordinate system and preserving the original spatial resolution were pivotal considerations for maintaining data integrity throughout subsequent analyses. The thematic mapping of raster data, with a specific focus on wetland classification categories, laid the foundation for a detailed and meaningful comparative evaluation. Table 2 provides essential parameters characterizing each RS product, offering readers a succinct reference to comprehend key attributes influencing subsequent analyses. This systematic approach to data pre-processing enhances the reliability and robustness of the comparative assessment, thereby contributing to the scientific rigor of our study.

2.4. Research Method

2.4.1. Descriptive Analysis of Wetland Areas and Types

(1) Wetland Category Selection

The objective of this study is to comprehensively assess the consistency and accuracy of eleven RS products in characterizing wetland features within the Irtysh River Basin, and the flowchart is shown in Figure 2. Additionally, we aim to discuss the potential and limitations of these RS products for effective wetland conservation and management. In pursuit of these goals, we identified wetland categories from the eleven RS products as the focal points of our analysis. These selected categories serve as representative indicators of the products' capabilities and characteristics in delineating wetland landscapes. Our primary research questions include: (1) How do different RS products exhibit variations in depicting the distribution and types of wetlands in the Irtysh River Basin? (2) What are the specific strengths and weaknesses associated with each RS product in describing wetlands? To address these questions, we initially curated a list of wetland categories for analysis, computed their respective areas within the study region, generated corresponding charts, and ultimately characterized the pattern of wetland distribution.

Table 2. Key information on the eleven RS-based LULC and wetland products.

Data Name	Primary Data Sources	Research and Development Organisations	Spatial Resolution/m	Vintages	Number of Classifications/pc	Data Sources
GlobeLand30 V2020	Landsat TM5/ETM+/OLI, HJ-1, GF-1	China National Centre for Basic Geographic Information	30	2020	10	http://www.globallandcover.com/ (accessed on 9 July 2023)
CGLS-LC100	External datasets such as PROBA-V, GeoWIKI, etc.	Copernicus Land Monitoring Service Global Team	100	2019	23	https://land.copernicus.eu/global/products/lc (accessed on 10 July 2023)
ESRI_Global-LULC_10m	Deep Learning Models and Training Datasets for ESA Sentinel-2, Impact Observatory	ESRI Corporation	10	2020	10	https://esri.maps.arcgis.com/ (accessed on 18 July 2023)
ESA WorldCover 10m v100	Sentinel-1/2 data, GeoWIKI samples	European Space Agency (ESA)	10	2020	11	https://viewer.esa-worldcover.org/worldcover/ (accessed on 10 July 2023)
GWL_FCS30	Landsat TM5/ETM+/OLI, Sentinel-1 SAR, ASTER GDEM	Institute of Space and Astronautical Information Innovation, Chinese Academy of Sciences	30	2020	9	https://zenodo.org/record/7340516 (accessed on 17 July 2023)
FROM_GLC10_2017	2015 30 m Landsat-8 data	Gong Peng Team, Tsinghua University	10	2017	10	http://data.starcloud.pcl.ac.cn/ (accessed on 16 August 2023)
GLC_FCS30-2020	Landsat time series imagery, CCI-LC products, MODIS NBAR data, 2014–2016	Institute of Space and Astronautical Information Innovation, Chinese Academy of Sciences	30	2020	30	https://data.casearth.cn/thematic/glc_fcs30 (accessed on 11 August 2023)
ESA CCI-LC	SPOT-VEGETATION images	European Space Agency (ESA)	300	2020	22	http://maps.elie.ucl.ac.be/CCI (accessed on 21 August 2023)
MCD12Q1 v061	MODIS data and other ground-based observations	NASA	500	2021	17	https://lpdaac.usgs.gov/products/mcd12q1v061/ (accessed on 17 August 2023)
GLC2000	SPOT-4 VEGETATION image	European Union Joint Research Centre (JRC)	1000	2000	28	https://forobs.jrc.ec.europa.eu/glc2000 (accessed on 21 August 2023)
AGLC-2015	Globeland30, FROM-GLC, and GLC-FCS30 products	Sun Yat-sen University	30	2015	10	https://doi.org/10.11834/jrs.20211261 (accessed on 18 August 2023)

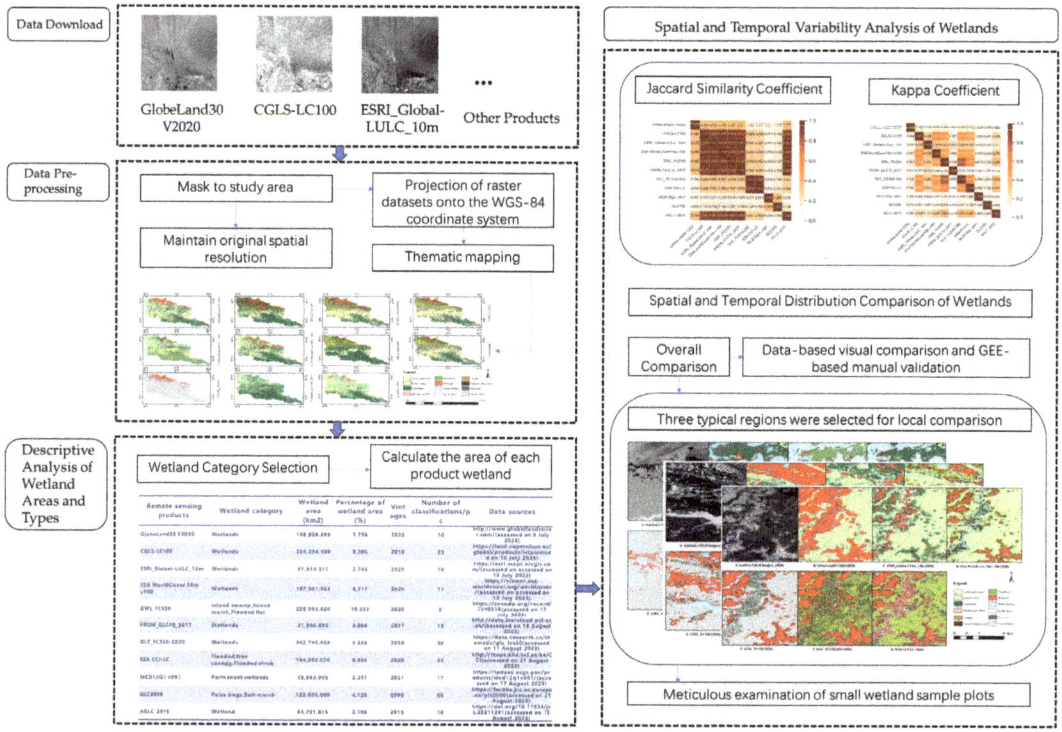

Figure 2. Research Flow Chart.

(2) Wetland Area Calculation

The area of each wetland category in the study area was determined using the following formula:

$$A_i = \frac{N_i \times {R_i}^2}{10^6} \quad (1)$$

where A_i is the wetland area of the i-th RS product (km^2), N_i is the number of pixels of the wetland category in the i-th product (pixel), and R_i is the spatial resolution of the i-th product (m).

The count of wetland pixels for each RS product was conducted using ArcGIS software. Subsequently, we multiplied this count by the square of the spatial resolution to obtain the area of each wetland category. The resulting data were organized into Table 3, and the percentage of each wetland category in the study area was calculated for further comparative analysis.

While this approach yields a reasonably precise estimation, it is not exempt from inherent limitations [46]. One notable constraint is the assumption that each pixel uniformly represents the same area. In reality, factors such as earth curvature and projection distortion can introduce variability in the area represented by each pixel, leading to potential errors in wetland area calculations. To remedy this limitation, future research could incorporate terrain correction models or pixel area normalization approaches to minimize variability introduced by earth curvature and projection distortion when calculating wetland areas.

Table 3. Area of each wetland category in the study area in eleven RS products.

Remote Sensing Product	Wetland Category	Wetland Area (km^2)	Percentage of Wetland Area (%)
GlobeLand30 V2020	Wetlands	139,809.699	7.756
CGLS-LC100	Wetlands	224,334.400	9.305
ESRI_Global-LULC_10m	Wetlands	81,814.511	2.783
ESA WorldCover 10 m v100	Wetlands	197,501.892	6.717
GWL_FCS30	Inland swamp, inland marsh, flooded flat	230,053.628	10.353
FROM_GLC10_2017	Wetlands	31,980.692	3.996
GLC_FCS30-2020	Wetlands	242,745.659	8.256
ESA CCI-LC	Flooded tree canopy, flooded shrub	164,252.070	5.935
MCD12Q1 v061	Permanent wetlands	43,993.000	2.207
GLC2000	Palsa bogs, salt-marsh	122,859.000	4.128
AGLC-2015	Wetland	64,281.615	2.186

(3) Wetland Distribution Description

Thematic maps of the eleven products within the study area were generated using ArcGIS software to facilitate the comparison and analysis of wetland distribution. Wetland-related categories were extracted from the raster data to focus specifically on wetland characteristics. Figure 3 illustrates the thematic maps of the eleven products in the Irtysh River Basin, where consistent colors denote identical land cover types, providing a visually accessible representation of their land cover distribution. These thematic maps serve as a means to observe the wetland classification outcomes of each RS product and discern the spatiotemporal distribution variations of wetlands among different RS products.

2.4.2. Spatial and Temporal Variability Analysis of Wetlands

To assess disparities in the spatial and temporal distribution of wetlands within the Irtysh River Basin across the eleven RS products, we employed two distinct methods.

(1) Jaccard Similarity Coefficient

The Jaccard similarity coefficient, a statistical metric quantifying the resemblance between two data sets, was utilized in this study [47]. It is defined as the proportion of the intersection over the union of two data sets:

$$J(A,B) = \frac{|A \cap B|}{|A \cup B|} \tag{2}$$

where A and B are two data sets, $|A \cap B|$ denotes the count of elements in their intersection, and $|A \cup B|$ indicates the count of elements in their union. The Jaccard coefficient ranges from 0 to 1, where 0 signifies no similarity whatsoever, and 1 denotes complete identity A higher value indicates a greater degree of similarity, whereas a lower value suggests less similarity.

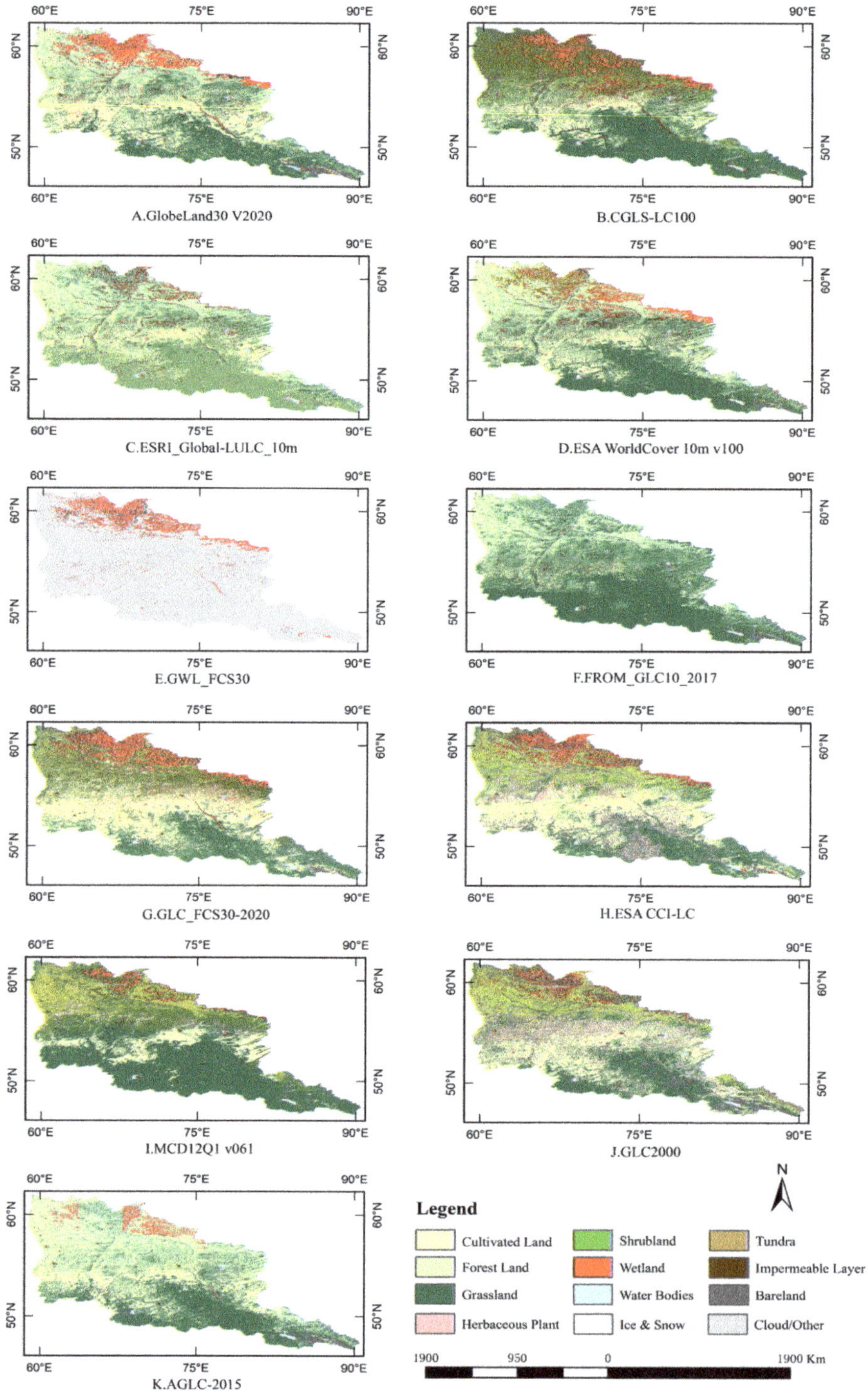

Figure 3. Thematic map of 11 categories of products in the Irtysh River Basin.

(2) Kappa Coefficient Analysis

The Kappa coefficient, a classification metric quantifying the agreement between two classification results [48], was employed in this study to compare the classification accuracy of each wetland category across the eleven RS products in the Irtysh River Basin. The methodology involves the following steps:

Firstly, the Overall Accuracy (*OA*), representing the ratio of matching pixels to the total number of pixels, was calculated from the confusion matrix using the formula:

$$OA = \frac{\sum_{i=1}^{n} X_{ii}}{\sum_{i=1}^{n}\sum_{j=1}^{n} X_{ij}} \tag{3}$$

where n denotes the number of wetland categories, and X_{ii} represents the element in row i and column i of the confusion matrix, indicating the number of matching pixels between category i in the reference data and category i in the comparison data. X_{ij} represents the element in row i and column j of the confusion matrix, indicating the number of matching pixels between class i in the reference data and class j in the comparison data. The order is determined by the rank order of the confusion matrix, not by the order of the reference and comparison data. The overall accuracy reflects the proportion of pixels of the same category in the two classification results, with a higher value indicating greater consistency.

Secondly, the Expected Accuracy (*EA*), representing the expected proportion of the number of pixels of the same category to the total number of pixels under the assumption of independence between the two categorization results, was computed based on the confusion matrix using the formula:

$$EA = \frac{\sum_{i=1}^{n}(\sum_{j=1}^{n} X_{ij})(\sum_{j=1}^{n} X_{ji})}{(\sum_{i=1}^{n}\sum_{j=1}^{n} X_{ij})^2} \tag{4}$$

where $\sum_{j=1}^{n} X_{ij}$ denotes the sum of elements in row i of the confusion matrix, i.e., the count of pixels of class i in the reference data. $\sum_{j=1}^{n} X_{ji}$ denotes the count of elements in column i of the confusion matrix, i.e., the count of pixels of class i in the comparison data. Expectation here refers to the theoretical value calculated according to the probabilities of random assignment, not the actual observed value.

Where $(\sum_{j=1}^{n} X_{ij})(\sum_{j=1}^{n} X_{ji})$ denotes the product of the counts of pixels in each class i in both reference and comparison data, i.e., the count of pixels that are identical in both according to the probability of random assignment. Summing such probabilities for all categories gives the expected accuracy. The expected accuracy reflects the proportion of pixels of the same category that two categorized results would have in a random situation, with smaller values indicating that they are less correlated.

Finally, the Kappa coefficient, the normalized difference between the overall accuracy and the expected accuracy, was derived from both values:

$$K = \frac{OA - EA}{1 - EA} \tag{5}$$

The Kappa coefficient ranges from -1 to 1, with -1 indicating complete disagreement, 0 indicating complete randomness, and 1 indicating complete agreement. It serves as a widely used measure of the level of agreement between two categorical outcomes after removing the random factor. Using this approach, Kappa coefficients were obtained between wetland categories in the eleven RS-based LULC and wetland products, quantifying their level of consistency in classifying wetlands in the Irtysh River Basin.

3. Results

3.1. Overview of Wetland Areas and Types

This section provides a comprehensive analysis and comparison of wetland area descriptions in the Irtysh River Basin across the eleven RS products. Table 3 presents data revealing substantial variations in the total wetland area, ranging from 64,281.615 km^2 to 230,053.628 km^2. These differences arise from distinct classification schemes, data sources, and production methods employed by the RS products. A visual representation of these disparities is illustrated in Figure 4a, depicted as a bar chart for enhanced clarity.

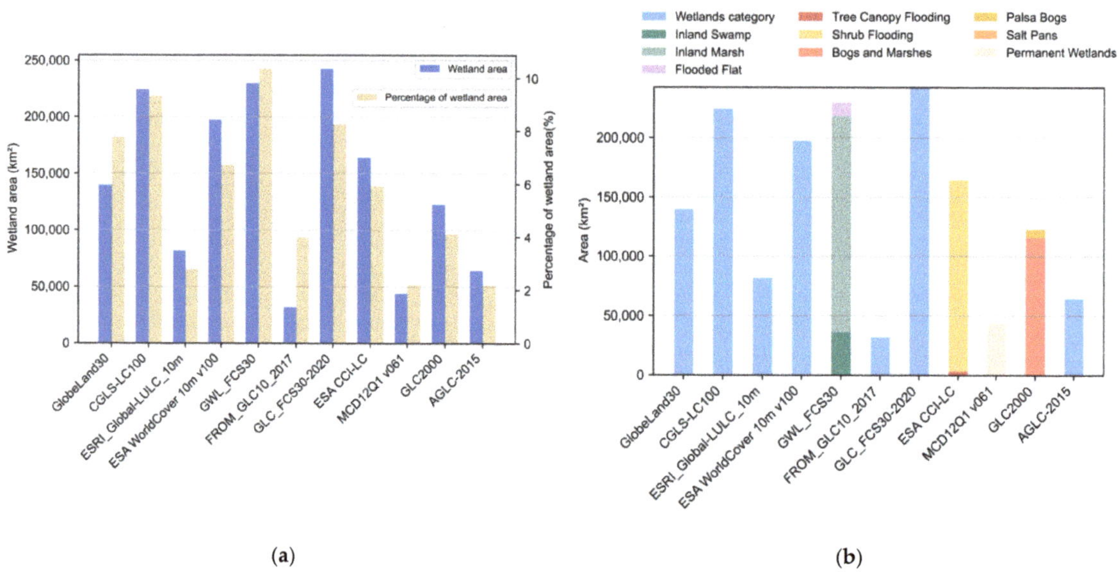

Figure 4. Overview of Wetland Areas and Types. (**a**) Bar chart illustrating the differences in total wetland area among eleven RS products in the Irtysh River Basin. (**b**) Comparative representation of the area and percentage distribution of various wetland types across the eleven RS products.

Furthermore, a detailed examination of different wetland types is presented in Figure 4b, showcasing the area and percentage covered by each wetland type among the eleven RS products. Notably, some products, such as GlobeLand30 V2020, CGLS-LC100, ESRI_Global-LULC_10m, ESA WorldCover 10 m v100, FROM_GLC10_2017, GLC_FCS30-2020, MCD12Q1 v061, and AGLC-2015, exclusively describe one wetland type, encompassing wetlands, or permanent wetlands. Additionally, datasets like GWL_FCS30 describe three wetland types—inland swamp, swamp marsh, and flooded flat. The ESA CCI-LC dataset encompasses two wetland types—tree canopy flooding and shrub flooding. The GLC2000 dataset classifies wetlands into three categories—bogs and marshes, palsa bogs, and salt pans. These discrepancies can be attributed to variations in wetland definitions, classification accuracy, and data quality employed by the diverse RS products.

The systematic presentation of these findings establishes a robust foundation for understanding the diversity in wetland areas and types, laying the groundwork for subsequent detailed analyses and discussions in the following sections.

3.2. *Assessment of Consistency and Accuracy of Wetland Products*

This section undertakes a comprehensive comparison and evaluation of the spatiotemporal distribution of wetlands in the Irtysh River Basin using eleven RS products. Quantitative indicators are employed to assess the performance of these RS products in wetland description and classification, aiming to identify similarities, differences, and potential areas for improvement.

To begin, we computed the Jaccard similarity coefficient matrix for wetland categories among the different RS products (Figure 5). Key observations from this matrix include:

- ESRI_Global-LULC_10m and FROM_GLC10_2017 exhibit the highest agreement in wetland identification, with a Jaccard coefficient of 0.971. This alignment is attributed to the shared utilization of 10-m-resolution Sentinel-2 data and the application of deep learning models such as random forest model and Impact Observatory's deep learning AI land classification model in their classification processes;
- GlobeLand30 V2020 and MCD12Q1 v061 demonstrate the lowest agreement in wetland identification, featuring a Jaccard coefficient of only 0.051. This discrepancy is likely due to variations in data sources and classification schemes. GlobeLand30 V2020 utilizes multi-source data with 10 land cover classes, while MCD12Q1 v061 relies on MODIS data with 17 land cover classes and a resolution of 500 m;
- AGLC-2015 showcases a high degree of consistency, with a mean Jaccard coefficient of 0.595 across all products. This robust agreement is attributed to the data fusion strategy of AGLC-2015, integrating information from GlobeLand30, FROM-GLC, and GLC-FCS30,
- MCD12Q1 v061 exhibits lower agreement, with a mean Jaccard coefficient of 0.102. This could be due to its utilization of MODIS data at a resolution of 500 m and the application of data mining and machine learning algorithms for classification.

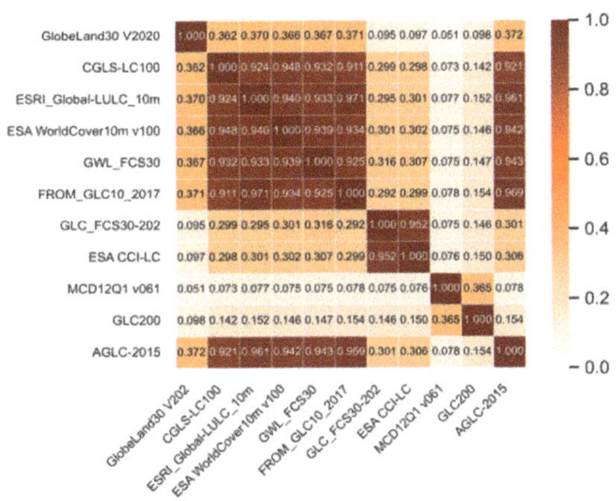

Figure 5. Jaccard similarity coefficient of wetland categories between different RS products.

The consistency of wetland RS products is influenced by factors such as data resolution, classification schemes, and methods [49–51]. These factors determine the applicability and accuracy of different products in defining, identifying, and describing wetlands. Selecting appropriate data sources and resolutions, along with designing rational classification schemes and methods, is crucial for improving the quality of wetland RS products. Considerations such as optical data susceptibility to clouds and fog, SAR data's ability to penetrate clouds, and the difficulty of terrain data in distinguishing vegetation types highlight the need for a balanced approach when choosing data sources and resolutions. Furthermore, the integration of multi-source data can enhance wetland identification results. Rational classification schemes and methods are pivotal for improving wetland description accuracy, necessitating a comprehensive evaluation of research scale and wetland types to strike a balance between detail and scope.

The use of the Jaccard similarity coefficient facilitates visual comparisons of the consistency in wetland identification among different RS products. This analysis, considering data resolution and methodology, serves as a valuable reference for the subsequent selection and enhancement of wetland RS products.

To further assess the consistency in wetland classification, Kappa coefficients between RS products were calculated (Figure 6). The Kappa coefficient, a key indicator measuring the level of agreement between classification results, provides insights into differences and reasons in wetland description.

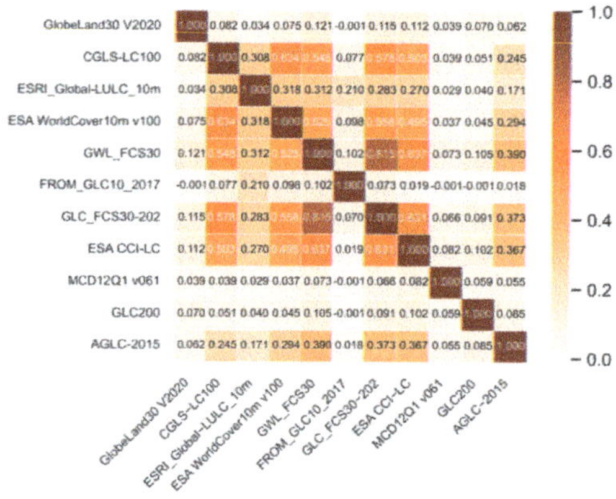

Figure 6. Kappa coefficients of wetland categories among RS products.

- Notably, the Kappa coefficient between GWL_FCS30 and GLC_FCS30-2020 is the highest at 0.815, indicating a robust consistency in wetland classification. This can be attributed to their detailed wetland classification system and hierarchical classification strategy, enabling the distinction of more wetland types and subtypes and capturing spatial details and changes,
- Conversely, the Kappa coefficients between FROM_GLC10_2017 and MCD12Q1 v061, FROM_GLC10_2017 and GLC2000, and FROM_GLC10_2017 and GlobeLand30 V2020 are the lowest, reaching only −0.001. This suggests a low consistency in wetland classification, potentially stemming from differences in data sources, resolutions, classification schemes, and methods, leading to deviations in wetland delineation and characterization.

Quantitative analysis of Kappa coefficients facilitates the evaluation of consistency among RS products in wetland classification, aiding in the identification of sources and factors contributing to errors. This analysis provides valuable guidance for the enhancement of wetland classification products, assisting both scientific and application fields in the selection and optimal use of RS products.

3.3. Spatial and Temporal Distribution Comparison of Wetlands

In this section, we rigorously assess the accuracy and consistency of various RS products in depicting wetlands within the Irtysh River Basin. Utilizing 3000 randomly generated points via the randomPoints function on the Google Earth Engine (GEE) platform, we conducted a representative and randomized assessment. Subsequently, we employed Google Earth high-resolution images to visually interpret each product's target year, following a common manual verification method. It is important to note that while effective, this approach may introduce subjective errors or uncertainties.

Upon conducting a comparative analysis of wetland distribution and types across the eleven RS products, we preliminarily assessed the degree of consistency between each product and the author's verification accuracy at the time of production. The observed distribution of each land cover closely aligned with the thematic map. Our comparative evaluation encompasses three aspects: overall, local, and detailed, offering methodological insights for subsequent wetland RS monitoring product selection, verification, and optimization.

(1) Overall Comparison

A comprehensive analysis of wetland area and type descriptions among the eleven RS products reveals significant diversity and discrepancies. Table 3 presents a range of total wetland area values, spanning from 64,281.615 km^2 to 230,053.628 km^2. These variations can be attributed to differences in data sources, classification schemes, and processing methods. Thematic maps in Figure 3 visually depict the concentration of wetlands extracted by different products in the middle and lower reaches of the Irtysh River Basin, with dense distribution along rivers and lakes.

Notably, the GWL_FCS30 product stands out for its proficiency in identifying small-scale wetlands, accurately delineating expansive wetland areas in the middle-eastern part, and successfully identifying scattered wetlands in the upstream mountainous area and southwestern part. This highlights the product's capability to capture diverse wetland features across different landscapes.

These findings underscore the importance of considering the specific strengths and limitations of each RS product in wetland analysis, as their performance can vary based on regional characteristics and the intended application. Figure 3 provides a visual representation of the spatial distribution, offering valuable insights for researchers and practitioners involved in wetland monitoring and management.

(2) Local Comparison

This refined comparative analysis, presented in Figure 7, aims to intricately evaluate wetland identification among six prominent remote sensing products (denoted as A to G), tailored for the year 2020 in the Irtysh River Basin, across three typical areas. The year 2020 is specifically emphasized because it is the most recent and comprehensive year for which data from multiple remote sensing products are available, and because it has important implications for wetland monitoring and conservation in the context of the United Nations Decades on Biodiversity and Ecosystem Restoration.

Among these products, the Sentinel-2 RGB Imagery (A) serves as a reference standard, providing high-resolution visual cues for wetland delineation. The remaining products, namely GlobeLand30 V2020 (B), ESRI_Global-LULC_10m (C), ESA WorldCover 10 m v100 (D), GWL_FCS30 (E), GLC_FCS30-2020 (F), and ESA CCI-LC (G), undergo a rigorous local comparison to discern their respective strengths and limitations within each of the three typical areas.

The comparative analysis of wetland localization in three typical areas of the Irtysh River Basin can effectively reveal subtle differences in wetland identification details between RS products. Through the analysis we can draw the following conclusions:

- GWL_FCS30 and GLC_FCS30-2020 products exhibit excellence in identifying small-scale and complex wetlands, showcasing their capability to capture spatial morphological details such as band or network distribution in swamp wetlands. However, GLC_FCS30-2020 may encounter instances of misclassification;
- The ESA WorldCover 10 m product excels in delineating wetland extension contours with higher accuracy;
- GlobeLand30 V2020 may exhibit misclassifications for certain land features,
- ESRI_Global-LULC_10m and ESA CCI-LC products may have difficulty identifying or misclassifying some fine wetland areas.

This localized comparative analysis, centered around the specifics of the Irtysh River Basin in 2020 and spanning three typical areas, refines our understanding of wetland dynamics. Furthermore, it offers valuable insights for the selection and optimization of remote sensing products in wetland monitoring applications, ensuring a more nuanced interpretation of wetland distribution and characteristics.

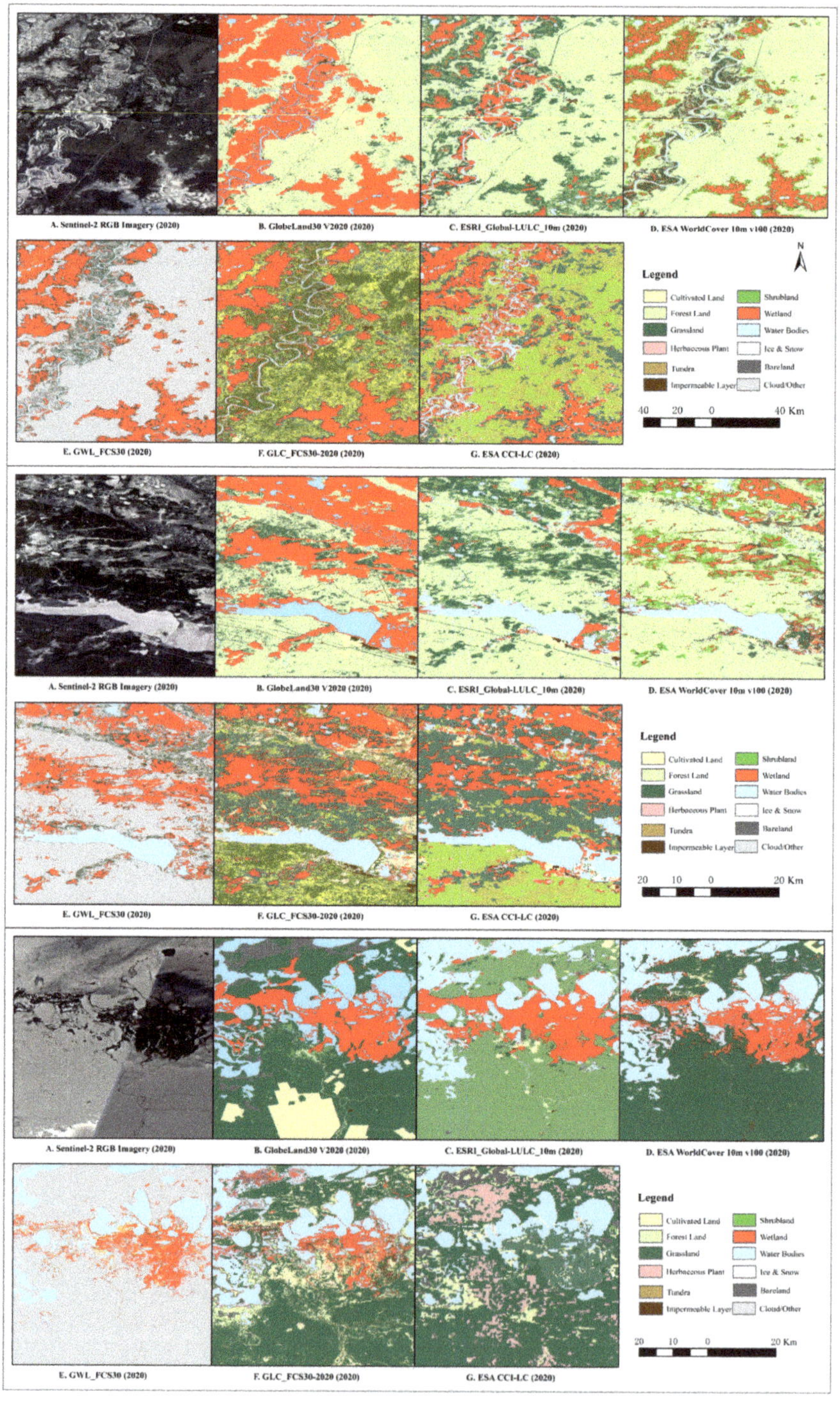

Figure 7. Comparative assessment of wetland identification in three typical areas of the Irtysh River Basin for the year 2020.

(3) Detailed Comparison

Upon meticulous examination of small wetland sample plots, distinct characteristics come to light. Wetland patches identified by the GWL_FCS30 and GLC_FCS30-2020 products exhibit smaller areas, greater numbers, and well-defined, continuous, and smooth boundaries. In contrast, wetland patches extracted by the ESRI-Global-LULC-10m and AGLC-2015 products feature relatively simplified boundaries, occasionally displaying omissions and misjudgments. These distinctions primarily arise from variations in wetland information classification and extraction algorithms between the two products.

The systematic assessment of spatial and temporal wetland distribution across the Irtysh River Basin contributes to a nuanced understanding of the strengths and weaknesses inherent in various RS products. This in-depth analysis serves as a foundation for informed decision-making in the selection and optimization of wetland RS monitoring products. The insights gained from this evaluation are invaluable for both scientific research and practical applications, providing a comprehensive perspective on the capabilities and limitations of different RS products in the context of wetland monitoring.

3.4. Factors Influencing Variances in Wetland Information Extraction

This section critically examines the principal factors contributing to disparities in wetland information extraction among the eleven RS products. The identified factors include data sources, spatial resolution, classification schemes, classification algorithms, and validation methods. These factors introduce variability in RS products' wetland descriptions, necessitating a thoughtful selection process aligned with research objectives while acknowledging inherent product characteristics and limitations. Simultaneously, enhancing the quality of RS products is crucial to support effective wetland conservation and management.

(1) Characteristics of Data Sources

The distinctiveness of land cover types is determined by the characteristics of data sources. Optical data susceptibility to clouds and fog, SAR data vulnerability to terrain influences, and terrain data sensitivity to source accuracy and processing methods underscore the need for preprocessing and correction measures.

(2) Spatial Resolution

Spatial resolution plays a pivotal role in delineating the level of detail and spatial variability in RS products. Higher-resolution data afford the capacity to distinguish more wetland subtypes and spatial patterns, while lower-resolution data can only capture broader wetland categories and large-scale changes. Selecting an appropriate spatial scale is therefore imperative.

(3) Classification Schemes and Algorithms

The core technologies underpinning RS product generation are classification schemes and algorithms, shaping the expression and granularity of land cover information. Variations in these components may lead to differences in naming, division, and attribution of identical land feature types. Intelligent classification methods such as deep learning, and optimization steps introduce further complexity. When utilizing RS products employing diverse classification schemes and algorithms, careful consideration is warranted to understand their impact on wetland type identification and differentiation, necessitating conversion and comparison procedures.

(4) Validation Methods

Validation methods represent a critical step in evaluating the quality and accuracy of RS products. Divergent validation methods can yield disparate accuracy assessment results for the same RS product. Considerations such as sample distribution, rigor, and cross-validation with other products or datasets are pivotal. When working with RS products employing distinct validation methods, it is essential to account for their influence on wetland accuracy assessment results, prompting calibration and adjustment efforts.

(5) Strengths and Weaknesses of Considered RS Products

Distinct RS products exhibit varying strengths and weaknesses in wetland description based on factors like data sources, spatial resolution, classification schemes, algorithms, and validation methods. Tailoring their usage to specific scenarios and wetland monitoring objectives is paramount. For instance, GlobeLand30 V2020, with its high resolution (30 m) and updated time range (2020), provides wetland extent and location in the Irtysh River Basin but lacks detailed wetland characteristics. Conversely, GWL_FCS30, boasting high resolution (30 m) and multiple wetland types, captures intricate wetland details and ecological features but may not be optimal for long-term changes and global comparisons. ESA CCI-LC, with lower resolution (300 m) and a broader temporal scope, facilitates long-term changes and global comparisons but sacrifices detail in wetland characteristics and ecological features. Understanding these nuances enables judicious selection and application of RS products in diverse wetland monitoring and management scenarios.

4. Discussion

This study meticulously assessed the accuracy and consistency of eleven RS wetland products in extracting wetland information within the Irtysh River Basin. Our findings elucidate substantial differences in both wetland area and type among the various products, primarily attributed to factors such as data sources, spatial resolution, classification schemes, and production processes.

For instance, the GWL_FCS30 product, leveraging multi-source data, markedly enhances the identification of small-area wetlands, resulting in a wetland area almost 3.5 times larger than that identified by the AGLC-2015 product. Noteworthy is GWL_FCS30's adoption of a detailed wetland subtype division, revealing richer internal differences within wetlands. Furthermore, spatial patterns and morphological characteristics exhibited heterogeneity across different products, influenced by factors such as data sources, classification methods, and processing techniques. The GWL_FCS30 product excels in depicting intricate internal morphological features, presenting clearer and continuous boundary information.

The examination of indicators such as the similarity coefficient and Kappa coefficient revealed consistency in the primary wetland distribution areas across different products. However, the presence of significant "false differences" highlighted the impact of classification errors, possibly intensified by data uncertainty during the classification process.

Our research carries significant implications. Firstly, we propose a comprehensive framework and method for selecting and validating wetland products, applicable beyond wetlands to other land surface elements. This contribution provides valuable support for RS data evaluation and application. Secondly, our study illuminates the profound influence of data sources and methods on the quality of wetland information extraction. This insight not only guides the technical trajectory and research focus of relevant production institutions but also serves as a reference for the ongoing innovation and development of RS technology. Lastly, our comparative results offer practical guidance for government departments and research institutions, aiding in the selection of suitable wetland products for monitoring regional changes. This, in turn, provides crucial information and data for informed scientific decision-making, wetland protection, and management.

Nevertheless, our study is not without limitations. Primarily, our comparative analyses were confined to a specific watershed region, potentially limiting the generalizability of our results to different geographic environments. Additionally, the use of a limited validation sample hinders a comprehensive assessment of absolute classification accuracy. Further, the examination of products for a single year limits insights into their performance across temporal changes. Future research avenues involve expanding samples across diverse topographic regions, establishing a standardized large-sample validation database for comprehensive accuracy assessment, and introducing time-series products for a more nuanced understanding of wetland change dynamics. These expansions aim to deepen our understanding and exploration within this field, addressing identified limitations and bolstering the robustness of our findings.

5. Conclusions

This study undertook a comprehensive comparison and evaluation of RS wetland products within the Irtysh River Basin, with the overarching goal of furnishing a scientific basis for wetland protection and management. Meanwhile, it aims to offer valuable insights for the ongoing development of wetland RS monitoring technology. The analysis incorporated eleven global-scale land cover or wetland RS products, dissecting their disparities, advantages, and drawbacks in delineating wetland area, type, distribution, and consistency in the Irtysh River Basin. The ensuing conclusions are as follows:

1. Significant Disparities in Wetland Description: Considerable differences were observed in the depiction of wetlands within the Irtysh River Basin among various RS products, attributed to factors such as data sources, spatial resolution, classification schemes, algorithms, and validation methods. Notably, the GWL_FCS30 product exhibited the highest wetland area extraction, while the AGLC-2015 product yielded the lowest. GWL_FCS30 demonstrated an enhanced capacity to portray internal morphological features, presenting clearer and more continuous boundary information. The ESRI_Global-LULC_10m and FROM_GLC10_2017 products exhibited superior classification consistency;
2. Key Influencing Factors on Information Quality: Central to the precision of wetland information extraction in the Irtysh River Basin are the influencing factors of data sources, classification methods, and validation schemes. Achieving high-quality wetland products necessitates obtaining raw data with elevated spatial and temporal resolution through the fusion of multiple sources. It is imperative to establish a detailed and scientifically grounded local wetland classification system, employ effective methods to unlock the full value of the data, and adhere to standardized and scientifically rigorous validation schemes to ensure result reliability,
3. Guidance for Future Endeavors: This study serves as a pivotal guide and reference for the selection and optimization of subsequent wetland products. It contributes valuable insights for the ongoing innovation and development of RS technology, providing a robust data foundation for informed wetland protection and management decisions. Nonetheless, certain limitations, including a restricted sample area, inadequate validation samples, and the absence of time-series products, underscore the necessity for future expansion and refinement in subsequent studies.

In summary, this research advances wetland monitoring technologies and underscores the critical considerations pivotal for accurate wetland information extraction. While acknowledging the limitations, the study's significance lies in its potential to inform strategic decision-making for wetland conservation and management, thus bridging the gap between RS technology and practical applications. To fortify the study's impact, future efforts should focus on addressing the identified limitations, expanding the scope of analysis, and incorporating temporal dimensions through the inclusion of time-series products. This would further amplify the relevance and applicability of the study's findings.

Author Contributions: Conceptualization, K.L. and A.S.; methodology, K.L. and A.S.; software, K.L.; validation, K.L., A.S., J.A. and W.L.; resources, A.S. and J.A.; data curation, K.L.; writing—original draft preparation, K.L.; writing—review and editing, K.L., A.S., J.A. and W.L.; visualization, K.L.; supervision, A.S.; funding acquisition, A.S. and J.A. All authors have read and agreed to the published version of the manuscript.

Funding: This work was partially supported by the Xinjiang Third Scientific Expedition (2022xjkk0702) and the Western Young Scholars Program of the Chinese Academy of Sciences (2022-XBQNXZ-001).

Data Availability Statement: The research data used in this study is available from corresponding authors on request.

Acknowledgments: We would like to extend our gratitude to the following organizations for providing the remote sensing products utilized in this study: China National Centre for Basic Geographic Information, Copernicus Land Monitoring Service Global Team, ESRI Corporation, European Space Agency, Institute of Space and Astronautical Information Innovation of the Chinese Academy of Sciences, Gong Peng Team at Tsinghua University, NASA, European Union Joint Research Centre, and Sun Yat-sen University. We also wish to acknowledge the Google Earth Engine platform for facilitating aspects of the analysis. Finally, we sincerely appreciate the valuable comments from reviewers and editors, which significantly helped us to improve the quality of this manuscript.

Conflicts of Interest: The authors declare no conflicts of interest.

References

1. McCarthy, M.J.; Radabaugh, K.R.; Moyer, R.P.; Muller-Karger, F.E. Enabling efficient, large-scale high-spatial resolution wetland mapping using satellites. *Remote Sens. Environ.* **2018**, *208*, 189–201. [CrossRef]
2. Xi, Y.; Peng, S.; Ciais, P.; Chen, Y. Future impacts of climate change on inland Ramsar wetlands. *Nat. Clim. Chang.* **2020**, *11*, 45–51. [CrossRef]
3. Mao, D.; Wang, Z.; Du, B.; Li, L.; Tian, Y.; Jia, M.; Zeng, Y.; Song, K.; Jiang, M.; Wang, Y. National wetland mapping in China: A new product resulting from object-based and hierarchical classification of Landsat 8 OLI images. *ISPRS J. Photogramm. Remote Sens.* **2020**, *164*, 11–25. [CrossRef]
4. Fluet-Chouinard, E.; Stocker, B.D.; Zhang, Z.; Malhotra, A.; Melton, J.R.; Poulter, B.; Kaplan, J.O.; Goldewijk, K.K.; Siebert, S.; Minayeva, T.; et al. Extensive global wetland loss over the past three centuries. *Nature* **2023**, *614*, 281–286. [CrossRef] [PubMed]
5. Ludwig, C.; Walli, A.; Schleicher, C.; Weichselbaum, J.; Riffler, M. A highly automated algorithm for wetland detection using multi-temporal optical satellite data. *Remote Sens. Environ.* **2019**, *224*, 333–351. [CrossRef]
6. Hu, S.; Niu, Z.; Chen, Y. Global Wetland Datasets: A Review. *Wetlands* **2017**, *37*, 807–817. [CrossRef]
7. Davidson, N.C. How much wetland has the world lost? Long-term and recent trends in global wetland area. *Mar. Freshw. Res.* **2014**, *65*, 934–941. [CrossRef]
8. Jiang, W.; Wang, W.; Chen, Y.; Liu, J.; Tang, H.; Hou, P.; Yang, Y. Quantifying driving forces of urban wetlands change in Beijing City. *J. Geogr. Sci.* **2012**, *22*, 301–314. [CrossRef]
9. Cui, L.; Li, G.; Ouyang, N.; Mu, F.; Yan, F.; Zhang, Y.; Huang, X. Analyzing Coastal Wetland Degradation and its Key Restoration Technologies in the Coastal Area of Jiangsu, China. *Wetlands* **2018**, *38*, 525–537. [CrossRef]
10. Davidson, N.C.; Finlayson, C.M. Extent, regional distribution and changes in area of different classes of wetland. *Mar. Freshw. Res.* **2018**, *69*, 1525–1533. [CrossRef]
11. Zhao, Z.; Zhang, Y.; Liu, L.; Liu, F.; Zhang, H. Recent changes in wetlands on the Tibetan Plateau: A review. *J. Geogr. Sci.* **2015**, *25*, 879–896. [CrossRef]
12. Kim, S.G. The evolution of coastal wetland policy in developed countries and Korea. *Ocean Coast. Manag.* **2010**, *53*, 562–569. [CrossRef]
13. Let, M.; Pal, S. Socio-ecological well-being perspectives of wetland loss scenario: A review. *J. Environ. Manag.* **2023**, *326*, 116692. [CrossRef] [PubMed]
14. Brinson, M.M.; Malvárez, A.I. Temperate freshwater wetlands: Types, status, and threats. *Environ. Conserv.* **2002**, *29*, 115–133. [CrossRef]
15. Junk, W.J. Long-term environmental trends and the future of tropical wetlands. *Environ. Conserv.* **2003**, *29*, 414–435. [CrossRef]
16. Seifollahi-Aghmiuni, S.; Kalantari, Z.; Land, M.; Destouni, G. Change Drivers and Impacts in Arctic Wetland Landscapes—Literature Review and Gap Analysis. *Water* **2019**, *11*, 722. [CrossRef]
17. Chakraborty, S.K.; Sanyal, P.; Ray, R. Basics of Remote Sensing Techniques Applicable in Wetlands Ecosystems. In *Wetlands Ecology: Eco-Biological Uniqueness of a Ramsar Site (East Kolkata Wetlands, India)*; Springer: Berlin/Heidelberg, Germany, 2023; pp. 303–377.
18. Guo, M.; Li, J.; Sheng, C.; Xu, J.; Wu, L. A review of wetland remote sensing. *Sensors* **2017**, *17*, 777. [CrossRef]
19. Huang, Y.; Peng, J.; Chen, N.; Sun, W.; Du, Q.; Ren, K.; Huang, K. Cross-scene wetland mapping on hyperspectral remote sensing images using adversarial domain adaptation network. *ISPRS J. Photogramm. Remote Sens.* **2023**, *203*, 37–54. [CrossRef]
20. Betbeder, J.; Gond, V.; Frappart, F.; Baghdadi, N.N.; Briant, G.; Bartholomé, E. Mapping of Central Africa forested wetlands using remote sensing. *IEEE J. Sel. Top. Appl. Earth Obs. Remote Sens.* **2013**, *7*, 531–542. [CrossRef]
21. Kuleli, T.; Guneroglu, A.; Karsli, F.; Dihkan, M. Automatic detection of shoreline change on coastal Ramsar wetlands of Turkey. *Ocean Eng.* **2011**, *38*, 1141–1149. [CrossRef]
22. Wang, H.; Ma, M. Impacts of Climate Change and Anthropogenic Activities on the Ecological Restoration of Wetlands in the Arid Regions of China. *Energies* **2016**, *9*, 166. [CrossRef]
23. Chen, L.; Ren, C.; Zhang, B.; Li, L.; Wang, Z.; Song, K. Spatiotemporal Dynamics of Coastal Wetlands and Reclamation in the Yangtze Estuary During Past 50 Years (1960s–2015). *Chin. Geogr. Sci.* **2018**, *28*, 386–399. [CrossRef]
24. Wang, D.; Zhang, S.; Wang, G.; Liu, Y.; Wang, H.; Gu, J. Reservoir Regulation for Ecological Protection and Remediation: A Case Study of the Irtysh River Basin, China. *Int. J. Environ. Res. Public Health* **2022**, *19*, 11582. [CrossRef] [PubMed]

25. Hanchu, L.; Jie, F.; Baoyin, L.; Li, W.; Qin, Q. Practical Exploration of Ecological Restoration and Management of the Mountains-Rivers-Forests-Farmlands-Lakes-Grasslands System in the Irtysh River Basin in Altay, Xinjiang. *J. Resour. Ecol.* **2021**, *12*, 766–776. [CrossRef]
26. Huang, F.; Xia, Z.; Li, F.; Guo, L.; Yang, F. Hydrological Changes of the Irtysh River and the Possible Causes. *Water Resour. Manag.* **2012**, *26*, 3195–3208. [CrossRef]
27. Kingsford, R.T.; Basset, A.; Jackson, L. Wetlands: Conservation's poor cousins. *Aquat. Conserv. Mar. Freshw. Ecosyst.* **2016**, *26*, 892–916. [CrossRef]
28. Fan, J.; Abudumanan, A.; Wang, L.; Zhou, D.; Wang, Z.; Liu, H. Dynamic Assessment and Sustainability Strategies of Ecological Security in the Irtysh River Basin of Xinjiang, China. *Chin. Geogr. Sci.* **2023**, *33*, 393–409. [CrossRef]
29. Lehner, B.; Döll, P. Development and validation of a global database of lakes, reservoirs and wetlands. *J. Hydrol.* **2004**, *296*, 1–22. [CrossRef]
30. Gumbricht, T.; Roman-Cuesta, R.M.; Verchot, L.; Herold, M.; Wittmann, F.; Householder, E.; Herold, N.; Murdiyarso, D. An expert system model for mapping tropical wetlands and peatlands reveals South America as the largest contributor. *Glob. Chang. Biol.* **2017**, *23*, 3581–3599. [CrossRef]
31. Chen, G.; Jin, R.; Ye, Z.; Li, Q.; Gu, J.; Luo, M.; Luo, Y.; Christakos, G.; Morris, J.; He, J.; et al. Spatiotemporal Mapping of Salt Marshes in the Intertidal Zone of China during 1985–2019. *J. Remote Sens.* **2022**, *2022*, 9793626. [CrossRef]
32. Radelyuk, I.; Zhang, L.; Assanov, D.; Maratova, G.; Tussupova, K. A state-of-the-art and future perspectives of transboundary rivers in the cold climate–a systematic review of Irtysh River. *J. Hydrol. Reg. Stud.* **2022**, *42*, 101173. [CrossRef]
33. Jiang, S.; Yuan, Y.; Chen, F.; Shang, H.; Zhang, T.; Yu, S.; Qin, L.; Zhang, R. A 291 year precipitation reconstruction in the upper Irtysh River basin based on tree-ring width. *Acta Ecol. Sin.* **2016**, *36*, 2866–2875.
34. Hrkal, Z.; Gadalia, A.; Rigaudiere, P. Will the river Irtysh survive the year 2030? Impact of long-term unsuitable land use and water management of the upper stretch of the river catchment (North Kazakhstan). *Environ. Geol.* **2006**, *50*, 717–723. [CrossRef]
35. Jun, C.; Ban, Y.; Li, S. Open access to Earth land-cover map. *Nature* **2014**, *514*, 434. [CrossRef] [PubMed]
36. Buchhorn, M.; Smets, B.; Bertels, L.; De Roo, B.; Lesiv, M.; Tsendbazar, N.-E.; Herold, M.; Fritz, S. *Copernicus Global Land Service: Land Cover 100m: Collection 3: Epoch 2019: Globe. Version V3.0.1*; Laboratory of Geo-information Science and Remote Sensing: Wageningen, The Netherlands, 2020.
37. Karra, K.; Kontgis, C.; Statman-Weil, Z.; Mazzariello, J.C.; Mathis, M.; Brumby, S.P. Global land use/land cover with Sentinel 2 and deep learning. In Proceedings of the 2021 IEEE International Geoscience and Remote Sensing Symposium IGARSS, Brussels, Belgium, 11–16 July 2021; pp. 4704–4707.
38. Zanaga, D.; Van De Kerchove, R.; Daems, D.; De Keersmaecker, W.; Brockmann, C.; Kirches, G.; Wevers, J.; Cartus, O.; Santoro, M.; Fritz, S. ESA WorldCover 10 m 2021 v200. Zenodo 2022. Available online: https://zenodo.org/records/7254221 (accessed on 10 July 2023).
39. Zhang, X.; Liu, L.; Zhao, T.; Chen, X.; Lin, S.; Wang, J.; Mi, J.; Liu, W. GWL_FCS30: A global 30 m wetland map with a fine classification system using multi-sourced and time-series remote sensing imagery in 2020. *Earth Syst. Sci. Data* **2023**, *15*, 265–293. [CrossRef]
40. Gong, P.; Liu, H.; Zhang, M.; Li, C.; Wang, J.; Huang, H.; Clinton, N.; Ji, L.; Li, W.; Bai, Y.; et al. Stable classification with limited sample: Transferring a 30-m resolution sample set collected in 2015 to mapping 10-m resolution global land cover in 2017. *Sci. Bull.* **2019**, *64*, 370–373. [CrossRef] [PubMed]
41. Zhang, X.; Liu, L.; Chen, X.; Gao, Y.; Xie, S.; Mi, J. GLC_FCS30: Global land-cover product with fine classification system at 30 m using time-series Landsat imagery. *Earth Syst. Sci. Data* **2021**, *13*, 2753–2776. [CrossRef]
42. Harper, K.L.; Lamarche, C.; Hartley, A.; Peylin, P.; Ottlé, C.; Bastrikov, V.; San Martín, R.; Bohnenstengel, S.I.; Kirches, G.; Boettcher, M. A 29-year time series of annual 300-metre resolution plant functional type maps for climate models. *Earth Syst. Sci. Data Discuss.* **2022**, *2022*, 1–37.
43. Friedl, M.; Sulla-Menashe, D. MCD12Q1 MODIS/Terra+ aqua land cover type yearly L3 global 500m SIN grid V006. *NASA EOSDIS Land Process. DAAC* **2019**, *10*, 200.
44. Roy, P.; Agrawal, S.; Joshi, P.; Shukla, Y. *The Land Cover Map for Southern Asia for the Year 2000. GLC2000 Database*; European Commission Joint Research Centre: Brussels, Belgium, 2003.
45. Xu, X.; Li, B.; Liu, X.; Li, X.; Shi, Q. Mapping annual global land cover changes at a 30 m resolution from 2000 to 2015. *Natl. Remote Sens. Bull.* **2021**, *25*, 1896–1916.
46. Lu, D.; Weng, Q.; Moran, E.; Li, G.; Hetrick, S. Remote sensing image classification. *Adv. Environ. Remote Sens. Sens. Algorithms Appl.* **2011**, *9*, 219–240.
47. Real, R.; Vargas, J.M. The probabilistic basis of Jaccard's index of similarity. *Syst. Biol.* **1996**, *45*, 380–385. [CrossRef]
48. Congalton, R.G. A review of assessing the accuracy of classifications of remotely sensed data. *Remote Sens. Environ.* **1991**, *37*, 35–46. [CrossRef]
49. Jensen, J. Thematic information extraction: Image classification. In *Introductory Digital Image Processing: A Remote Sensing Perspective*; Pearson Education: London, UK, 1996; pp. 197–256.

50. Herold, M.; Mayaux, P.; Woodcock, C.; Baccini, A.; Schmullius, C. Some challenges in global land cover mapping: An assessment of agreement and accuracy in existing 1 km datasets. *Remote Sens. Environ.* **2008**, *112*, 2538–2556. [CrossRef]
51. Maxwell, A.E.; Warner, T.A.; Fang, F. Implementation of machine-learning classification in remote sensing: An applied review. *Int. J. Remote Sens.* **2018**, *39*, 2784–2817. [CrossRef]

geosciences

MDPI

Article

Assessment of Unmanned Aerial System Flight Plans for Data Acquisition from Erosional Terrain

Valentina Nikolova [1,*], Veselina Gospodinova [2] and Asparuh Kamburov [2]

[1] Department of Geology and Geoinformatics, University of Mining and Geology "St. Ivan Rilski", 1700 Sofia, Bulgaria

[2] Department of Mine Surveying and Geodesy, University of Mining and Geology "St. Ivan Rilski", 1700 Sofia, Bulgaria; veselina.gospodinova@mgu.bg (V.G.); asparuh.kamburov@mgu.bg (A.K.)

* Correspondence: v.nikolova@mgu.bg

Abstract: Accurate data mapping and visualization are of crucial importance for the detection and monitoring of slope morphodynamics, including erosion processes and studying small erosional landforms (rills and gullies). The purpose of the current research is to examine how the flight geometry of unmanned aerial systems (UASs) could affect the accuracy of photogrammetric processing products, concerning small erosion landforms that are a result of slope wash and temporary small streams formed by rain. In October 2021, three UAS flights with a different geometry were carried out in a hilly to a low-mountain area with an average altitude of about 650 m where erosion processes are observed. UAS imagery processing was carried out using structure-from-motion (SfM) photogrammetry. High-resolution products such as photogrammetric-based point clouds, digital surface models (DSMs) and orthophotos were generated. The obtained data were compared and evaluated by the root mean square error (RMSE), length measurement, cloud-to-cloud comparison, and 3D spatial GIS analysis of DSMs. The results show small differences between the considered photogrammetric products generated by nadir-viewing and oblique-viewing (45°—single strip and 60°—cross strips) geometry. The complex analysis of the obtained photogrammetric products gives an advantage to the 60°—cross strips imagery, in studying erosional terrains with slow slope morphodynamics.

Keywords: unmanned aerial system; flight plan; point clouds; modelling

1. Introduction

The availability of high-quality data about the terrain, including elevation models, is of great importance for the studying of the topographic surface, and particularly the studying of geomorphological changes, and various land-forming processes on a local scale. Photogrammetry has proven to be an accurate and reliable method for the mapping and monitoring of landslides and soil erosion by providing precise data about the terrain and soil cover. Many studies present the application of close-range photogrammetry using digital cameras for such purposes [1,2]. With the development of modern technologies and systems for data acquisition, such as unmanned aerial systems (UASs), the capabilities of generating high-precision models are increasing and they are being used more and more to study geomorphic processes [3–6]. These advances have been facilitated by the development of structure-from-motion (SfM) [7], a technique that combines well-established photogrammetric principles with modern computational methods [8–10]. Its use expands the possibilities of obtaining quality images from different angles and distances to the studied object, which allows the simultaneous processing of nadir and oblique photos [11]. The need for terrain data with high resolution and accuracy, and the large volume of data pose the question of the methods of obtaining the data and the efficiency of photogrammetric processing. In recent years, studies that evaluate how UAS data acquisition affects the accuracy of the obtained photogrammetric products of areas with different relief characteristics have been elaborated. Most of the publications analyze the impact of the number, distribution, and quality of the control points [12–14] and flight height [15–17]. The availability of a UAS with RTK (Real-Time Kinematic) positioning would facilitate the solution to such issues related to the generation of highly accurate digital surface models and orthophoto mosaics [14,18,19]. Despite the increasing attention to the flight plans and particularly to the impact of camera tilt on photogrammetric products, there is still a need for deepening the research in this direction, particularly on the effectiveness of oblique imagery [18]. Many publications show that the most accurate and detailed terrain models are obtained in the case of combining nadir and oblique images [11,18,20–24]. On the other hand, combining nadir and oblique images for large areas leads to a large volume of data, the processing of which is time-consuming. This determines the need for the assessment of the impact of flight geometry on the quality of the products used in geomorphological studies to find the most appropriate way to solve the tasks.

Regarding the above, the current study aims to evaluate the different flight plan geometries using UASs with RTK and ground control points (GCPs) in obtaining centimeter-level precision data on the erosional terrain. The literature review shows that there is a very limited number of publications that give special attention to flight geometry in studying small erosional landforms, particularly for data acquisition from terrain with low vertical dissection of the relief. Thus, the results of this research can contribute to an increase in the effectiveness of the study of erosion landforms as well as to receiving a more precise assessment of erosion processes. The UAS survey was carried out by three flight plans—one in nadir mode—a single strip and two in oblique mode (at 45°—a single strip and 60°—cross strips). The effectiveness of flight plans is evaluated by comparison and assessment of the respective dense point clouds, digital elevation models, and orthophoto mosaics. The comparison approach is widely used in many publications for determining the differences and similarities between objects or phenomena. In contrast to the previous publications [18], we used the Multiscale Model to Model Cloud Comparison (M3C2) [25] and a theoretical plane to analyze the clouds. In addition to these methods, we analyzed the point density, which is an indicator that impacts the quality of the digital surface models. The specific indicators for the comparison of the photogrammetric products are described in Section 2.3. The results of the study provide contemporary quantitative data about an erosional terrain that is characterized by low values of vertical dissection. Taking into account published material [11,18,21], the results could have a further methodological contribution regarding the use of nadir and oblique imagery for the geomorphological study of different landscape conditions. The application of the common methodology to different landscapes is of importance for a further assessment of the transferability of the methodology.

2. Materials and Methods

2.1. Study Area

The area of interest is located in the Eastern Rhodopes mountain (southern Bulgaria), near the village of Kostino, Kardzhali district. It covers a mountainous area of approximately 3.5 ha, with an altitude of around 650 m a.s.l. (Figure 1). The area is made up of sandstone (a breccia-conglomerate sandstone formation) while some parts are composed of volcanic rocks—latite to andesite lava breccias [26]. The highly weathered rocks together with rare vegetation and high slope gradients (15–30° are predominant) facilitate the development of erosion processes. They are triggered by rainfalls that follow a period of dry weather. As a result, rills and gullies have been formed. The field observations and measurements show a prevailing rate of vertical erosion of around 2 cm per year. The length of the gullies is between 50 and 110 m. The area is characterized by low vertical dissection of the relief (around 4 m per 100 m^2). The area is partly covered by grass, single trees, and shrubs, which makes it difficult to evaluate the changes in the topographic surface and requires high-resolution terrain models to determine areas of erosion and accumulation.

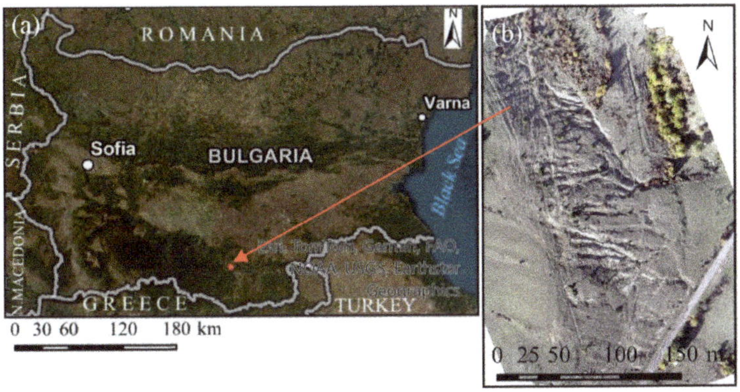

Figure 1. Study area: (**a**) geographical location; (**b**) orthophoto.

2.2. Equipment and Data Collection

A DJI Phantom 4 RTK was used for the acquisition of UAS imagery. The UAS has a 1-inch CMOS sensor with effective pixels of 20 M and a 24 mm wide-angle lens. It is equipped with a multi-frequency multi-system high-precision RTK GNSS. The RTK module provides centimeter-level positioning data for improved absolute accuracy on the image metadata. A DJI GS RTK app was used to create flight plans. The application provides the planning and execution of a flight along an optimal trajectory to create maps or three-dimensional models. In the current study, two application modes (a single strip and cross strips) were used for the creation of flight plans. As a result, an s-shaped flight route is received in 2D mode, which can be used for a digital orthophoto map. In 3D (double grid) mode, s-shaped crisscrossed routes make up the flight route which allows the development of more detailed 3D models [27].

Three UAS flights (one in a nadir—single strip and two in oblique mode at 45°—a single strip and a 60°—cross strip) were completed in one day, using identical flight parameters (Table 1). The flight altitude was set as low as possible to receive the most precise data, taking into account the various terrain obstacles—the slope, trees, electric poles, and powerlines. The image overlap is of high importance for vertical and horizontal accuracy, among the other influencing factors, while the quantity of GCPs has a high significance for the vertical accuracy [28]. In this regard, we performed the flights with 80% overlap (along and across the track). Highly overlapping images increase the accuracy of the generated 3D point clouds, avoiding false matches, and providing significant 3D point reliability at centimeter-level vertical accuracy [29]. A high overlap of 70–90% front overlaps and 60–80% side overlaps are recommended for topographic surveys and DTM generation [19]. The following three types of UAS imagery were obtained: nadir imagery—a single strip with a gimbal pitch angle at 90°; oblique imagery—a single strip with a gimbal pitch angle at 45°; and oblique imagery—cross strips with a gimbal pitch angle at 60° (Figure 2).

Table 1. Parameters of UAS flights.

Indicator	Value
UAV altitude (m)	60
Ground sample distance (cm)	$1.67 \approx 2$
Along the track overlap (%)	80
Across the track overlap (%)	80

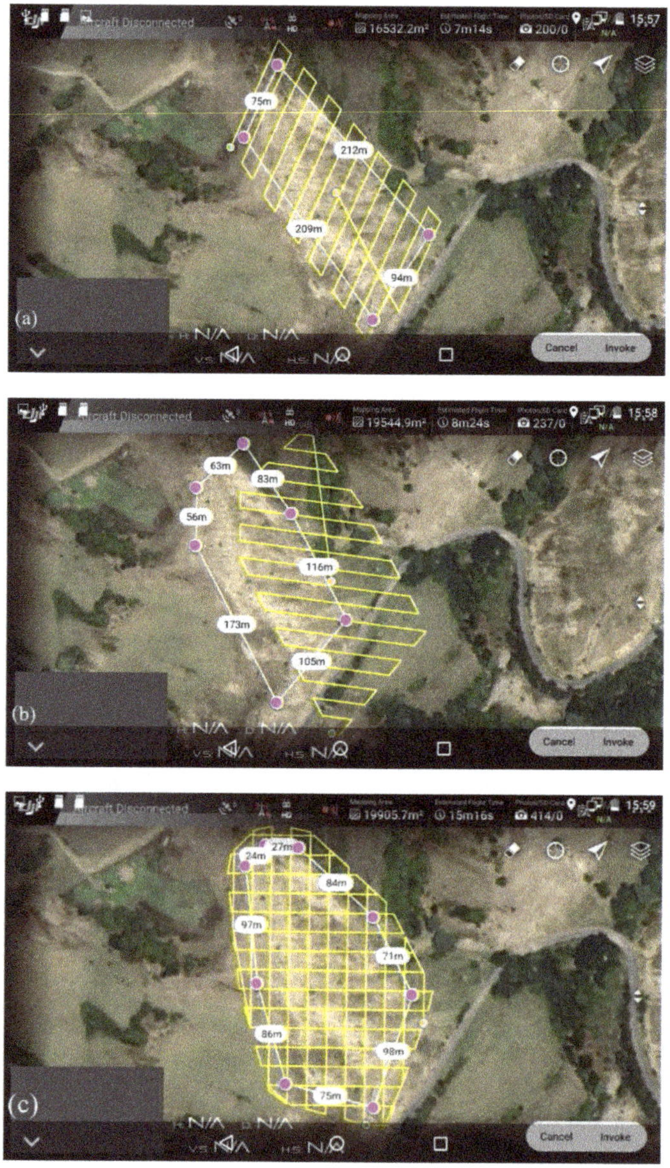

Figure 2. UAS trajectories of data acquisition designed by DJI GS RTK app for Android, version: v. 2.2.5-GSP: (**a**) nadir—single strip with gimbal pitch angle at 90°; (**b**) oblique—single strip with gimbal pitch angle at 45°; (**c**) oblique—cross strip with gimbal pitch angle at 60°.

The georeferencing of the model required the establishment of 20 ground points, of which 7 were used as control points (GCPs) and 13 as check points (CPs). They were distributed over the area with a separation of about 20–30 m from each other (Figure 3). These points were marked with 20-cm wooden pegs hammered into the ground and a 30 × 30 cm black-and-white paper/plastic target, and subsequently measured via network RTK observations using a virtual reference station (VRS) correction service from the Bulgarian RTK Network "1 Yocto" (https://1yocto.bg/, accessed on 25 October 2021). A multi-system multi-frequency GNSS receiver CHCNAV i50, operated via Landstar, v. 7 field surveying software was used—all measurements were performed in fixed GNSS mode, with estimated RMS accuracies around 2 cm horizontally and 2.5 cm vertically. The results are given in Table 2.

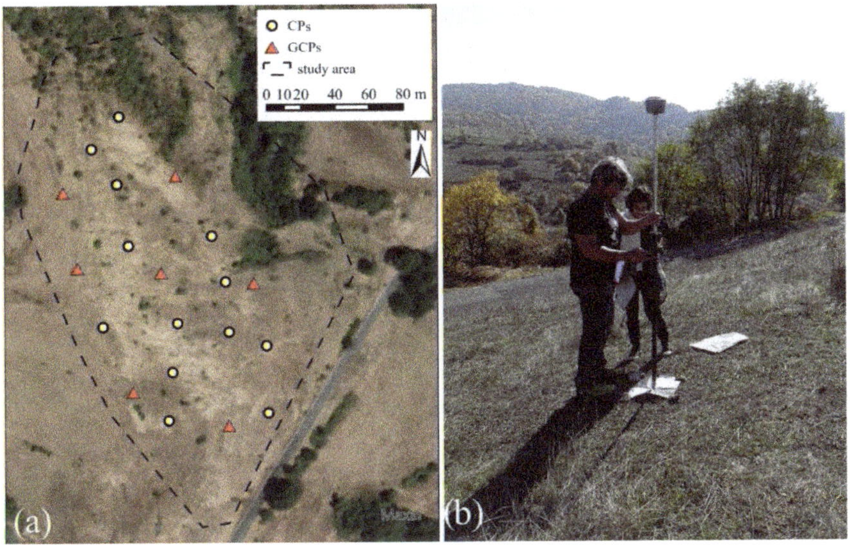

Figure 3. Control points: (**a**) location of GCPs and CPs; (**b**) measurements of the points.

Table 2. RTK GNSS observations of the ground points in WGS 84/UTM 35N (EPSG: 32635) spatial reference system (points from 1 to 7 are GCPs, and these from 100 to 112 are CPs).

Point	Northing	Easting	Height (m)	Hz. RMS (m)	Vert. RMS (m)	Satellites Used
1	4618381.220	359135.010	613.960	0.018	0.026	31
2	4618371.366	359067.864	642.235	0.017	0.023	35
3	4618326.176	359076.314	637.090	0.018	0.024	35
4	4618252.511	359108.964	623.773	0.019	0.024	37
5	4618232.469	359165.857	605.703	0.018	0.025	34
6	4618317.131	359180.885	603.648	0.020	0.026	32
7	4618323.493	359126.056	619.714	0.020	0.027	31
100	4618416.033	359101.018	625.236	0.019	0.026	32
101	4618396.905	359084.878	631.483	0.019	0.026	32
102	4618290.638	359091.014	630.283	0.018	0.024	35
103	4618234.896	359130.596	616.081	0.018	0.024	36
104	4618239.270	359189.416	600.595	0.019	0.025	34
105	4618279.774	359188.594	601.805	0.019	0.026	35
106	4618344.579	359155.869	607.370	0.020	0.026	30
107	4618338.956	359106.313	623.936	0.021	0.027	32
108	4618375.844	359099.596	628.198	0.021	0.028	32
109	4618263.361	359132.658	615.944	0.019	0.025	35
110	4618287.738	359166.025	608.150	0.020	0.025	35
111	4618292.918	359135.854	615.921	0.020	0.027	36
112	4618317.514	359163.893	607.896	0.019	0.025	30

The analysis of the published materials shows that using only RTK data for georeferencing makes the construction of the high-accuracy digital products of UAS photogrammetry easier, but systematic elevation errors are obtained [30,31]. In this regard, for our study, we used 20 ground points, 7 of which were GCPs. The number of GCPs and CPs was determined by taking into account the published research [19], horizontal and vertical dissection of our study area, and aiming to receive high-resolution models.

2.3. Photogrammetric Processing and Analysis of the Photogrammetric Products

The UAS data were processed through Agisoft Metashape Professional software, version 2.0.2. This is a fully automated platform with an intuitive workflow that produces high-quality digital models from imagery [32]. The software uses the structure-from-motion approach [9,10,33]. It allows the processing of imagery acquired at different distances from the objects (from any position), capture angles, and focal lengths. The first stage of the SfM approach comprises detecting identical points in all overlapping images and applying the scale invariant feature transform (SIFT) algorithm. This algorithm transforms the input images into a set of local specific features that are invariant to the scaling and rotation of the underlying image, and partially invariant to the 3D camera viewpoint and to changes in illumination. The key points are automatically identified in each image at all scales and locations. After that, a feature descriptor computed by transforming local image gradients is created. These descriptors allow features to be matched in large datasets [8,34]. The second stage of the SfM approach is a bundle adjustment. This iterative procedure defines simultaneously the geometry of the scene, camera positions and orientations. In this way, the bundle adjustment leads to minimizing the reprojection error between the locations of the observed and predicted points in the images. This algorithm is based on multiple iterative procedures through which a sparse point cloud is created. The three-dimensional coordinates of the object are obtained in a random coordinate system based on the homological points, detected in the images. The coordinates of the GCP and check points are then entered for georeferencing and validation of the internal quality of the model. Obtainment of the object geometry is followed by a calculation of the 3D point coordinates and the consequent production of a dense point cloud, DSMs, and orthomosaics.

In the current study, the photogrammetric data processing was carried out in Agisoft Metashape Professional with the following parameters: alignment accuracy—high; optimization—all parameters. For the generation of the dense data cloud, we used: dense point cloud generation quality—high; filtering mode—mild; source date—depth maps. Camera calibration parameters and RMSE of GCPs and CPs are presented in Table 3.

Table 3. Camera location errors and RMSE at the different flight plans, according to the SfM Agisoft report.

Capture Type	Average Camera Location Error Total Error (cm)	RMSE (m) GCPs	RMSE (m) CPs
Nadir—single strip	1.5	0.020	0.025
Oblique—single strip with gimbal pitch angle at 45°	1.7	0.014	0.021
Oblique—cross strip with gimbal pitch angle at 60°	1.8	0.007	0.013

As a result of the photogrammetric processing of the data, three dense point clouds were obtained from each of the three flight plans, as well as DSMs and orthophotos. These products were compared to evaluate the impact of the flight plan parameters on the products' characteristics and in regard to their use in studying small erosional landforms. For this purpose, the following aspects were considered:

- Point clouds comparison—cloud to plane and cloud to cloud distances

The three obtained point clouds of 45°—single strip, 60°—cross strips, and nadir—single strip geometry were processed with open-source software Cloud Compare, v. 2.13 beta [35]. They were compared with a flat plane and by a Multiscale Model to Model Cloud Comparison (M3C2) plug-in [25]. The clouds were filtered using a Cloth Simulation Filter (CSF) plug-in [36], with a cloth resolution of 0.1 and a Relief type of scene. This provided a useful approach to removing vegetation and artificial objects (e.g., electric poles). For the purpose of the comparison, a theoretical plane was automatically generated with reference to the

60° point cloud using the Fit Plane tool. The orientation of the fitted plane is a 16° dip angle at a 93° dip direction, centered in the middle of the Z values of the oblique 60° point cloud (613.925 m). For further processing (e.g., rasterization of the plane in ArcGIS Pro), it was subsampled into a flat 100 mln point cloud using the random method. The three point clouds were compared to the fitted plane using the Cloud/Primitive Distance tool, with the results stored in the relevant layer's scalar field attribute (Figure 4). The method computes the exact distance from the cloud to the plane.

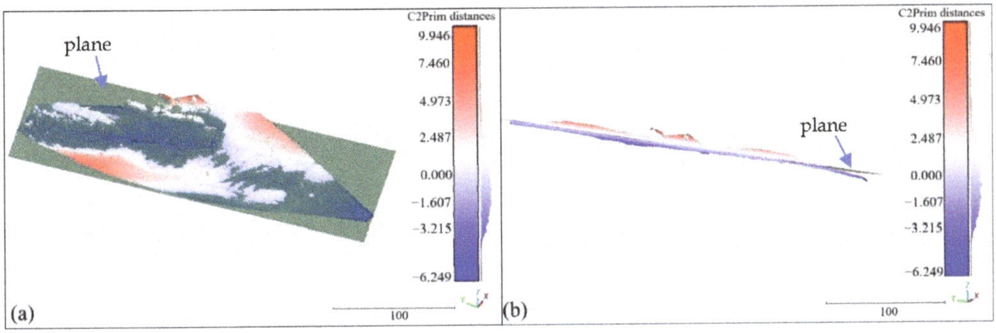

Figure 4. Cloud to primitive entity comparison: (**a**) plan view; (**b**) profile view.

The oblique-viewing point cloud obtained at 45°—a single strip mode and the cloud of a nadir single strip were compared to the reference 60°—a cross strips mode cloud using M3C2 [25]. This algorithm allows direct comparison of point clouds without requiring gridding or meshing. The distance is calculated along a normal vector for core points. In the current study, all the points of the first cloud are used for core points. For normal calculation, a multi-scale mode was applied. The cloud with the highest number of points was used as a reference cloud.

Another indicator for cloud comparison is the number of points and their spatial distribution. These characteristics of the clouds are of importance when creating DSMs and choosing an interpolation method. For the purpose of cloud comparison and evaluation, rasters that present the number of points per square meter were created by LAS Point Statistics As Raster tool in ArcGIS Pro 3.2.1 [37].

- Evaluation of the DSMs

The derived DSMs were evaluated for accuracy through the computation of the root mean square error (RMSE) of check points determined on the models, and those that were measured on the terrain. As a result of the photogrammetric processing of the data in Agisoft software, v.2.0.2, DSMs with a different horizontal resolution were received due to the different flight geometry. When the alignment of the images was performed with a high-quality setting, the following cell size DEMs were obtained: at nadir-viewing imagery—0.035 m; at oblique imagery 60°—cross strip—0.032 m, and at oblique imagery 45°—single strip—0.045 m. For comparison purposes, DSMs with one and the same resolution (0.05 and 0.1 m) from the three point clouds were created in ArcGIS Pro 3.2.1 [37] using a binning interpolation type, IDW for cell assignment, and a natural neighbor for the void fill method. Cell sizes of 0.05 and 0.1 m were set taking into account the peculiarities of the topographic surface and keeping the size as low as possible but not less than the ground sampling distance, received at the photogrammetric processing of the UAS data.

Additionally, the DSMs are compared with a flat surface—the plane that was used for clouds comparison. The application of the CutFill tool of ArcGIS Pro 3.2.1 allowed us to compute the volumes above and under the surface. This gives relative information about a possible calculation of the erosion and deposition on the base of the data, obtained by different flight geometry.

- Orthophotos

For the assessment of the UAS flight plan impact on the resulting orthophotos, the length of the lines calculated using the coordinates of the two points defining the line, and measured by GNSS, were compared with the length of the same lines which were digitized in the ArcGIS Pro 3.2.1 software, according to a visual identification on the three orthophotos of each subcategory. For this purpose, six lines were considered. They are selected regarding the morphometric features of the terrain—in the direction of the maximal slope gradient, following nearly a gully bed and across the slope, and perpendicular to the line of the maximal slope gradient. The lines are presented and analyzed in Section 3.3.

3. Results

3.1. Point Clouds

The results of the cloud to primitive distance (C2Prim) calculations show that there are minimal differences between the clouds of the three types of flight geometry (Table 4, Figure 5). The patterns of the spatial distribution of the C2Prm distances are very similar. Despite the larger range of the differences (between—6.457 and 10.180 m), the values close to 0 take most of the cases, and the extreme values are single cases and are due to some outliers. However, it should be taken into account that these differences are the differences between the theoretical surface and the cloud. If we compare the maximal values of C2Prim distances received for the three point clouds, the larger difference is obtained between the nadir and 45° cloud—38.0 cm and the lowest—between the nadir and 60° cloud—14.7 cm. The comparison of the minimal distances from C2Prim shows the same pattern—the largest is the difference between the nadir and the 45° cloud—7.5 cm, and the lowest is between the nadir and the 60° cloud—2.8 cm. In this regard, we can conclude that the clouds of oblique 60°—cross strips mode and nadir viewing—single strip mode are closer to each other.

Table 4. Distribution of positive and negative values of cloud to primitive distances (% of the total cases).

Oblique Imagery 60°—Cross Strips		Oblique Imagery 45°—Single Strip		Nadir Imagery 90°—Single Strip	
Positive Values	Negative Values	Positive Values	Negative Values	Positive Values	Negative Values
40.99	59.01	41.91	58.09	40.96	59.04

The M3C2 distance computation shows similar results The range of the distance differences between the 60°—cross strips mode cloud and nadir—a single strip mode cloud is a bit larger but the extreme values are single and can be neglected. Distances close to 0 are the predominant cases in the calculated distances using M3C2, which is visible on the histograms (Figure 6). This is observed in the comparison of the nadir—single strip mode cloud to the 60°—cross strips mode cloud where values between 0.027 and −0.027 m take 82% of the computed distances. In the comparison of the 45°—single strip mode cloud to the 60°—cross strips mode, 80% of the cases are between −0.031 and 0.046 m, and the values from −0.021 to 0.025 m take 36% of the computed distances. This distribution of the distance values confirms a greater uniformity between the 60° and nadir clouds than between the clouds of 60° and 45° imageries.

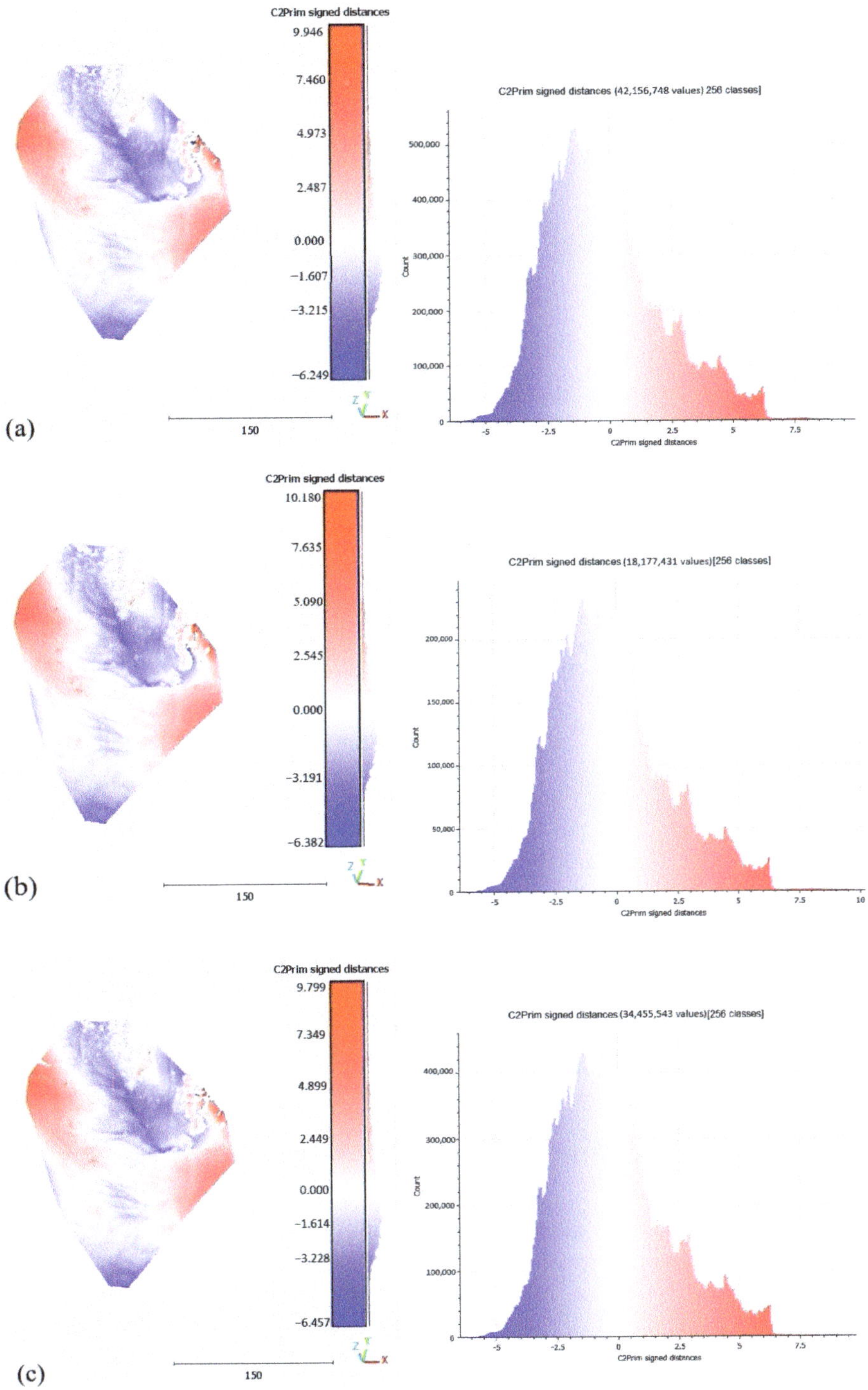

Figure 5. Cloud to primitive distances in meters: (**a**) cloud of oblique 60°—cross strips mode to plane; (**b**) cloud of oblique 45°—single strip mode to plane; (**c**) cloud of nadir—single strip mode to plane.

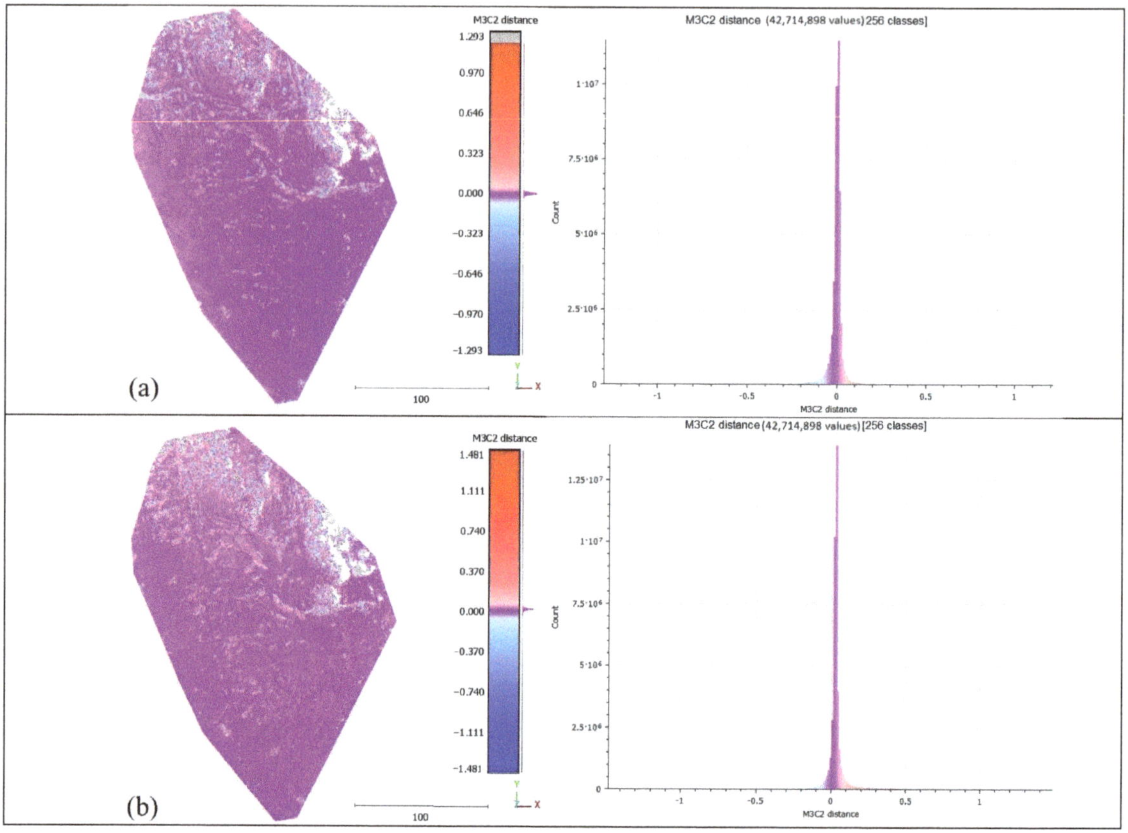

Figure 6. M3C2 distances in meters computed between the reference point cloud of 60°—cross strips mode and: (**a**) nadir—single strip cloud; (**b**) cloud of 45°—single strip mode.

Regarding the number of points in the clouds, the maximum number is contained in the 60°—cross strips cloud—42,714,898, in the cloud of 45°—single strip mode the number of points is the least—18,454,515, and in the nadir cloud, the points are 34,770,923. For the majority of the studied area (51%), the density of points is from 1000 to 1500 per square meter in the 60°—cross strips mode cloud and from 500 to 1000 per square meter in the nadir clouds (in 65% of the studied area), while for the 45°—single strip mode cloud most of the UAS captured area (58%) has 200 to 500 points/m^2. The point clouds statistics are given in Table 5. The results of the analysis of the point density show similar patterns of the spatial distribution of the points in the three clouds but that they are closer between the clouds of the 60°—cross strips mode and nadir imagery (Figure 7).

Table 5. Point counts per m^2. (ArcGIS Pro statistics of rasters whose cell values reflect statistical information about LAS points).

Capture Type	Mean	Maximum	Minimum
Oblique imagery 45°—single strip	510	1483	1
Oblique imagery 60°—cross strips	1183	3460	1
Nadir imagery—single strip	965	2858	1

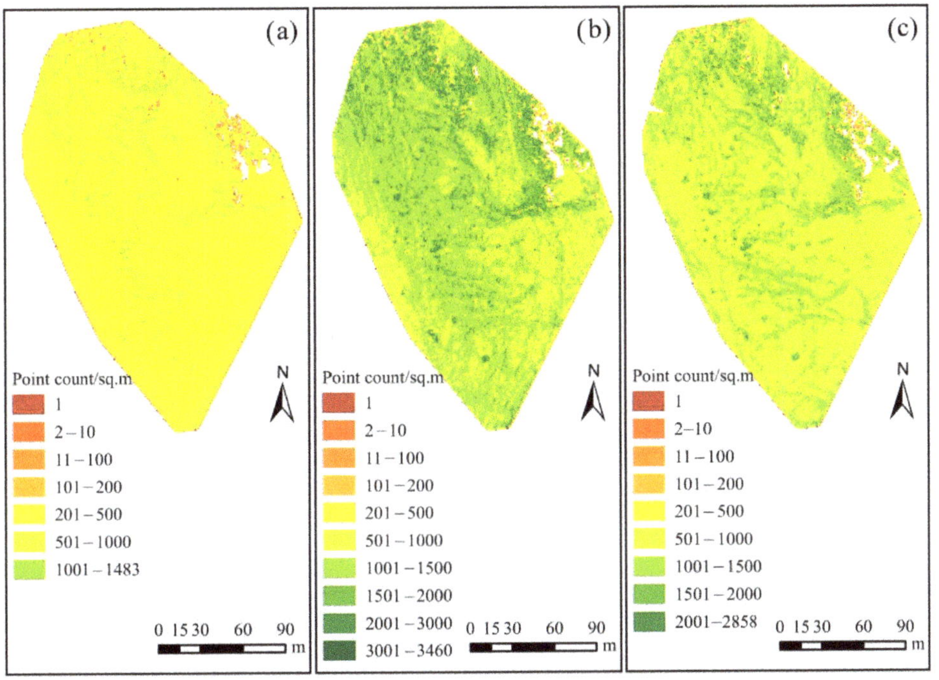

Figure 7. Spatial distribution of the points in the clouds: (**a**) oblique imagery 45°—single strip mode; (**b**) oblique imagery 60°—cross strips mode; (**c**) nadir viewing imagery—single strip mode.

There are single cases for the considered clouds (some areas where high vegetation was filtered) for which we received 1 point/m^2. Areas with less than 10 points/m^2 take 0.2% of the study area for nadir cloud, 0.17%—for 60° cloud and 0.3%—for 45° cloud. A large number of points in the clouds and high point density allow the creation of high-resolution DSMs.

3.2. Digital Surface Models

One of the most often used metrics for the assessment of the geometric accuracy of the photogrammetric products and geospatial models is the RMSE. For the study of erosion and deposition, the changes in the elevation are of great importance. In this regard, the accuracy of the generated DSMs was evaluated by RMSE of the calculated elevation of the respective locations in the model according to the measured check points. The results are given in Table 6.

Table 6. Vertical RMSE (m) of the DSMs, generated from the clouds according to the respective control points.

	Point Cloud for the Creation of the DSM		
DSM	**60°—Cross Strips**	**45°—Single Strip**	**Nadir—Single Strip**
DSM—automatically generated in Agisoft in cloud processing	0.024 (DSM cell size 0.032 m)	0.040 (DSM cell size 0.049 m)	0.022 (DSM cell size 0.35 m)
DSM—cell size 0.05 m	0.024	0.037	0.022
DSM—cell size 0.10 m	0.023	0.037	0.022

DSMs based on nadir viewing imagery point cloud have the highest vertical accuracy, but the same can be said for the DSMs based on 60°—cross strips mode cloud. The differences in the RMSE are of millimeters. The results are closely related not only to the UAS flight plan but also to the method of interpolation in the creation of the DSMs, as well as to the resolution of the models. It has to be noted also that the vertical RMS accuracy of the measured control points is around 0.025 m.

Comparison of the DSMs with a theoretical plane, using the CutFill tool of ArcGIS Pro 3.2.1., shows smaller differences in the calculated volumes above and below the surface for the models created of the point clouds from the 60°—cross strips mode and nadir—single strip mode imagery than the models of 45°—single strip mode cloud (Table 7).

Table 7. CutFill comparison with a theoretical plane, volumes in m^3.

	60°—Cross Strips Mode Cloud	**45°—Single Strip Mode Cloud**		**Nadir—Single Strip Mode Cloud**	
Above the surface	31,389.58	32,255.84		31,462.17	
Below the surface	35,194.55	34,027.38		35,083.88	
Difference: between 60° and 45° clouds		above the surface −866.26	below the surface 1167.16		
Difference: between 60° and 90° clouds				above the surface −72.59	below the surface 110.67

Distributed over the whole test area, the differences in the volumes between the clouds from 60°—cross strips mode and 45°—single strip mode, above and below the reference plane, are about 6 cm, and those between 60°—cross strips mode and nadir—single strip mode are 0.5 cm. It can be concluded that for areas with small values of vertical dissection of the relief, the use of UAS data obtained from nadir imagery instead of 60°—cross strips mode will not significantly affect the erosion and accumulation assessment. On the other hand, using DEMs based on 45°—single strip cloud data can lead to significant overestimation or underestimation of erosion and accumulation rates.

3.3. *Orthophotos*

Three orthophotos were generated from the three image subcategories. A purely visual difference between them could not be seen as they look quite similar. For the purpose of the assessment, six lines were considered (Figure 8). They were selected regarding the morphometric features of the terrain. L1, L2, and L3 are located across the slope, perpendicular or nearly perpendicular to the line of maximal slope gradient. L4 and L5 are in the direction of the maximal slope gradient, and L6 follows nearly the gully bed in the lower part and cuts across the slope in the upper part. The differences between the lengths of these lines are determined by coordinate calculation and measured digitized lines are presented in Table 8.

In general, the differences are fairly small, which is associated with the high-accuracy and high-resolution products of UAS photogrammetry. Oblique imagery with a gimbal pitch angle at 60°—cross strips works better in measuring the length of L1, L2, and L6 where the percentage difference reached almost zero in three orthophotos of each subcategory. The differences between the length of reference lines L1, L2, and L3, and the corresponding three subcategories of lines are slightly larger than the corresponding differences for L4 and L5. This can be explained by the greater fragmentation of the relief in the cross-slope direction. In this case, oblique imagery (60°—cross strips mode) shows the best results, despite the fact that length differences in both other cases (nadir and 45° imagery) are also small.

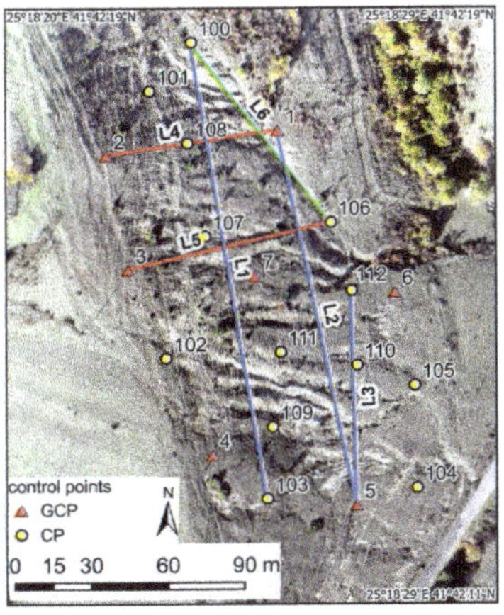

Figure 8. Spatial distribution of the control points and the location of lines (L1–L6) throughout the study area (blue lines are across the line of maximal slope gradients; red lines are in the direction of maximal slope gradients, and the green line follows the gully cross-section in the upper part and nearly gully bed in the lower part).

Table 8. Lengths of the reference lines and the corresponding digitized lines (the number of lines corresponds to the one given in Figure 8).

UAS Data	Distance, m	Difference with the Reference Line, m *	Difference, %
	Line 1 (L1)		
Reference lines from X and Y coordinates	183.536		
Nadir imagery (90°—single strip)	183.582	−0.046	−0.03
Oblique imagery (45°—single strip)	183.585	−0.049	−0.03
Oblique imagery (60°—cross strips)	183.580	**−0.044**	−0.02
	Line 2 (L2)		
Reference lines from X and Y coordinates	151.916		
Nadir imagery (90°—single strip)	151.951	−0.035	−0.02
Oblique imagery (45°—single strip)	151.937	−0.021	−0.01
Oblique imagery (60°—cross strips)	151.936	**−0.020**	−0.01
	Line 3 (L3)		
Reference lines from X and Y coordinates	85.068		
Nadir imagery (90°—single strip)	85.118	−0.05	−0.06
Oblique imagery (45°—single strip)	85.111	−0.043	−0.05
Oblique imagery (60°—cross strips)	85.107	**−0.039**	−0.05
	Line 4 (L4)		
Reference lines from X and Y coordinates	67.865		
Nadir imagery (90°—single strip)	67.867	**−0.002**	0.00
Oblique imagery (45°—single strip)	67.921	−0.056	−0.08
Oblique imagery (60°—cross strips)	67.881	−0.016	−0.02
	Line 5 (L5)		
Reference lines from X and Y coordinates	81.656		
Nadir imagery (90°—single strip)	81.660	−0.004	−0.01
Oblique imagery (45°—single strip)	81.658	**−0.002**	0.00
Oblique imagery (60°—cross strips)	81.661	−0.005	−0.01
	Line 6 (L6)		
Reference lines from X and Y coordinates	90.080		
Nadir imagery (90°—single strip)	90.099	−0.019	−0.02
Oblique imagery (45°—single strip)	90.069	0.011	0.01
Oblique imagery (60°—cross strips)	90.083	**−0.003**	0.00

* the smallest differences are presented in bold.

Regarding the length of L4 and L5, which nearly follow the longitudinal slope profile, the results are very good using either nadir-viewing geometry. The differences in the length of the lines at the three subcategories compared to the reference line are negligible, except for L4, oblique imagery (45°—single strip). This can be explained by the relatively small length of the lines and the smaller variation of the terrain down the slope in comparison to the terrain variation in the direction across the slope. Similar results are presented by Kyriou et al. [18]. The authors state that nadir-viewing geometry gave the best results when the line intersects a smoother surface, while for a more complex profile line, a combination of nadir and oblique imagery shows better results in measuring the distances.

The difference in the measured distances in a gully longitudinal profile is insignificant (in a range of millimeters), in contrast to the measured distances in a direction across the slope, which reach up to 4–5 cm.

4. Discussion

Considering the point clouds, DSMs, and orthophotos, the results of the current study show that the differences in the properties of the photogrammetric products of 60°—cross strips oblique imagery and nadir—single strip mode are very small and can even be neglected. This can be explained by the features of the test area and the low vertical dissection of the relief. With regard to our results and the published materials on this topic, it can be concluded that in the delineating of small erosional landforms (rills, gullies) the flight geometry does not have a significant impact. However, the analysis of the differences in the length of the lines digitized on the orthophoto and reference length calculated by coordinates shows that oblique 60°—cross strips imagery is more reliable. The results of the study are in line with the findings of previous publications [18,20,22,38] that underline the advantage of the oblique imagery or a combination of nadir and oblique ones in the modelling of complex terrain. James and Robson [39] consider a single oblique camera angle as an option for reducing systematic deformation of the digital elevation model. According to Nesbit and Hugenholtz [22], supplementing nadir image blocks with oblique images in the UAV–SfM workflow consistently improves spatial accuracy. The tilt angle of the camera, combined with the slope of the terrain, has a significant impact on the detection of tie-points [40], which ensures the best fit of the images and impacts the quality of the results. Considering the DSM evaluations and differences in the values received from CutFill analysis, we can say that the flight geometry has a higher influence on volume calculation.

The spatial resolution of the photogrammetric products depends on the UAS flight parameters, characteristics of the images, camera parameters, and software capabilities. Śledź and Ewertowski [41] emphasize the level of accuracy set for dense point cloud generation. According to the cited authors, at the lowest level of processing, the resolution was ~28 cm, at the low level ~14 cm, at the medium level ~7 cm, at the high level ~3.5 cm, and at the ultra-high level ~1.75 cm. For the purpose of the current study, we performed the alignment of the images with a high-quality setting. At this option, the software works with photos of the original size [32]. Apart from the processing setting, the resolution of the derived models also depends on the camera tilt, if all other conditions are equal. In our study, we received at nadir-viewing imagery—3.5 cm; at oblique imagery 60°—cross strips—3.2 cm, and at oblique imagery 45°—single strip—4.5 cm for the pixel size. At the medium level of data processing, the results are, respectively: 7.0 cm -at nadir-viewing imagery and 9.7 cm—oblique imagery 45°—single strip. The above results direct attention to the parameters of photogrammetric processing and to the use of high or ultra-high levels of dense point cloud generation if the task concerns the evaluation of the geomorphic change due to slow geomorphological processes.

Regarding the evaluation of the DSMs and the computed vertical RMSE, the products based on the data of oblique imagery 60°—cross strips and nadir-viewing imagery are very similar; even the RMSE for the DSM derived from the nadir cloud is less compared to that of 60°—cross strips mode. Using GCPs significantly decreases high z errors for RTK surveys.

Nota et al. [14] found that GCPs minimize the z errors to below 2 cm. The larger number of GCPs and their uniform distribution over the area can provide high-quality digital surface models, even with a medium option of the quality parameter of the photogrammetric processing [13]. GCPs are used not only for georeferencing but also for improving internal and external orientation parameters in photogrammetric processing [19]. In our study, we used a total of 20 points at a distance of about 20–30 m from each other and obtained GCPs and CPs RMSE around 2 cm and even less (Table 3).

According to Meinen and Robinson [42], the accuracy of nadir orientation images with a dense distribution of GCPs is comparable and quite close to that obtained from a terrestrial laser scanner. We can say that for a terrain with developed small erosional landforms and low values of vertical dissection of the relief, there are no significant differences between the DSMs of oblique imagery 60°—cross strips mode and nadir-viewing imagery, and the high number and density of the points in the clouds determine the small differences.

5. Conclusions

Based on the comparison of the point clouds, the lengths of the lines measured on the orthophoto mosaics, and the RMSE of the DSMs generated by the data of the three types of flight geometry used in the current study, the following conclusions can be drawn.

The M3C2 between the reference 60°—cross strip mode cloud and both compared clouds (nadir—single strip and 45°—single strip) shows small differences. The values near 0 are predominant. Closer are oblique 60°—cross strips and nadir—single strip imageries.

UAS camera tilt and plan geometry have a greater influence on volume calculation. To avoid over- and underestimation, RTK UAS is to be used with ground control points.

The comparison of the lengths of the lines digitized on the orthophotos against the reference line obtained from the X and Y coordinates shows larger differences in the case of the lines across the slope than of the lines drawn in the longitudinal direction of the slope (following the maximal slope gradient). In this particular study, oblique imagery 60°—cross strips mode shows greater accuracy. However, for this kind of erosional terrain, if the slope gradient is changing slowly, the nadir viewing geometry also produces good results.

In general, the comparisons of the photogrammetric products of the nadir single strip, oblique 45°—single strip, and oblique 60°—cross strips imagery reveal small differences that do not decisively determine the choice of the optimal UAS flight geometry in mapping small erosional landforms.

The small differences between the DSMs of the three flight geometries can be explained by the small size of the studied area, the large count of points in the clouds, and their high density. The results of the current study are closely related to the terrain features and the particular landscape conditions. Further studies of other areas with a relatively small vertical dissection of the relief are needed to evaluate the efficacy of the oblique geometry in studying erosional slopes, as well as to assess the transferability of the methodology.

Author Contributions: The authors contributed equally to elaborating the paper. Particular contributions are: conceptualization, V.N. and V.G.; methodology data acquisition, V.G. and A.K.; data processing and analyses, V.G., V.N. and A.K.; discussion and conclusion, V.N. and V.G., visualizations, V.N. All authors have read and agreed to the published version of the manuscript.

Funding: This research was carried out in the frame of the project "Application of geoinformation technologies in erosion research in mountain areas—Case studies of Eastern Rhodopes (Bulgaria)", funded by the National Science Fund, Ministry of Education and Science (Bulgaria), Competition for financial support for basic research projects-2019 (Contract No КП-06-Н34/3, signed on 5 December 2019).

Data Availability Statement: The data presented in this study are available on request from the corresponding author, upon reasonable request.

Acknowledgments: The authors thank the anonymous reviewers for their time and comments.

Conflicts of Interest: The authors declare no conflict of interest.

References

1. Liu, W.-C.; Huanga, W.-C. Close range digital photogrammetry applied to topography and landslide measurements. *Int. Arch. Photogramm. Remote Sens. Spat. Inf. Sci.* **2016**, *41*, 875–880. [CrossRef]
2. Mokhtar, M.R.M.; Wahab, S.N.A.; Bin, M.N.; Hashim, H.H.; Kasim, A.C. Landslide monitoring using close range photogrammetry. *Plan. Malays. J. Malays. Inst. Plan.* **2021**, *19*, 160–168. [CrossRef]
3. Bazzoffi, P. Measurement of rill erosion through a new UAV-GIS methodology. *Ital. J. Agron.* **2015**, *10* (Suppl. S1), 708. [CrossRef]
4. Krenz, J.; Greenwood, P.; Kuhn, N.J. Soil degradation mapping in drylands using unmanned aerial vehicle (UAV) data. *Soil Syst.* **2019**, *3*, 33. [CrossRef]
5. Liao, K.; Song, Y.; Xie, S.; Zheng, H. Monitoring of Benggang erosion based on UAV photogrammetry technology. *IOP Conf. Ser. Earth Environ. Sci.* **2019**, *330*, 052003. [CrossRef]
6. Urban, R.; Štroner, M.; Blistan, P.; Kovanič, L.; Patera, M.; Jacko, S.; Ďuriška, I.; Kelemen, M.; Szabo, S. The suitability of UAS for mass movement monitoring caused by torrential rainfall—A study on the talus cones in the Alpine terrain in High Tatras, Slovakia. *ISPRS Int. J. Geo.-Inf.* **2019**, *8*, 317. [CrossRef]
7. Ullman, S. The interpretation of structure from motion. *Proc. R. Soc. Lond.* **1979**, *203*, 405–426.
8. Westoby, M.J.; Brasington, J.; Glasser, N.F.; Hambrey, M.J.; Reynolds, J.M. 'Structure-from-motion' photogrammetry: A low-cost, effective tool for geoscience applications. *Geomorphology* **2012**, *179*, 300–314. [CrossRef]
9. Eltner, A.; Sofia, G. Structure from motion photogrammetric technique. *Dev. Earth Surf. Process.* **2020**, *23*, 1–24. [CrossRef]
10. Rosas, M.A.; Clapuyt, F.; Viveen, W.; Vanacker, V. Quantifying geomorphic change in Andean river valleys using UAV-PPK-SfM techniques: An example from the western Peruvian Andes. *Geomorphology* **2023**, *435*, 108766. [CrossRef]
11. Nikolakopoulos, K.G.; Kyriou, A.; Koukouvelas, I.K. Developing a Guideline of Unmanned Aerial Vehicle's Acquisition Geometry for Landslide Mapping and Monitoring. *Appl. Sci.* **2022**, *12*, 4598. [CrossRef]
12. James, M.R.; Robson, S.; D'Oleire-Oltmanns, S.; Niethammer, U. Optimising UAV Topographic Surveys Processed with Structurefrom-Motion: Ground Control Quality, Quantity and Bundle Adjustment. *Geomorphology* **2017**, *280*, 51–66. [CrossRef]
13. Forlani, G.; Dall'Asta, E.; Diotri, F.; Cella, U.M.d.; Roncella, R.; Santise, M. Quality Assessment of DSMs Produced from UAV Flights Georeferenced with On-Board RTK Positioning. *Remote Sens.* **2018**, *10*, 311. [CrossRef]
14. Nota, E.W.; Nijland, W.; De Haas, T. Improving UAV-SfM time-series accuracy by co-alignment and contributions of ground control or RTK positioning. *Int. J. Appl. Earth Obs. Geoinf.* **2022**, *109*, 102772. [CrossRef]
15. Zimmerman, T.; Jansen, K.; Miller, J. Analysis of UAS Flight Altitude and Ground Control Point Parameters on DEM Accuracy along a Complex, Developed Coastline. *Remote Sens.* **2020**, *12*, 2305. [CrossRef]
16. Lee, S.; Park, J.; Choi, E.; Kim, D. Factors Influencing the Accuracy of Shallow Snow Depth Measured Using UAV-based Photogrammetry. *Remote Sens.* **2021**, *13*, 828. [CrossRef]
17. Seifert, E.; Seifert, S.; Vogt, H.; Drew, D.; Van Aardt, J.; Kunneke, A.; Seifert, T. Influence of Drone Altitude, Image Overlap, and Optical Sensor Resolution on Multi-view Reconstruction of Forest Images. *Remote Sens.* **2019**, *11*, 1252. [CrossRef]
18. Kyriou, A.; Nikolakopoulos, K.; Koukouvelas, I. How image acquisition geometry of UAV campaigns affects the derived products and their accuracy in areas with complex geomorphology. *ISPRS Int. J. Geo.-Inf.* **2021**, *10*, 408. [CrossRef]
19. Jiménez-Jiménez, S.I.; Ojeda-Bustamante, W.; de Jesús Marcial-Pablo, M.; Enciso, J. Digital Terrain Models Generated with Low-Cost UAV Photogrammetry: Methodology and Accuracy. *ISPRS Int. J. Geo.-Inf.* **2021**, *10*, 285. [CrossRef]
20. Amrullah, C.; Suwardhi, D.; Meilano, I. Product accuracy effect of oblique and vertical non-metric digital camera utilization in UAV-photogrammetry to determine fault plane. *ISPRS Ann. Photogramm. Remote Sens. Spat. Inf. Sci.* **2016**, *III-6*, 41–48. [CrossRef]
21. Rossi, P.; Mancini, F.; Dubbini, M.; Mazzone, F.; Capra, A. Combining nadir and oblique UAV imagery to reconstruct quarry topography: Methodology and feasibility analysis. *Eur. J. Remote Sens.* **2017**, *50*, 211–221. [CrossRef]
22. Nesbit, P.R.; Hugenholtz, C.H. Enhancing UAV–SfM 3D model accuracy in high-relief landscapes by incorporating oblique images. *Remote Sens.* **2019**, *11*, 239. [CrossRef]
23. Nesbit, P.R.; Hubbard, S.M.; Hugenholtz, C.H. Direct Georeferencing UAV-SfM in High-Relief Topography: Accuracy Assessment and Alternative Ground Control Strategies along Steep Inaccessible Rock Slopes. *Remote Sens.* **2022**, *14*, 490. [CrossRef]
24. Parmehr, E.G.; Savadkouhi, M.; Nopour, M. The impact of oblique images and flight-planning scenarios on the accuracy of UAV 3D mapping. *Photogramm. Rec.* **2023**, *38*, 563–580. [CrossRef]
25. Lague, D.; Brodu, N.; Leroux, J. Accurate 3D comparison of complex topography with terrestrial laser scanner: Application to the Rangitikei canyon (N-Z). *ISPRS J. Photogramm. Remote Sens.* **2013**, *82*, 10–26. [CrossRef]
26. Jordanov, B.; Sarov, S.; Georgiev, S.; Valkov, V.; Balkanska, E.; Grozdev, V.; Marinova, R.; Markov, N. *Explanatory Note to Geological Map of Bulgaria, 1:50,000, Map Sheet K-35-75-G (Nikolovo)*; Ministry of Environment and Water, Bulgarian Geological Survey, Uniscorp Ltd.: Sofia, Bulgaria, 2008; p. 92. (In Bulgarian)
27. DJI. Phantom 4 RTK User Manual, v.2.4. 2021. Available online: https://dl.djicdn.com/downloads/phantom_4_rtk/20210716/Phantom_4_RTK_User_Manual_v2.4_EN.pdf (accessed on 20 July 2023).
28. Liu, Y.; Han, K.; Rasdorf, W. Assessment and Prediction of Impact of Flight Configuration Factors on UAS-Based Photogrammetric Survey Accuracy. *Remote Sens.* **2022**, *14*, 4119. [CrossRef]
29. Haala, N.; Cramer, M.; Rothermel, M. Quality of 3D point clouds from highly overlapping UAV imagery. *Int. Arch. Photogramm. Remote Sens. Spat. Inf. Sci.* **2013**, *40*, 183–188. [CrossRef]

30. Losè, T.L.; Chiabrando, F.; Tonolo, F.G. Boosting the timeliness of UAV large scale mapping. Direct georeferencing approaches: Operational strategies and best practices. *ISPRS Int. J. Geo.-Inf.* **2020**, *9*, 578. [CrossRef]
31. Štroner, M.; Urban, R.; Seidl, J.; Reindl, T.; Brouček, J. Photogrammetry using UAV-mounted GNSS RTK: Georeferencing strategies without GCPs. *Remote Sens.* **2021**, *13*, 1336. [CrossRef]
32. Agisoft LLC. Agisoft Metashape Professional. 2020. Available online: https://www.agisoft.com/pdf/metashape-pro_2_0_en.pdf (accessed on 1 December 2023).
33. Özyeşil, O.; Voroninski, V.; Basri, R.; Singer, A. A survey of structure from motion. *Acta Numer.* **2017**, *26*, 305–364. [CrossRef]
34. Lowe, D. Distinctive image features from scale-invariant keypoints. *Int. J. Comput. Vis.* **2004**, *60*, 91–110. [CrossRef]
35. CloudCompare (Version 2.13 Beta) [GPL Software]. 2023. Available online: https://www.danielgm.net/cc/ (accessed on 28 November 2023).
36. Zhang, W.; Qi, J.; Wan, P.; Wang, H.; Xie, D.; Wang, X.; Yan, G. An Easy-to-Use Airborne LiDAR Data Filtering Method Based on Cloth Simulation. *Remote Sens.* **2016**, *8*, 501. [CrossRef]
37. ESRI Inc. *ArcGIS Pro Software*; version 3.2.1; ESRI Inc.: Redlands, CA, USA, 2023.
38. Sammartano, G.; Chiabrando, F.; Spanò, A. Oblique images and direct photogrammetry with a fixed wing platform: First test and results in Hierapolis of Phrygia (TK). *ISPRS Int. Arch. Photogramm. Remote Sens. Spat. Inf. Sci.* **2020**, *43*, 75–82. [CrossRef]
39. James, M.R.; Robson, S. Mitigating systematic error in topographic models derived from UAV and ground-based image networks. *Earth Surf. Process. Landf.* **2014**, *39*, 1413–1420. [CrossRef]
40. Jaud, M.; Letortu, P.; Théry, C.; Grandjean, P.; Costa, S.; Maquaire, O.; Davidson, R.; Le Dantec, N. UAV survey of a coastal cliff face—Selection of the best imaging angle. *Measurement* **2019**, *139*, 10–20. [CrossRef]
41. Śledź, S.; Ewertowski, M.W. Evaluation of the Influence of Processing Parameters in Structure-from-Motion Software on the Quality of Digital Elevation Models and Orthomosaics in the Context of Studies on Earth Surface Dynamics. *Remote Sens.* **2022**, *14*, 1312. [CrossRef]
42. Meinen, B.U.; Robinson, D.T. Mapping erosion and deposition in an agricultural landscape: Optimization of UAV image acquisition schemes for SfM-MVS. *Remote Sens. Environ.* **2020**, *239*, 111666. [CrossRef]

geosciences

MDPI

Article

Assessments of Gravity Data Gridding Using Various Interpolation Approaches for High-Resolution Geoid Computations

Onur Karaca *, Bihter Erol and Serdar Erol

Gravity Research Group, Geomatics Engineering Department, Istanbul Technical University, Istanbul 34469, Turkey; bihter@itu.edu.tr (B.E.); erol@itu.edu.tr (S.E.)
* Correspondence: karacaonu@itu.edu.tr

Abstract: This article investigates the role of different approaches and interpolation methods in gridding terrestrial gravity anomalies. In this regard, first of all, simple and complete Bouguer anomalies are considered in gravity data gridding. In the comparison results of gridding these two Bouguer anomaly datasets, the effect of the high-frequency contribution of topographic gravitation (by means of the terrain correction) is clarified. After that, the role of the used interpolation algorithm on the resulting grid of mean gravity anomalies and hence on the geoid modeling accuracy is inspected. For this purpose, four different interpolation methods including geostatistical Kriging, nearest neighbor, inverse distance to a power (IDP), and artificial neural networks (ANNs) are applied. Here, the IDP and nearest neighbor methods represent simple-structured algorithms among the interpolation methods tested in this study. The ANN method, on the other hand, is preferred as a complex, optimization-based soft computing method that has been applied in recent years. In addition, the geostatistical Kriging method is one of the conventional methods that is mostly applied for gridding gravity data in geodesy and geophysics. The calculated gravity anomalies in grids are employed in high-resolution geoid model computations using the least squares modifications of Stokes formula with additive corrections (LSMSA) technique. The investigations are carried out using the test datasets of Auvergne, France that are provided by the International Service for the Geoid for scientific research. It is concluded that the interpolation algorithms affect the gravity gridding results and hence the geoid model determination. The ANN method does not provide superior results compared to the conventional algorithms in gravity gridding. The geoid model with 4.1 cm accuracy is computed in the test area.

Keywords: gravity anomalies; data interpolation; terrain correction; gravity reduction; gridding; geoid computation; GNSS/leveling

1. Introduction

In many applications of geodesy, geophysics, geodynamics, and engineering disciplines, the height information of a point relative to a reference surface is essential. Ellipsoidal height, which is the vertical distance between a point on the Earth's surface and an ellipsoid along the ellipsoidal normal, can be obtained anywhere and anytime, regardless of weather conditions, using global navigation satellite systems (GNSS). However, this kind of height cannot be utilized for the majority of engineering and surveying applications, including drainage projects, building engineering structures, managing natural disasters, etc., as it only has geometrical meaning and does not refer to the Earth's gravitational field [1–3]. Instead, the orthometric height is utilized, which indicates a natural flow of the fluids, as it refers to the Earth's gravitational field and is described as the distance between the geoid and a point on the Earth's surface [4–6]. Leveling is the conventional measurement technique for precise determination of the orthometric heights. However, this method is a time-consuming and expensive task in practice. In addition, it is prone to error accumulation, especially in large areas, and is more difficult to carry out in rough terrain. Deformations, especially occurring in active tectonic regions, require updating the vertical control networks that rely on leveling measurements, which is a significant disadvantage for vertical control networks. Therefore, acquiring the orthometric height (H) of a topography point using its GNSS-ellipsoidal height (h) and geoid height (N), which is the distance between the geoid and ellipsoid along the ellipsoid normal at the point ($H = h - N$), is an efficient method in surveying applications. In this method, the geoid height parameter is obtained using a geoid model, and the accuracy of the orthometric height in this transformation is directly related to the accuracy of the geoid model. In order to obtain the orthometric heights within a few centimeters of accuracy, high-resolution regional geoid models determined using dense and accurate terrestrial gravity data are used [5,7–9].

The geoid is a level surface with a constant gravity potential, and it coincides with the mean sea level after removing the effects of sea surface topography over the oceans [10]. The determination of the geoid surface relies on a boundary value problem solution represented by the Stokes formula and adopts an assumption of no masses outside of the geoid surface. In order to fulfill this condition, the topographical masses outside of the geoid are mathematically excluded in geoid computation theory. The implementation of the Stokes formula requires a global coverage of terrestrial gravity observations, but having dense terrestrial gravity observations over the entire Earth is not practically possible. The consequence of this data deficiency in the results is minimized by integrating locally available dense terrestrial gravity data and globally available satellite gravity observations in a modified Stokes integral formula. A stochastic modification of the Stokes integral formula is utilized in the least squares modification of Stokes integral with additive corrections (LSMSA) approach developed by the Royal Institute of Technology in Sweden (or shortly the KTH approach). This approach has been used in the determination of the regional geoid models in many studies so far, and it has yield accurate results in these studies (see [3,7,11–31]). In addition to the KTH method, there are also different gravimetric geoid determination algorithms used for regional gravimetric geoid modeling in the literature (see [32–37]).

In geoid model computations using the LSMSA method, the free air gravity anomalies in the grid form on the surface of the topography are used. These free air anomalies are restored from gridded Bouguer anomalies on the geoid. For gridding terrestrial gravity data, observed gravity values (g) are transformed into gravity anomalies (Δg) and are reduced to the reference surface "geoid" using the Bouguer reduction procedure. In the reduction of gravity anomalies, the gravitational attraction effects caused by topographic masses above the geoid are simply calculated and removed from gravity observations [38,39]. According to implemented reduction components, either simple or complete (refined) Bouguer gravity anomalies can be calculated and applied. In simple Bouguer reduction, the topography is identified just as the Bouguer plate (or shell), and the masses described by this plate above the geoid are removed. Even though the simple Bouguer anomalies (SBAs) do not include the effects of residual topographical masses deviating from the plate, complete Bouguer anomalies consider these excess terrain effects. Thus, the CBAs are acquired by adding the terrain corrections (TC) to the simple Bouguer anomalies. Simple or complete Bouguer gravity anomalies are preferred to be used for gravity data interpolation and gridding due to their low-frequency characteristics with less dependency on topographic variations [40,41]. Several studies have investigated whether SBAs may be utilized in gravity data interpolation and gridding instead of CBAs. While free-air anomalies (FAAs) are affected by the aliasing effect more than simple Bouguer anomalies, complete Bouguer anomalies are smoother and less sensitive to the negative effects of aliasing than both anomalies, in theory. According to Abbak et al. [38], Goos and Featherstone [42], and Kuhn et al. [43], the use of CBAs is crucial when topography is rough due to extreme elevation variations, but SBAs can also be used in smooth regions despite the omission of the high-frequency effect of residual topographical masses. Carrying out data gridding using appropriate types of gravity anomalies is crucial for the geoid determination. Inappropriately gridded gravity dataset may lead to a loss of the gravity signal in the input data and hence reduce the quality of the preformed regional geoid model. As stated above, removing high-frequency information obtained via terrain corrections prior to the gridding can solve the aliasing problem in the input gravity grid data. On another hand, densifying the gravity observations can also minimize this error, but this option may not always be possible, especially in rough topographies and restricted observational areas for. Digital elevation model (DEM) data are used for calculating the gravitational effects of topographic masses in Bouguer reduction process [2,41,44]. Also, the geoid computation using the LSMSA method employs DEM for calculating the terrain and downward continuation (DWC) corrections, which are restored onto the approximate geoid height parameter obtained from a global geopotential model (GGM) as the low-frequency component of the geoid [44].

This research is carried out using gravity observations in the Auvergne test area in France. It aims to investigate the role of interpolation methods in gridding the terrestrial gravity anomalies and their consequences on geoid model determination. Within the scope of the study, firstly, we test and clarify the role of the used Bouguer gravity anomaly type in the gridding process and hence its consequence on the geoid model determination. For this purpose, we grid the simple (SBAs) and complete (CBAs) Bouguer anomalies, employing the geostatistical Kriging interpolation algorithm, which is a commonly used technique in gravity field determination studies. Since the LSMSA method uses free air anomalies on the Earth's surface, free air anomalies in grid form are obtained by applying the Bouguer plate and terrain correction values to the gridded Bouguer anomalies [7]. The statistics obtained from the areal comparison of the SBA and CBA grids at grid nodes and validations of the geoid models in the study area are used to clarify the significance of the Bouguer anomaly type in the data gridding process.

Secondly, the effects of the interpolation techniques used in gridding the CBAs on geoid determination are investigated in this study. Four interpolation algorithms (Kriging, inverse distance to a power (IDP), nearest neighbor, and artificial neural network (ANN) methods) are employed. Here, it is worth emphasizing that this investigation does not aim to find the best interpolator but rather to clarify the role of employing a different interpolator in gravity gridding. Among these methods, ordinary Kriging is one of the most common two methods (the other is the Least Squares Collocation method) used in gravity field and geoid modeling studies. These methods are generally employed in data interpolation modules of the geoid computation software suits as in the KTH-Geolab software for precise geoid determination [18–20], as well as GRAVSOFT geodetic gravity field modeling programs [28,35,36]. The IDP and nearest neighbor algorithms are two widely used interpolation methods not only in the field of gravity research but also in various application areas where spatial data interpolation is required. Both algorithms dominate the local characteristics in the interpolation results via weighting the data or limiting the interpolation boundaries. Additionally, the simplicity of their formulations is an advantage for these two algorithms from a computational practicality point of view. In addition to these three widely used interpolation methods, we also apply an artificial neural network for gridding the complete Bouguer anomalies as the fourth method in this study. The ANN is a learning-based soft computing algorithm, and due to this aspect, it is quite different from the other three algorithms tested here. We included the ANN in this study to verify the performance of a recent generation interpolation algorithm in gravity gridding and to compare it with the widely used conventional techniques. Subsequently, new geoid models are acquired using free air anomaly grid datasets. A comparison of the calculated geoid models at the grid nodes provides the areal differences among the models. The validation of the geoid models using the GPS/leveling benchmarks in the area reveals an absolute accuracy value for the calculated geoid models.

The conclusions drawn from this investigation indicate that in regions with plain topography, since the terrain correction parameters remain within certain limits, either simple or complete Bouguer gravity anomalies can be used in gravity gridding. However, in rough terrain, the differences between simple and complete Bouguer anomalies are considerable; therefore, complete Bouguer anomalies should be used. In gravity gridding, the used interpolation algorithm affects both the magnitude and distribution pattern of the gravity values at the grid nodes and influences the geoid model computations. Using the conventional interpolation algorithms (Kriging, the IDP, and the nearest neighbor), the gravity value at a grid node is estimated directly using the data at the observation points, and the contribution of the data point is formulated in the algorithm. The Kriging method also takes the stochastic properties of the observations, employing covariances in its algorithm. Since they have similarities in their mathematical backgrounds, their output grids have smaller differences in magnitude. Contrary to expectations, the ANN, a learning-based soft computing method, does not provide superiority over the conventional methods in this study. As a result of the tests, the grid datasets obtained using the Kriging, IDP and nearest neighbor interpolation algorithms were found to be the most compatible. As a result of the calculations, the accuracy of the geoid models computed in the region is 4.1 cm.

The methods used in the computations are introduced in Section 2, together with explanatory visuals and formulas. Then, the data, study area, and numerical results are provided in Section 3 under the title of "Case Study". In Section 4, interpretations of the results and conclusions are provided.

2. Materials and Methods

Basic definitions and formulas of gravity anomalies and terrain reductions are given here within the context of gravimetric geoid computations. Then interpolation techniques, which play an important part in the gridding procedure for Bouguer anomalies, are discussed. Thereafter, least squares modifications of Stokes formula with additive corrections (LSMSA) method is expressed using detailed formulas.

2.1. Gravity Anomalies and Terrain Correction

The observed gravity values on the topography are not suitable for interpolation since they contain high-frequency components of Earth's gravitational field. Thus, randomly distributed gravity data on topography has to be reduced to the geoid surface for the purposes of geoid determination, interpolation, and extrapolation of gravity data and geophysical exploration studies. The geoid can be computed using Stokes integral and gravity anomalies on the geoid are its input data in computations. The geoid height parameter (N) is the output in this computation [7,24,38,44,45]. In other words, surface gravity anomalies have to be downward continued to the geoid surface using a suitable reduction schema before the geoid computation. The difference between the gravity value at a point on the geoid surface (g_{P_0}) and the normal gravity at its corresponding point on the ellipsoid (γ_{Q_0}) is described as the gravity anomaly on the geoid (Δg); see Equation (1) below [10,39,41]:

$$\Delta g = g_{P_0} - \gamma_{Q_0} \tag{1}$$

The need for reductions in gravity anomalies to the geoid stems from the requirement of regular grid usage of geoid computation algorithms such as the LSMSA method. Therefore, pointwise gravity anomalies on the Earth's surface are transformed into simple or complete (refined) Bouguer anomalies that have smooth characteristics for interpolation processes [11,13]. Although the interpolation process directly affects the accuracy of the geoid computation, the choice of the optimum interpolation techniques plays an important role [7,46].

In geodesy, Bouguer anomalies are important to use for gridding process and data interpolation purpose [11,13]. These anomalies are also necessary because many geoid modeling algorithms evaluate the gravity data in grid form. Additionally, Bouguer anomalies are also used to explain geological structures in geophysics disciplines [43]. In Bouguer reduction, the topography is represented as plate (or shell), and the gravitation of all the masses within this plate above the geoid are calculated and removed from the observed gravity. Here, the thickness of the Bouguer plate is H_P, and it is equal to the height of topographic point P [40,41]. An illustration of a Bouguer plate is given in Figure 1 below [10,47]:

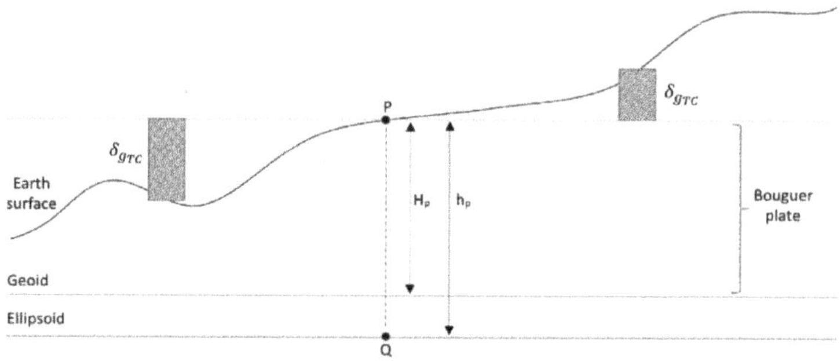

Figure 1. Illustration of a Bouguer plate having H_P thickness and residual topographical masses that deviate from the plate and are subject to calculating terrain corrections (δg_{TC}) in the Bouguer reduction.

Simple Bouguer anomalies (Δg_{SB}) can be computed as shown in Equation (2) [39,44]:

$$\begin{aligned}\Delta g_{SB} &= g_{P_0} - \gamma_{Q_0} + \delta g_F - \delta g_B \\ &= g_{P_0} - \gamma_{Q_0} + 0.1967 H_P\end{aligned} \tag{2}$$

where g_{P_0} is the observed gravity at point P having orthometric height H_P, γ_{Q_0} is the normal gravity value calculated at point Q on the ellipsoid surface, δg_B is the Bouguer plate reduction parameter, and δg_F is the free air reduction parameter. The planar Bouguer reduction parameter is given in Equation (3):

$$\delta g_B = 2\pi G \rho H_P \tag{3}$$

where G is the gravitational constant (G = 6.672585 $\times$ 10^{-11} m^3 kg^{-1} s^{-2}) and ρ is the topographical mass density. In the computations, the mass density is generally assumed to be constant as ρ = 2670 kg m^{-3}. Hence, after substituting the constants into the above formula, it is rewritten as in Equation (4):

$$\delta g_B = 0.1119 H_P \tag{4}$$

in the equation, δg_B is in mGal and H_P is in meters. The free air correction that takes the observed gravity in free air down to the geoid using the vertical gradient of the gravity is formulated as in Equation (5) [5,41]. However, in the equation, there is an assumption because the normal gradient of gravity (associated with the ellipsoidal height h) $\partial\gamma/\partial h$ is used instead of $\partial g/\partial H$. But, in Hofmann-Wellenhof and Moritz [5], it is confirmed that this equation can be used, and it gives sufficient precision for many practical applications (see p. 134 in [5]).

$$\delta g_F = -\frac{\partial g}{\partial H} H \cong -\frac{\partial \gamma}{\partial h} H_P = 0.3086 H_P \tag{5}$$

Complete Bouguer anomalies (Δg_{CB}) are calculated by adding another correction term named terrain correction (δg_{TC}) to simple Bouguer anomalies, which removes the gravitational effect of residual topographical masses deviating from the Bouguer plate (see Figure 1), as formulated in Equation (6) [39,47]:

$$\Delta g_{CB} = g_{P_0} - \gamma_{Q_0} + \delta g_F - \delta g_B + \delta g_{TC} = \Delta g_{FAA} - \delta g_B + \delta g_{TC} \tag{6}$$

In Equation (6), Δg_{FAA} is the free air anomalies and is expressed as in Equation (7):

$$\Delta g_{FAA} = g_{P_0} - \gamma_{Q_0} + \delta g_F \tag{7}$$

The summation of the Bouguer reduction (δg_B) and terrain correction (δg_{TC}) parameters is called terrain reduction.

The normal gravity (γ_0) computed on the reference ellipsoid is calculated using Somigliana formula in Equation (8) [8,43]:

$$\gamma_0 = \gamma_e \frac{1 + k sin^2 \varphi}{\sqrt{1 - e^2 sin^2 \varphi}} \quad (8)$$

where φ is geocentric latitude, γ_e is the normal gravity at the equator ($\varphi = 0$), k is the normal gravity constant, and the eccentricity e is formulated as $e^2 = 1 - \frac{b^2}{a^2}$, where a and b are the semi-major and semi-minor axes of the reference ellipsoid, respectively. The GRS80 (Geodetic Reference System 1980) ellipsoid parameters are used in this study for the computation of normal gravity; see Table 1 for the parameters [48]:

Table 1. GRS80 (Geodetic Reference System 1980) reference ellipsoid parameters.

Constant Parameters	Definitions	GRS80
a	Semi-major axis	6,378,137.0000 m
b	Semi-minor axis	6,356,752.3141 m
γ_e	Normal gravity at equator	9.7803267715 ms^{-2}
γ_p	Normal gravity at pole	9.8321863685 ms^{-2}
GM	Geocentric gravitational constant	3,986,005 × 10^8 m^3 s^{-2}
e^2	First eccentricity	0.00669438002290
k	$(\frac{b\gamma_p}{a\gamma_e} - 1)$	0.001931851353
f	Flattening $(\frac{a-b}{a})$	0.00335281068118

The formula of terrain correction ($\delta_{g_{TC}}$) is given in Equation (9) [2,49]:

$$\delta_{g_{TC}} = G \iint_{\sigma_0} \int_{H_P}^{z} \frac{\rho(x, y, z)(z - H_P)}{\ell^3(x_P - x, y_P - y, z - H_P)} dz d\sigma \quad (9)$$

where H_P is the topographic height at the computation point, z corresponds to the height of the running point, σ_o is the integration area, and ρ is the density of the topographical masses. The distance ℓ between the computation point and the running point can be calculated as shown in Equation (10):

$$\ell = \sqrt{(x_P - x)^2 + (y_P - y)^2 + (H_P - z)^2} \quad (10)$$

In this study, terrain corrections are computed using the TC module of the GRAVSOFT geodetic gravity field computation software [50,51], and the geoid models are calculated using the ITU-LSMSA geoid computation software produced by Istanbul Technical University Gravity Research Group (ITU-GRG).

We utilize the planar Bouguer reduction formulas in the tests since the computation area has a reasonable size and planar approach, satisfying the objectives in this research. However, instead of a planar approach, the gravitation of the topographic masses can also be formulated in a spherical approximation using Bouguer reduction. The Bouguer shell correction (δg_B^s) in the spherical approach is formulated as shown in Equation (11):

$$\delta g_B^s = 4\pi G \rho H_P \quad (11)$$

The terrain corrections in the spherical Bouguer reduction are calculated using global topography data, which is a significant burden. Detailed formulation of the spherical Bouguer reduction schema can be found in Kuhn et al. [43]. Comparative results of planar and spherical approaches in the Bouguer reduction with discussions are given in

Abbak et al. [38], Vanicek et al. [52,53], Erol et al. [54], Hackney and Featherstone [8], and Tziavos and Sideris [41].

Digital elevation model (DEM) data is used in terrain reduction. The height information used in the computation of the Bouguer plate correction and terrain correction parameters are obtained from a high-resolution digital elevation model of the study area. Hence, the resolution and accuracy of the used DEM data affect the accuracy of the gridding process [40–42,55]. In addition, the DEM data are used in the computation of the downward continuation effect in geoid modeling using the LSMSA method (see in Section 2.3) [44,56]. In this study, we use high-resolution global shuttle radar topography mission (SRTM) digital elevation model data with 3 arc-second spatial resolution [57]. The SRTM DEM data are freely available from U. S. Geological Survey [58]. The horizontal position of the DEM grids is in the WGS84 datum, and its height data bases are on the EGM96 global geoid surface. The vertical accuracy of the SRTM3 DEM data is reported as ~9 m over the Earth [57,58]. This accuracy is confirmed with the local validation results from different countries such as the Himalayan site of India [59], Croatia [60], and Turkey [61].

2.2. Interpolation Algorithms

As stated in the previous sections, Bouguer anomalies are employed in the gridding process due to their smooth features, whereas free air anomalies have a higher correlation with the topographical changes [39]. For the gridding process of simple or complete Bouguer anomalies, interpolation algorithms including geostatistical Kriging, inverse distance to a power (IDP—also called IDW), nearest neighbor, and artificial neural network (ANNs) are carried out in this study [62,63]. Thus, the basic characteristics of these methods are described here [64,65].

2.2.1. Nearest Neighbor

The nearest neighbor interpolation algorithm (also known as point-sampling in some cases) is quite simple to implement and is commonly used in spatial data gridding problems. Rather than calculating an average value based on some weighting criteria as in the IDP method, this algorithm selects the value of the nearest point and does not consider the values of the other neighboring points. In data gridding using the nearest neighbor method, the nearest data point to the grid node to be estimated is searched for and used [66–68].

2.2.2. Inverse Distance to a Power (IDP)

This deterministic method is based on a weighted average of the observed values (data points) considering the closeness. The observations closer to the points to be interpolated have higher weights. This method is very fast and useful when the observed values are distributed irregularly or sparse; however, the formation of bull's eye can occur [63,69]. Modifying the smoothing parameter can resolve this problem. IDP can be formulated as shown in Equations (12) and (13) [66,70]:

$$\hat{Z}_j = \frac{\sum\limits_{i=1}^{n} Z_i P_i}{\sum\limits_{i=1}^{n} P_i} \tag{12}$$

$$P_i = \frac{1}{h_{ij}^k} \; and \; h_{ij} = \sqrt{{d_{ij}}^2 + \delta^2} \tag{13}$$

where $\hat{Z}_j$ is the interpolated data, Z_i is the observed data, P_i is the weight of Z_i, n is the number of observations, d_{ij} is the distance between the interpolated and observed value, h_{ij} is the effective separation distance between the interpolated point (j) and the observed point (i), δ is the smoothing parameter, and k is the power parameter (k = 1, 2 or 3).

The power parameter controls how fast the weights decrease as they approach the interpolated value. In other words, the influence of locations far from the estimation points during interpolation decreases with increasing weighting power. Usually, the power parameters are chosen within a range of one to three [63,71]. In our study, the power parameter is chosen as 2. The uncertainty factor can be applied to sample data using the smooth option on the software. By increasing the smoothing parameter, the effects of specific observations on gridding process can be decreased [66].

2.2.3. Kriging

This commonly used geostatistical gridding method, which can be exact or smooth, is based on spatial variance as a distance-dependent function and is applicable to all kinds of datasets. In contrast to IDP and nearest neighbor, Kriging has the slowest computation speed and performs with a smoother display [66,71].

The observed point's spatial continuity or roughness can be described using the variogram. There are numerous variogram models on the Surfer that identify the spatial relationship mathematically, such as exponential, Gaussian, linear, logarithmic, nugget effect, power, quadratic, rational quadratic, spherical, wave, cubic, and penta-spherical. The choice of the drift type identifies the discrimination of ordinary and universal Kriging. If the drift type is chosen as linear or quadratic, the Kriging is termed universal. Otherwise, no drift indicates an ordinary Kriging that presumes a constant unknown mean over the specified area [66].

Another separation can be practiced as point and block Kriging that are applicable to both ordinary and universal Kriging approaches. The points at the grid nodes are calculated using point Kriging, whereas the averages of the blocks are computed using block Kriging by using blocks that are centered on the nodes [65,66,72]. The basic equation of ordinary Kriging is as follows [73,74]:

$$\hat{Z}_j = \sum_{i=1}^{n} Z_i P_i \tag{14}$$

where P_i is the weight, Z_i is the observed data, and $\hat{Z}_j$ is the interpolated data. In the computations, we use ordinary Kriging with a linear variogram model. The Kriging type is the point Kriging.

2.2.4. Artificial Neural Network (ANN)

The artificial neural network (ANN) is a widely used soft computing algorithm that was designed by imitating the human nervous system. The algorithm can be trained using sample data and estimates the target value using network connections based on weights.

The algorithm can be used in many fields such as the automotive industry, banking, amusement sector, aerospace, and so on. Moreover, it can also be used for the calculation of geodetic deformations, prediction of sea level changes, and estimation of the orientation parameters of Earth [75–80].

The algorithm runs with the "nnstart" command on MATLAB, and "nntool" is not available for versions later than R2022a [81]. There are various ANN techniques built into the MATLAB software such as radial basis, generalized regression, NARX, hopfield, feed-forward backpropagation, Elman backpropagation, etc. [75,82,83]. Here, the feed-forward backpropagation technique is used with the "trainlm" function, which specifies the Levenberg–Marquardt approach. Additionally, the "tansig" transfer function and "learngdm" adaption learning function with 2 layers and 200 neurons are chosen to specialize the method (see Figure 2). In contrast to the Levenberg–Marquardt approach, which is fast but requires a high memory capacity, there are also two other common methods named Bayesian regularization (trainbr) and scaled conjugate gradient (trainscg). See Demuth et al. [75] for further information.

The network geometry of feed-forward backpropagation consists of neurons within parallel layers that are classed as input, hidden, and output layers. The data are presented to the network via the input layer, whereas the desired output values to be obtained are stored in the output layer. The hidden layer is where all the calculations are performed [77,84].

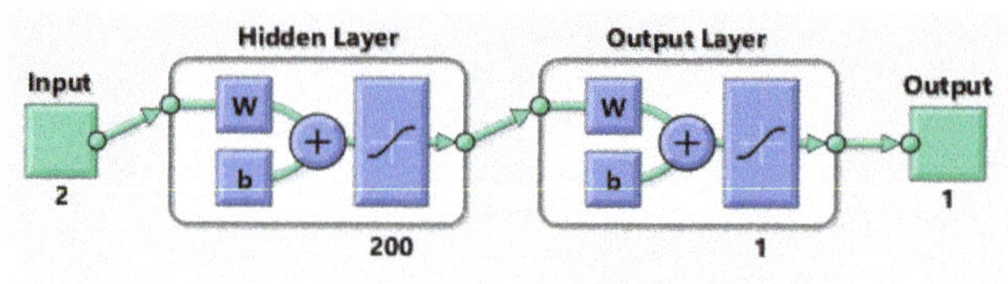

Figure 2. Architecture of the artificial neural network algorithm in MATLAB (w: weights, b: bias).

In this study, the ANN tool is applied to pointwise complete Bouguer anomalies to obtain gridded data to generate free air anomalies, which are the main component of the gravimetric geoid computation. Using 244,009 pointwise complete Bouguer anomalies (CBAs), the training process is performed on 15% test and 15% validation points. Then, the CBAs on 173,641 grid nodes are estimated by simulating the trained network. The illustrations that reflect the regression and performance of the training based on Levenberg–Marquardt method are given in Figure 3 (Network: 200, max fail: 6, epoch: 300).

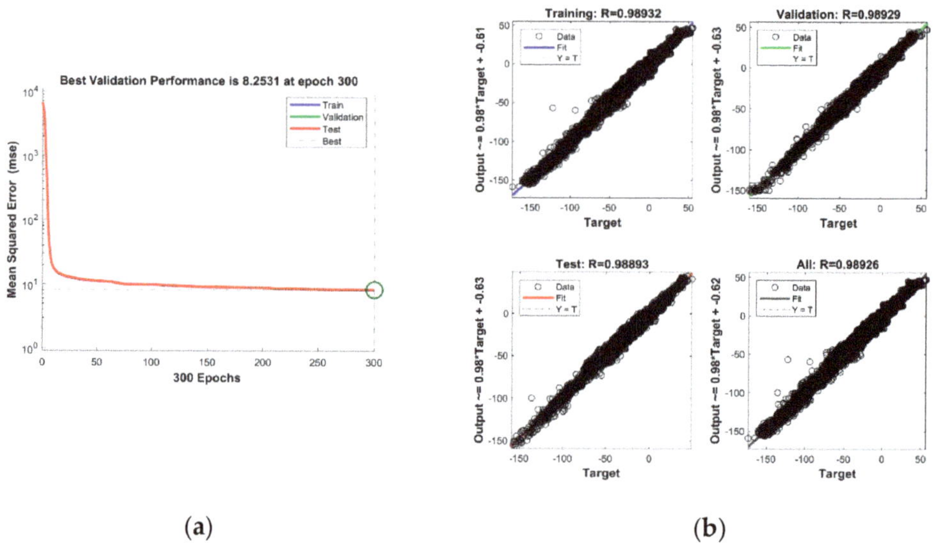

Figure 3. Neural network training performance (**a**) and regression (**b**).

2.3. Gravimetric Geoid Determination Using the LSMSA Method

The geoid undulation can be computed based on the gravity anomaly using the Stokes formula [2,10,13]:

$$N = \frac{R}{4\pi\gamma}\iint_{\sigma} S(\Psi)\Delta g d\sigma \tag{15}$$

where $S(\Psi)$ is the Stokes function, Ψ is the geocentric angle, Δg is the gravity anomaly, $d\sigma$ is the infinitesimal surface element of the unit sphere, R is the mean Earth radius, and γ is the normal gravity on the reference ellipsoid. The aforementioned equation requires gravity information on the entire globe that is not practically possible. Therefore, the method needs to be used in a limited region by neglecting gravity data in remote zones [7,13,22]. This limitation causes a truncation error, and a stochastic solution called the modified Stokes formula with additive corrections (LSMSA) is applied to overcome the truncation problem. The least squares modification of Stokes integral with additive corrections (LSMSA) method (also called the KTH method) was developed by the Royal Institute of Technology (KTH) in Sweden. In this method, the predicted global mean square error of the formula is minimized using the least squares concept with the contributions of GGM and terrestrial gravity data (see equation below) [13–20,23,85]:

$$\hat{N} = \tilde{N} + \delta N_{com}^{top} + \delta N_{DWC} + \delta N_{com}^{atm} + \delta N_{ell} \quad (16)$$

where $\hat{N}$ is the geoid undulation and $\tilde{N}$ is the approximate geoid height. In the equation, δN_{com}^{top} is the combined topographic correction, δN_{DWC} is the downward continuation effect, δN_{com}^{atm} is the combined atmospheric correction, and δN_{ell} is the ellipsoidal correction. All together, these correction parameters are called additive corrections. Further information regarding their formulations can be found in references [13,14,86–88]. The approximate geoid height $\tilde{N}$ can be computed as follows:

$$\tilde{N} = \frac{R}{4\pi\gamma} \iint_{\sigma_0} S^L(\Psi)\Delta g d\sigma + \frac{R}{2\gamma} \sum_{n=2}^{M} b_n \Delta g_n \quad (17)$$

where $S^L(\Psi)$ is the modified Stokes function, L is the maximum degree of modified harmonics (GGM expansion degree), M is the upper limit of the GGM, σ_0 is the spherical cap, b_n is a modification parameter, and Δg_n is the gravity anomaly obtained from the GGM. For the computation of $S^L(\Psi)$, the following formula can be used:

$$S^L(\Psi) = \sum_{n=2}^{\infty} \frac{2n+1}{n-1} P_n(cos\Psi) - \sum_{n=2}^{L} \frac{2n+1}{2} s_n P_n(cos\Psi) \quad (18)$$

where $P_n(cos\Psi)$ is a Legendre polynomial and s_n is the stochastic modification parameter. According to Abbak et al. [22], Abbak and Ustun [11], and Sjöberg [14], the b_n parameter can be formulized in three ways: biased ($b_n = s_n$), unbiased ($b_n = s_n + Q_n^L$), and optimum ($b_n = \frac{(s_n + Q_n^L)c_n}{(c_n + dc_n)}$). In this above formula, dc_n is the error degree variance, Q_n^L is truncation coefficient, and c_n is signal degree variance [13].

The LSMSA method described so far is applied in our study area using ITU-LSMSA software. The choice of the optimum parameters for the computation of the geoid models using the LSMSA method is crucial. These parameters are integration cap size (Ψ_0), the global geopotential model expansion degree (M), and the upper bound of the Stokes function (L) [1,7,89]. The lack of gravity data can be resolved using high-degree GGMs; however, this increases the harmonic coefficient errors in the resultant geoid height parameter. Therefore, special attention should be paid to select an optimal expansion degree of GGM [11,13]. Here, the optimum integration radii (Ψ_0), expansion degree of GGM (M), error variance (C), and biased/unbiased/optimum b_n parameter choices are investigated for the study area. The decision regarding the optimal computation parameters in geoid modeling is based on a trial-and-error based test procedure. Therefore, all the geoid models computed with varying parameters are compared with each other. As a GGM, which is the demonstration of Earth's gravitational field using spherical harmonic coefficients, XGM2019e is preferred [8]. The XGM2019e model coefficients are freely available from ICGEM [90], and an application using the XGM2019e model can be found in Tocho et al. [91]. The XGM2019e global geopotential model is up to degree and order 2190. Zingerle et al. [92] provides a comparison of the XGM2019e with its high-resolution counterparts: the EIGEN-6C4 and EGM2008 models [93].

3. Case Study

In this section, firstly, we introduce the study area and test data. Then, the results of the numerical studies are presented. In the numerical test results section, firstly, the effects of simple and complete Bouguer anomalies on the geoid computation are investigated. The optimum parameters used in the LSMSA gravimetric geoid calculation are determined afterwards. As the last topic of this section, the influences of interpolation methods on the geoid determination when gridding complete Bouguer anomalies are analyzed.

3.1. Study Area and Data

The Auvergne test region covers a 6° × 8° area. The dense and high-accuracy gravity observations measured in Auvergne are shared as the test dataset for gravity field modeling studies by the Institut Geographique National (IGN) in Bureau Gravimetrique International (BGI) database. These data are freely available to researchers in order to test geoid computation techniques. In the dataset, there are 244,009 gravity points within the coordinates in latitudes of 43° N–49° N and longitudes of 1° W–7° E. The approximate spatial density of the gravity observations corresponds to 1 point in every 1.3 km. This approximate distance between the gravity observations in the area is considered when determining the grid spatial resolution during the data gridding process. The accuracy of the gravity data is reported as ~2 mGal [94].

In addition to the gravity observations in the Auvergne test area, 75 GPS/leveling points are also available for validations of the calculated geoid models. However, the GPS/leveling points are distributed in a limited area in the center of Auvergne, and these control benchmarks cover the area between the coordinates in latitudes 45° N–47° N, and longitudes 1.5° E–4.5° E. Their orthometric heights in the National Leveling Network (NGF-IGN69) datum and ellipsoidal heights in the ITRF datum are provided with a ~2 cm accuracy in the dataset. The topographical heights of the GPS/leveling benchmarks change from 206.8 m to 1235.5 m [28,46,95]. Figure 4a shows the topography of the Auvergne test area and the distribution of the 75 GPS/leveling benchmarks in the area [57,96]. The distribution of gravity points in the area is shown in Figure 4b [96].

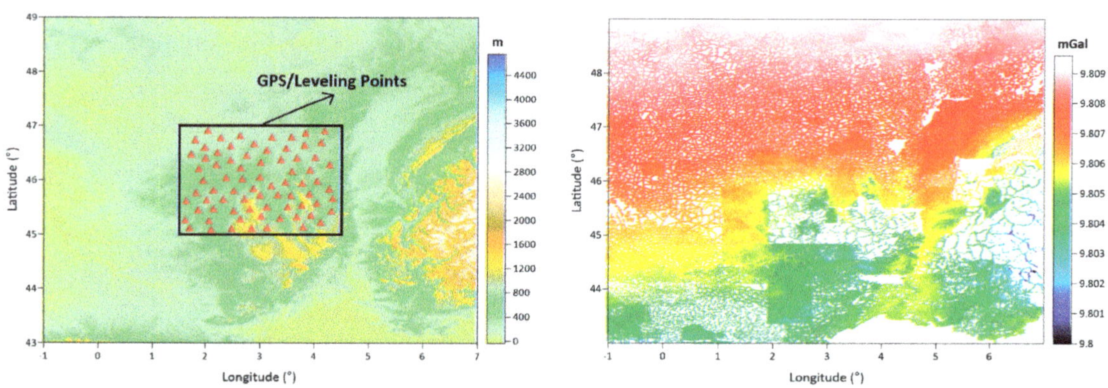

Figure 4. Topography in the Auvergne test area and distribution of 75 GPS/leveling benchmarks (red triangles) in the area (**a**) and distribution of the points with gravity observations in the area (**b**).

Figure 5 shows the steps that were followed in the numerical tests in this study.

3.2. Simple and Complete Bouguer Anomalies in Gravity Gridding

In the first part of the numerical tests, the role of the Bouguer anomaly type in gravity gridding is investigated. Both simple and complete Bouguer anomaly datasets in the grid form are also examined in the geoid determination in the test area, and the differences between the gridded gravity datasets and calculated geoid models using them are compared and interpreted using the derived statistics and maps.

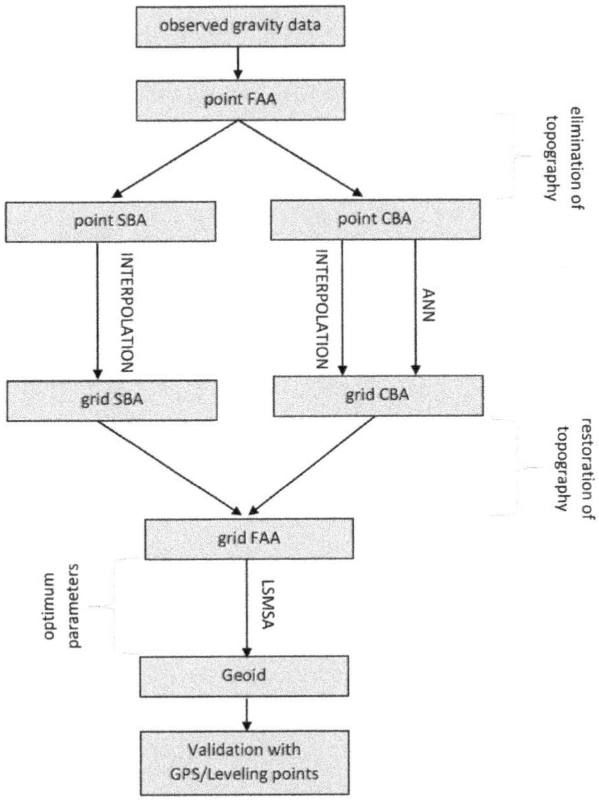

Figure 5. Followed data processing steps in numerical tests.

Complete Bouguer anomalies (CBAs) are computed by adding the terrain correction to the simple Bouguer anomalies (SBAs) according to Equation (6). For the computation of terrain corrections (TCs) using Fortran-based GRAVSOFT geodetic gravity field computation software, 3″ and 30″ resolution SRTM DEMs are used, and the adopted inner and outer radii of integration are r_1 = 40 km and r_2 = 100 km, respectively. The most time-consuming part in obtaining the CBAs is terrain correction computations, which take 8 h using a Razer Blade 14 Laptop that has Intel (R) Core™ i7-6700HQ 2.60 GHz Processor, 16 GB RAM, and a 64-bit Windows 10 operating system. In order to acquire the gridded free air anomalies (FAAs) for the geoid computation, terrain corrections are estimated once again on the grid nodes using the same computation parameters for the Bouguer reduction process ($\Delta g_{FAA}^{GRID} = \Delta g_{CBA}^{GRID} + \delta g_{B}^{GRID} - \delta g_{TC}^{GRID}$). Figure 6a maps the terrain corrections at the gravity points that are used for computing the CBAs before gridding the data. Then, the calculated CBAs are gridded with a 1 arc-minute spatial resolution. When deciding the spatial resolution of the grid, the density of the original gravity observations is considered. This is because, in the original dataset, the gravity points given every 1.3 km correspond to a distance of approximately 1 arc minute. After gridding the CBAs, the free air anomalies in grid form are calculated. In order to restore the free air anomalies from the CBAs, the terrain corrections are calculated for the grid nodes. Figure 6b shows the grid terrain corrections with a 1 arc-minute resolution. Table 2 gives the statistics of the terrain corrections both at the gravity points (Point TC) and at the grid nodes (Grid TC) as well as the change in the topographic heights in the area based on the SRTM 3 DEM data.

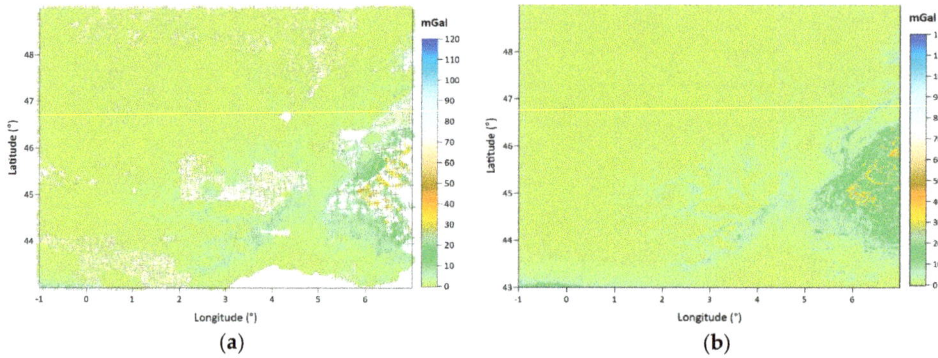

Figure 6. Terrain corrections at gravity points used for computing the complete Bouguer anomalies before gridding the gravity data (**a**) and terrain corrections at grid nodes having 1 arcminute grid spacing used to restore free air anomalies from the complete Bouguer anomalies (**b**).

Table 2. Statistics of terrain corrections at the gravity points (point TC) and at the grid nodes (grid TC), and the topographic heights in the area based on the SRTM 3″ DEM data.

	Max	Min	Mean	Median	Std	RMSE
Point TC (mGal)	56.533	−0.317	1.043	0.476	2.153	2.392
Grid TC (mGal)	118.221	−0.327	1.718	0.371	4.086	4.432
3″ DEM (m)	4745.649	−33.506	378.900	230.454	433.149	575.485

Considering the statistics given in Table 2, the topographic heights reach ~4750 m from sea level. The terrain corrections in the area vary between −0.317 mGal and 56.533 mGal at the pointwise data. When the TC values are calculated for the grid nodes, they are between −0.327 mGal and 118.221 mGal, with a 4.086 mGal standard deviation. In the table, the root mean square error (RMSE) values are also given (RMSE = $\sqrt{\mu^2 + \sigma^2}$, where μ: mean, and σ: standard deviation).

Figure 7a,b show the maps of the SBAs and CBAs that are calculated at the gravity observation points and then gridded using the Kriging interpolation algorithm with 1 arc-minute spacing.

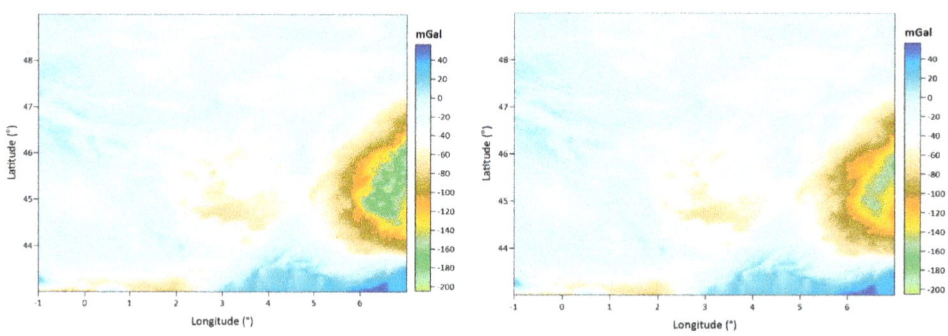

Figure 7. Simple Bouguer anomalies (SBAs) calculated at gravity points and interpolated at 1 arc-minute spacing grid nodes using Kriging (**a**). Complete Bouguer anomalies (CBAs) calculated at gravity points and interpolated at 1 arc minute spacing grid nodes using Kriging (**b**) (unit: mGal).

The difference map of the SBA and CBA values is also given in Figure 8, corresponding to the terrain corrections between the CBA and SBA values as given in Equation (6). The surface pattern of the difference map is naturally similar to the given terrain correction maps in Figure 6.

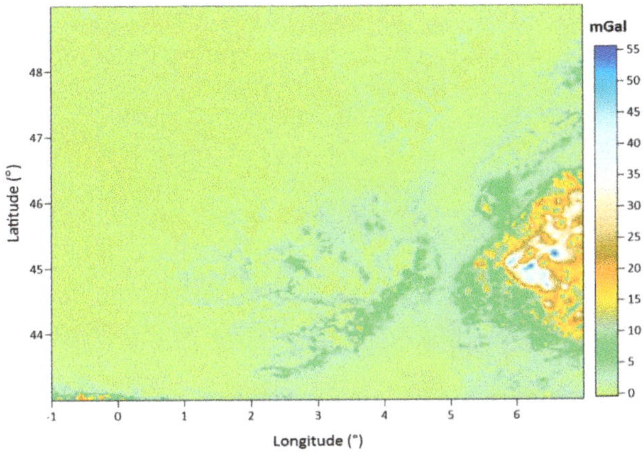

Figure 8. Difference map of the SBAs and CBAs at 1 arc-minute spacing grid nodes (unit: mGal).

Table 3 gives the statistics of the SBA, CBA, and their differences. Considering the table, it is seen that the SBAs change between −204.830 mGal and 56.373 mGal with a 31.640 mGal standard deviation, whereas the CBAs are between −172.656 mGal and 57.271 mGal with a 27.905 mGal standard deviation. Their differences are between −0.439 mGal and 55.590 mGal, with a 4.879 mGal standard deviation. These statistics are consistent with the terrain correction statistics at the points given in Table 2.

Table 3. Statistics of the SBAs and CBAs calculated at gravity points and interpolated at 1 arc-minute spacing grid nodes using the Kriging method, and their differences.

Unit: mGal	Max	Min	Mean	Median	Std	RMSE
SBA	56.373	−204.830	−34.170	−28.652	31.640	46.569
CBA	57.271	−172.656	−32.238	−28.097	27.905	42.638
Difference	55.590	−0.439	1.932	0.396	4.879	5.247

According to these comparison statistics, a significant difference between the SBAs and CBAs is found. On the other hand, when the difference map is considered, the differences between the SBAs and CBAs are clear at the mountainous part of the test area. However, in plain topography, their differences seem to be ignorable. This result arises from the fact that the terrain correction parameter increases significantly with topographic heights. Obtaining the difference between the calculated terrain corrections at the grid nodes and the derived terrain corrections by subtracting the interpolated SBAs from the CBAs emphasizes the role of the interpolation process in gravity field mapping. In the following step, we inspect whether the choice of Bouguer anomaly type affects the geoid modeling or not. Thus, we calculate the free air anomalies from the gridded SBA values and gridded CBA values, respectively. The free air anomaly datasets in grid form are used as inputs in the geoid model computations using the LSMSA approach.

Figure 9a,b show the free air anomalies in the grids calculated from the SBA map (in Figure 7a) and CBA map (in Figure 7b), respectively. In Figure 10, a difference map of these two free air anomaly grids is given. The descriptive statistics of these grid free air anomaly datasets and their differences are given in Table 4. The table shows that the difference between two free air anomaly grids is 37.167 mGal and −86.721 mGal extremums. Figure 10 shows the locations where these extreme differences occur between the two grid datasets.

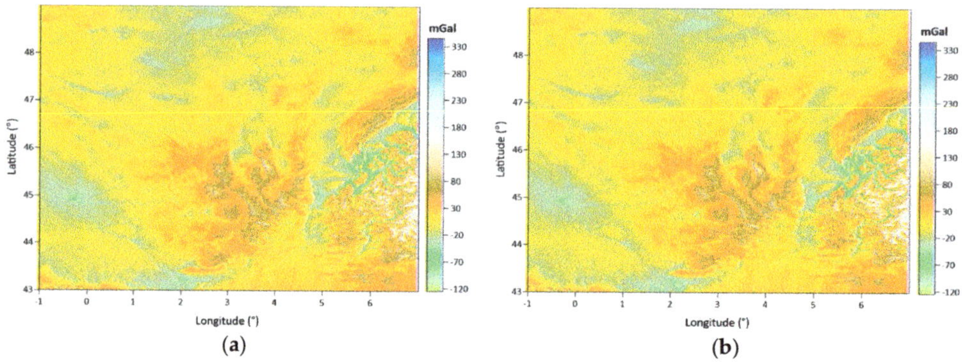

Figure 9. Free air anomalies in grid form with a 1 arc-minute resolution, obtained from: the SBAs in the grids (**a**) and the CBAs in the grids (**b**) (unit: mGal).

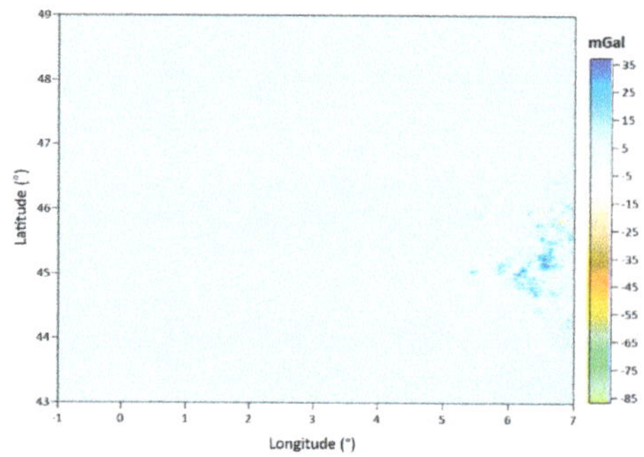

Figure 10. Difference map of free air anomaly grids produced from the SBAs and CBAs in the grids (unit: mGal).

Table 4. FAA gridding statistics of simple and complete Bouguer anomalies.

Unit: mGal	Max	Min	Mean	Median	Std	RMSE
FAA (via SBAs)	345.683	−126.313	8.245	1.888	29.463	30.595
FAA (via CBAs)	258.962	−129.942	8.458	1.948	29.867	31.042
Difference	37.167	−86.721	0.213	0.016	2.103	2.114

Thereafter, two different gravimetric geoid models are computed using these free air anomaly grids. In the geoid model computations using the LSMSA method, the adopted computation parameters are as follows: the spherical harmonic expansion degree of the geopotential model is $M = 780$, the error variance value is $C = 4$ mGal^2, the type of modification is used as b_n = biased, and the radius of integration is $\Psi_0 = 0.25°$. In the following text, a section (Section 3.3) explaining the selection process of the optimum parameters for the LSMSA geoid model computation method is given. Figure 11a,b shows the geoid maps, which are calculated using the free air anomalies derived from the SBA's grid and CBA's grid, respectively. The geoid models are calculated in 1 arc-minute spacing grid form. In Figure 12, the map shows the differences between these two geoid models. As seen from the difference maps given in Figures 10 and 12, the geoid height differences exhibit a similar distribution pattern to the free air anomaly differences, and the differences increase with topographical heights. Table 5 gives the statistics of the calculated geoid models as well as the statistics of their differences. In the first two lines of the table, the statistics of the two models seem to be similar, having the same mean and standard deviation. Although in this view, the difference statistics show that the two models actually differ from each other with −19.5 cm and 27.3 cm minimum and maximum values, respectively.

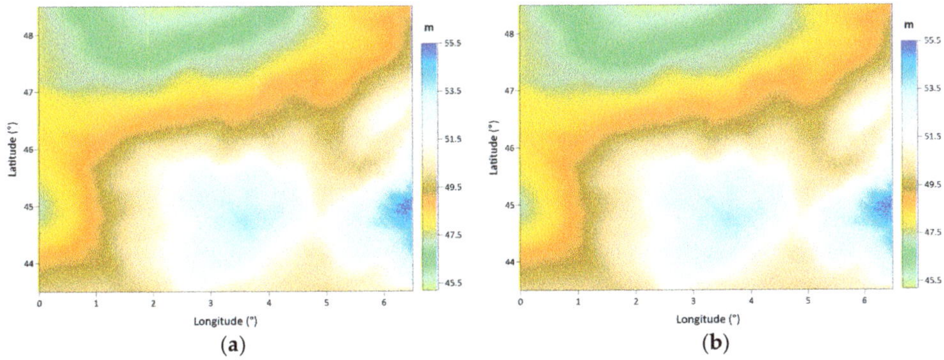

Figure 11. Geoid models in 1 arc-minute grid resolution that are calculated with free air anomalies derived from: the SBAs (**a**) and the CBAs grids (**b**) (unit: meter).

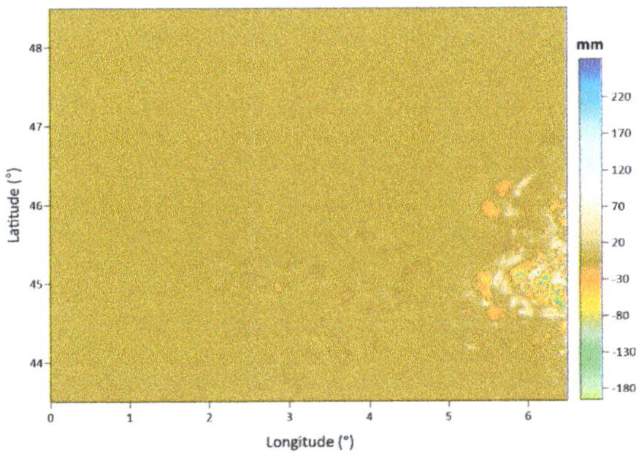

Figure 12. Difference map of geoid models given in Figure 11a,b (unit: mm).

Table 5. Statistics of geoid models calculated using free air anomalies derived from the SBAs and CBAs grid data.

Unit: m	Max	Min	Mean	Median	Std	RMSE
SBA based GEOID	55.743	45.268	49.616	49.567	2.021	49.657
CBA based GEOID	55.470	45.267	49.616	49.567	2.021	49.657
Difference	0.273	−0.195	0.000	0.000	0.008	0.008

In addition to the comparison of the two geoid models, they are also individually validated at GPS/leveling benchmarks in the area. The validation of the geoids relies on Equation (19):

$$\Delta N = N^{GPS/Lev} - N^{Geoid} \tag{19}$$

In the equation, the geoid height ($N^{GPS/Lev}$) obtained from GPS-height (h) and orthometric height (H) at a benchmark ($N^{GPS/Lev} = h - H$) is compared with the geoid height derived from the geoid model (N^{Geoid}). In the validations, the N^{Geoid} parameter is interpolated from the geoid model using the inverse distance to a power algorithm. Table 6 gives the validation statistics of the geoid models. According to these statistics, both models have the same absolute accuracy based on the standard deviation of the geoid height differences at the benchmarks, and this accuracy is 4.1 cm. The difference between the two geoid models is not visible in the GPS/leveling validation results. This is natural because the GPS/leveling control benchmarks are distributed in a limited area in the region and this area does not represent the topographical changes. Because of this disadvantage of the independent validation dataset, the differences between the two geoid models at the high topographical parts do not affect the validation statistics.

Table 6. Validation statistics of performed geoid via SBA and CBA grids.

Unit: m	Max	Min	Mean	Median	Std	RMSE
ΔN SBA	1.161	0.973	1.059	1.064	0.041	1.060
ΔN CBA	1.162	0.972	1.059	1.064	0.041	1.060

3.3. Optimum Parameters for the LSMSA Geoid Calculation Method

The geoid model computation results using the LSMSA method depend on the used computation parameters and determining the optimum parameter set, which gives the most accurate geoid model bases on a trial-and-error procedure. In this study, in order to determine the optimum computation parameters, we run the trial-and-error process using free air anomaly grids derived from the complete Bouguer anomalies using the Kriging interpolation approach. Each trial is made using a different parameter set, and the computed geoid model is validated at the GPS/leveling benchmarks. The comparison of the absolute accuracies of the solutions eventually leads to a determination of the optimum parameter set.

At first, the geoid models are calculated using varying harmonic expansion degrees of geopotential model (M) from 180° to 780° and changing the cap size for the assigned values of Ψ_0 = 0.10°, 0.25°, 0.50°, 0.75°, and 1.00°, respectively. In this first attempt, the error variance (C = 4 mGal^2) and b_n parameters (b_n = optimum) are kept unchanged. The validation results of the calculated geoid models are given in Table 7.

Table 7. Standard deviations of the geoid heights differences from the validations of geoid models calculated using varying M and Ψ_0 computation parameters.

Unit: m	M = 180	M = 240	M = 300	M = 360	M = 420	M = 480	M = 540	M = 600	M = 660	M = 720	M = 780
Ψ = 0.10°	0.261	0.109	0.096	0.083	0.078	0.076	0.076	0.076	0.076	0.076	0.075
Ψ = 0.25°	0.174	0.146	0.142	0.139	0.131	0.112	0.102	0.085	0.072	0.067	0.065
Ψ = 0.50°	0.183	0.188	0.134	0.100	0.092	0.091	0.091	0.091	0.087	0.082	0.079
Ψ = 0.75°	0.197	0.142	0.117	0.107	0.121	0.115	0.110	0.106	0.109	0.111	0.112
Ψ = 1.00°	0.197	0.158	0.153	0.172	0.175	0.166	0.167	0.183	0.188	0.195	0.192

The test results given in Table 7 reveal the optimum values for a harmonic expansion degree of M = 780 and a cap size (integration radius) of Ψ_0 = 0.25°. With these two parameters and b_n = optimum acceptance, the most appropriate error variance (C) value is inspected. The inspection results, reported as the means and standard deviations of the geoid height differences at the GPS/leveling points, are given in Table 8.

Table 8. Standard deviations of the geoid heights differences based on the validations of the geoid models calculated with varying error variance C parameters.

Unit: m	Max	Min	Mean	Std	RMSE
C = 1 mGal^2	1.113	0.803	0.987	0.065	0.989
C = 4 mGal^2	1.113	0.802	0.987	0.065	0.989
C = 9 mGal^2	1.113	0.801	0.986	0.066	0.989
C = 16 mGal^2	1.113	0.801	0.986	0.066	0.988

The validation statistics show that the effects of the error variance parameter on the geoid model accuracy are not as significant as the previous two parameters. Since all the solutions have similar performances, the error variance C = 4 mGal^2 is chosen as suitable based on the reported accuracy of the terrestrial gravity data.

Finally, the modification type (b_n) as biased, unbiased, or optimum is decided. Table 9 gives the validation results of the calculated geoid models. In the obtained results, the biased solution is selected for use. Thus, in the following tests, to clarify the role of the interpolation algorithm in gravity data gridding, geoid computations are carried out using computation parameters including a harmonic expansion degree of $M = 780$, a cap size of $\Psi_0 = 0.25°$, an error variance $C = 4$ mGal^2, and a biased solution.

Table 9. Standard deviations of the geoid heights differences from the validations of geoid models calculated for different modification choices (b_n).

Unit: m	Max	Min	Mean	Std	RMSE
biased	1.162	0.972	1.059	0.041	1.060
unbiased	1.113	0.815	0.991	0.063	0.993
optimum	1.113	0.802	0.987	0.065	0.989

3.4. *Effect of Gravity Data Interpolation Method in Geoid Model Computation*

The last part of the numerical tests in this study aims to explain the role of the interpolation algorithm in gravity gridding for preparing the Bouguer gravity anomaly maps and input gravity grid for geoid model computations. Based on previous tests, we agreed on the type of Bouguer gravity anomalies to use in this section. Based on the obtained results in Section 3.2, we use complete Bouguer gravity anomalies in this section, since the test area has a miscellaneous topographical pattern, and in the mountainous part, the differences between the SBAs and CBAs are significant. Figure 13a shows the complete Bouguer gravity anomalies at the gravity points reduced to the geoid surface. This map shows the CBAs before gridding with any interpolation algorithm. In the tests, these data are gridded with 1 arc-minute spacing (corresponding to the approximate spatial density of the gravity observations on the topography) using the ordinary Kriging, inverse distance to a power (IDP), nearest neighbor, and artificial neural network (ANN) algorithms. Among these methods, Kriging is a commonly used method in gravity field and geoid determination and is provided to users as an integrated computation module of the geoid model computation software. The gravity field and geoid modeling are based on measurements, and errors are unavoidable in the content of these measurements. These errors need to be minimized in the modeling process by employing an appropriate stochastic strategy. Kriging predicts the value of a function at a given point by computing the weighted average of the data in the neighborhood of a given point. However, the Kriging algorithm makes use of the Gauss–Markov theorem to consider the estimated value and its error independently and to provide a best linear unbiased estimator at an unsampled location (interpolation point) based on the adopted assumption of covariances. This advantage of the Kriging algorithm makes it a commonly used method in gravity field prediction studies.

The inverse distance to a power and nearest neighbor algorithms are also the most widely used approaches in spatial data-based applications because they are practical and fast algorithms that provide an adequate accuracy for most cases. In terms of formulation, these two methods are similar, but IDP is an advanced version of the nearest neighbor method. IDP allows for the inclusion of more observations than only the nearest observation. The value at the grid node is calculated from a linear combination of the neighboring stations. The weight of each data point is determined by the distance, which may not be linear depending on the assigned power.

The artificial neural network (ANN) algorithm is the last method that is tested in this section. This algorithm is based on a mathematical architecture and adopts a "feed forward–back propagation" optimization-based processing strategy. Due to these aspects, it is quite different from the first three methods tested here. The mathematical base of this method makes it more complicated and includes a number of computational parameters that must be determined in order to obtain optimal results. In the ANN interpolation in this study, we use the Levenberg–Marquardt training method since it is highly recommended in the neural network literature. We utilize trial-and-error tests in order to decide the optimal numbers of neurons and iterations for training. In the result, 200 neurons and 300 iterations are found to be optimal in these computations.

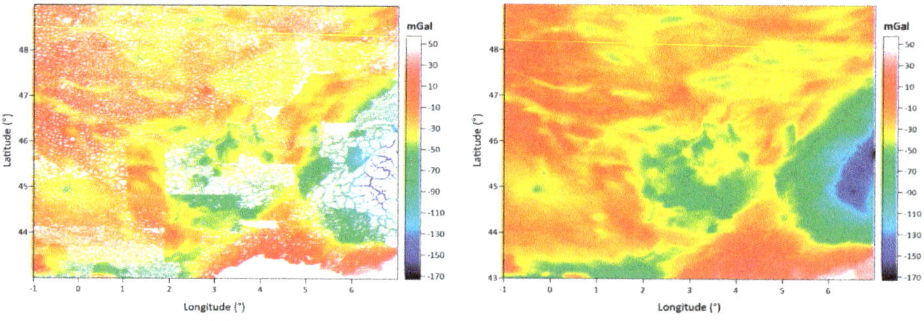

(**a**) (**b**)

Figure 13. Complete Bouguer gravity anomalies at gravity points reduced to the geoid surface (before gridding) (**a**). Map drawn with a 1 arc-minute resolution CBA grid calculated using the Kriging interpolation algorithm (**b**).

Figure 13b shows the map drawn using the 1 arc-minute resolution CBA grid calculated using the Kriging interpolation algorithm.

Figure 14a–c shows the Bouguer gravity anomaly maps, which are drawn using the CBA grid data calculated using inverse distance to a power, nearest neighbor, and artificial neural network interpolation algorithms, respectively.

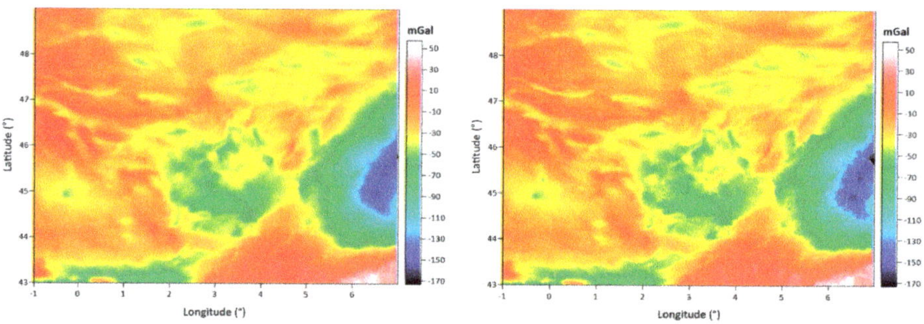

(**a**) (**b**)

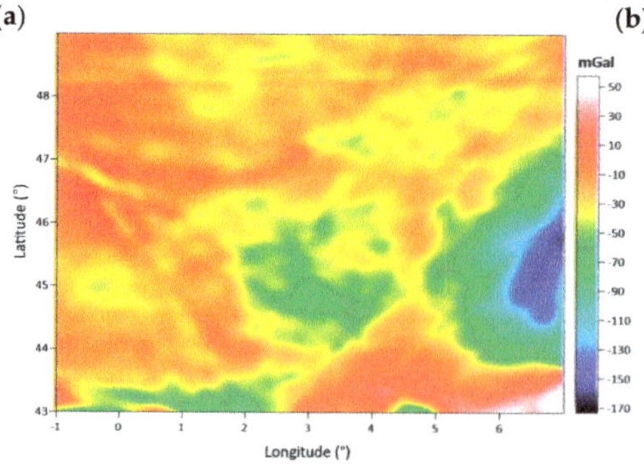

(**c**)

Figure 14. Bouguer gravity anomaly maps drawn using 1 arc-minute resolution CBA grids calculated using: inverse distance to a power (**a**), nearest neighbor (**b**), and artificial neural network (**c**).

Table 10 gives the statistics of the CBAs grid datasets calculated using each interpolation algorithm. In terms of the basic statistics given in the table, the interpolation methods seem to generate similar grids, and there are no significant differences among the grids. However, to verify whether the generated grids are really identical, we calculated the differences in the IDP, nearest neighbor, and ANN grids from the Kriging grid, which is assumed as the control dataset. Since Kriging is a widely used algorithm in gravity field calculation studies, we assign this algorithm as the control method.

Table 10. Grid statistics belonging to the complete Bouguer anomalies (CBAs) based on the used interpolation algorithms.

Unit: mGal	Max	Min	Mean	Median	Std	RMSE
Kriging	57.282	−172.616	−32.234	−28.095	27.901	42.632
IDP	55.503	−172.575	−32.310	−28.028	27.676	42.543
Nearest Neighbor	56.650	−173.173	−32.290	−28.087	27.866	42.652
ANN	55.196	−159.404	−32.538	−27.853	27.779	42.783

Figure 15a–c show the differences in the Kriging grid from the IDP, nearest neighbor and, ANN grids, respectively. When the grid difference maps are considered, it is seen that although the basic statistics are close, the generated grids with each algorithm represent different surface patterns. Additionally, the distribution of the grid differences between the datasets varies. Regarding this situation, the Kriging grid and nearest neighbor grid have the maximum consistency, both dataset fit better in the plain topography, and their differences increase as the topography rises. In Figure 15a, the grid differences between the Kriging and the IDP methods exhibit a seemingly homogeneous distribution over the area. The magnitude of their differences increases throughout the mountainous part of the area. In Figure 15c, the grid differences between the Kriging and the ANN methods represent a wavy pattern and an almost homogeneous distribution over the entire area. Based on this map, we do not recognize any correlation between the distribution of grid differences and the topography. Regarding the distribution pattern of grid differences in this map, we can say that the Kriging grid has a minimum consistency with the ANN grid among the datasets. Table 11 gives the statistics of the grid differences. According to these statistics, the Kriging grid fits the IDP and nearest neighbor grids with 1.3 mGal and 1.6 mGal standard deviations, and regarding the mean values of grid differences, there is no significant offset between these datasets. However, when the grid differences in Kriging and the ANN are considered, these two datasets deviate from each other, with a 5.3 mGal standard deviation and a 0.3 mGal offset between them. Therefore, these statistics confirm the visual interpretation of the grid difference maps. In each comparison, the grid difference maps exhibit a large difference localized in the southeast corner of the map sheets. The reason for this local discrepancy may be the low performance of the Kriging algorithm in that local area. That part corresponds to a coastline and lack of data at the sea part (southeast of the Auvergne test area), which may not be handled properly by the Kriging algorithm.

Table 11. Statistics of CBA grid differences between the Kriging grid compared to the IDP grid, nearest neighbor grid, and the ANN grid.

	Unit: mGal	Max	Min	Mean	Median	Std	RMSE
KRIGING vs.	IDP	13.646	−26.880	0.076	0.009	1.314	1.316
	Nearest Neighbor	45.066	−34.711	0.056	0.006	1.618	1.619
	ANN	100.999	−37.245	0.304	0.082	5.310	5.318

In the following tests, we analyze the consequences of the differences between the CBA grids on the geoid model determination using the LSMSA method. In order to carry out the geoid model computations with free air anomaly grids on the Earth's surface, we continue the CBA grids upward based on Equations (6) and (7). In the result, the free air anomaly grids, which are respectively obtained from the Kriging, IDP, nearest neighbor, the ANN complete Bouguer anomaly grids, are given in Figures 16b and 17a–c. In order to provide a comparison, the free air anomaly values at the gravity points (without any gridding) are also provided in Figure 16a. Table 12 provides the statistics of the free air anomaly grid datasets.

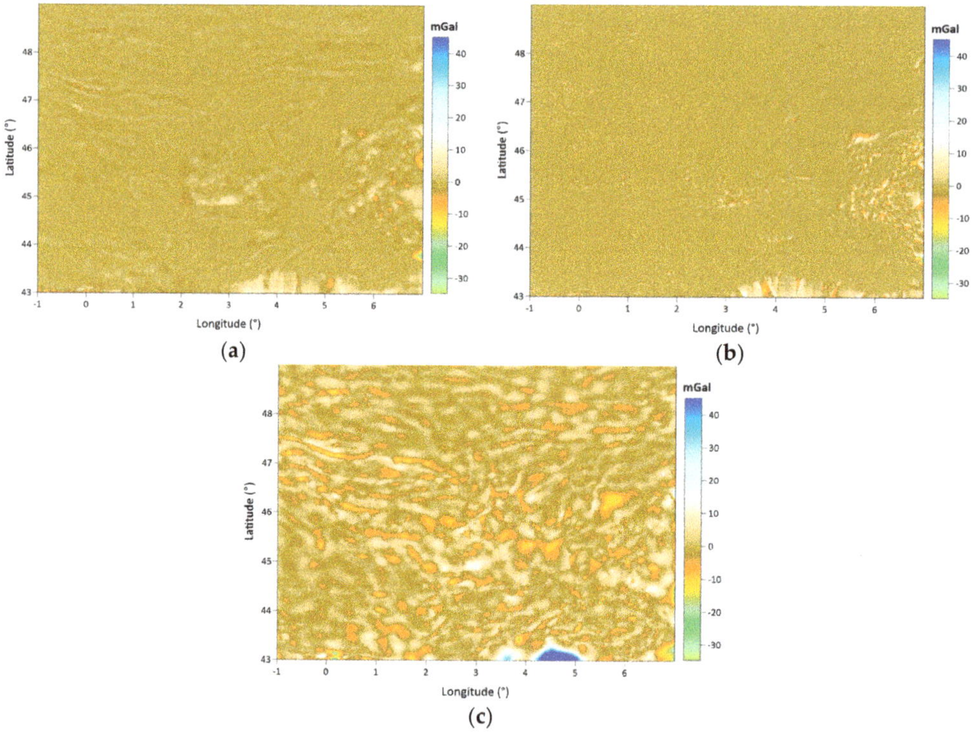

Figure 15. Difference maps of the Kriging grid compared to: the IDP grid (**a**), nearest neighbor grid (**b**), and the ANN grid (**c**) (complete Bouguer anomaly differences at 1 arc-minute resolution grid nodes).

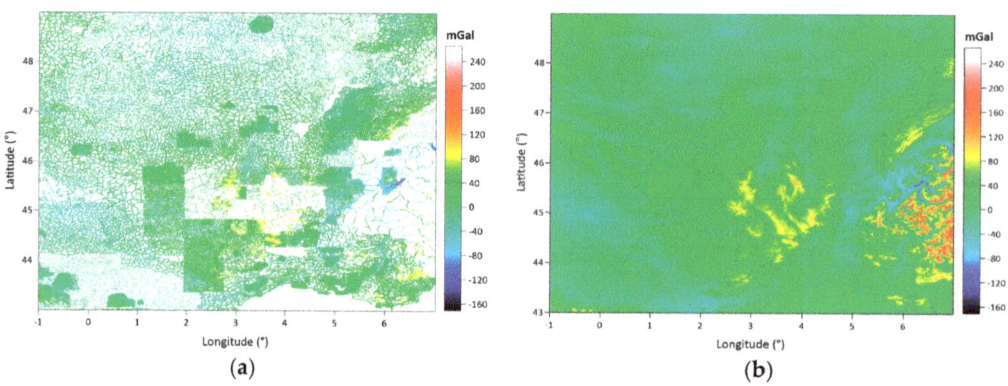

Figure 16. Free air anomalies at gravity points (without any gridding) (**a**), free air anomaly grid map obtained from Kriging CBAs grid with 1 arc-minute spacing (**b**).

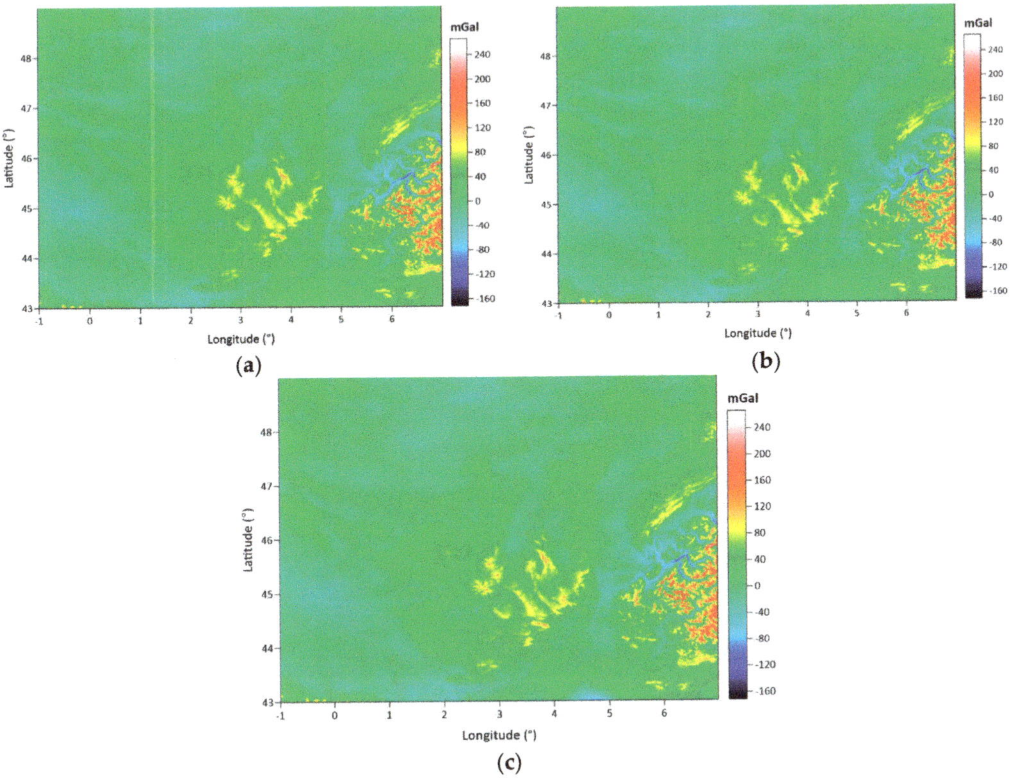

Figure 17. Free air anomaly grid maps obtained from: the IDP CBAs grid (**a**), nearest neighbor CBAs grid (**b**), and the ANN CBAs grid (**c**) with 1 arc-minute grid spacing.

Table 12. Statistics of free air anomaly grids obtained from Kriging, the IDP, nearest neighbor and the ANN CBAs grids, respectively.

Unit: mGal	Max	Min	Mean	Std	RMSE
Kriging	258.962	−129.942	8.458	29.867	31.042
IDP	266.128	−128.788	8.382	29.822	30.978
Nearest Neighbor	266.037	−132.073	8.402	29.884	31.042
ANN	256.155	−128.766	8.154	30.018	31.106

In order to investigate the relationships between the free air anomaly grid datasets in further detail, their differences are calculated. To keep the inspection and interpretation part concise, again, one dataset (Kriging-based grid dataset) is assumed as the control data, and the grid differences between it and the other sets are generated. Figure 18 shows the free air anomaly grid difference maps. As a result of a careful inspection of the maps, the similarities between these maps and their corresponding CBA grid difference maps in Figure 15 are recognized. The statistics of the free air anomaly differences between the grid datasets are given in Table 13. Also, these statistics are identical to the statistics given in Table 11. This conclusion is explained by Equation (6). While calculating the free air grid differences between the two datasets, the restored Bouguer reduction (δg_B) and terrain correction (δg_{TC}) parameters that are considered when calculating the free air anomalies according to Equation (6) are canceled because of the subtraction operation. Thus, the free air anomaly difference becomes equal to the complete gravity anomaly difference at a grid node in this comparison. Due to this result, we conclude that although the interpolation process is applied to the CBAs in gridding, it affects the free air anomaly grids as well. Thus, we expect to see the consequences of differences in free air anomaly grids on the geoid model determination results.

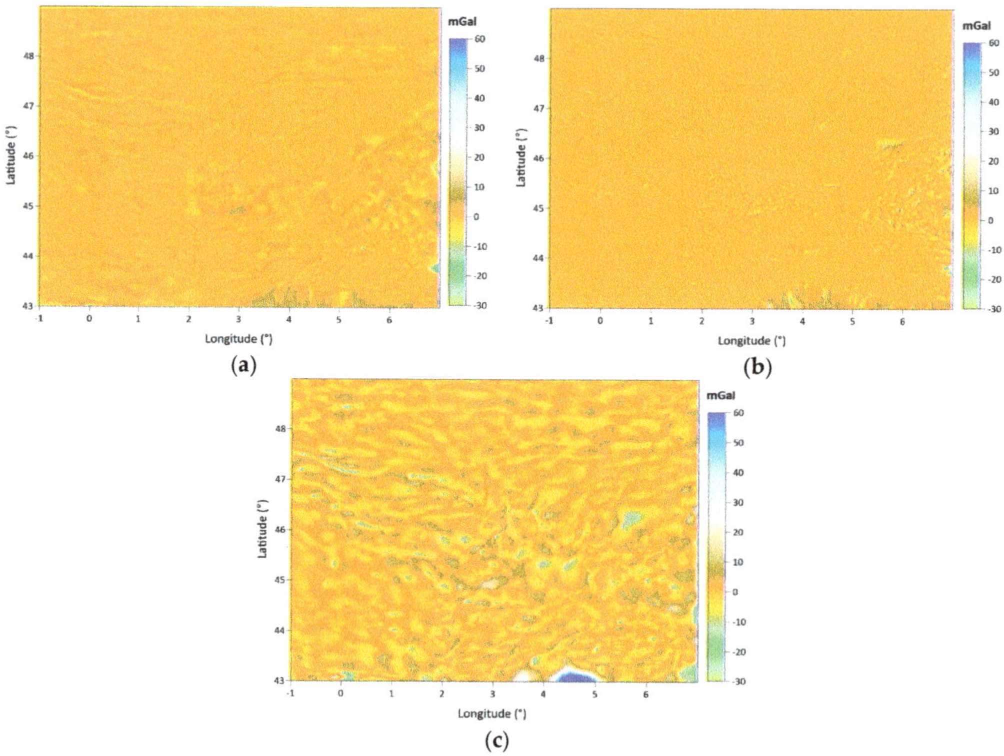

Figure 18. Free air anomaly differences between the Kriging-based grid and the IDP-based grid (**a**), nearest neighbor-based grid (**b**), and ANN-based grid (**c**) using 1 arc-minute grid spacing.

Table 13. Statistics of free air anomaly differences between the Kriging-based grid and the IDP-based grid, nearest neighbor-based grid, and the ANN based-grid, respectively.

	Unit: mGal	Max	Min	Mean	Median	Std	RMSE
	IDP	13.646	−26.880	0.076	0.009	1.314	1.316
KRIGING vs.	Nearest neighbor	45.066	−34.711	0.056	0.006	1.618	1.619
	ANN	100.999	−37.245	0.304	0.082	5.310	5.318

In the final step of the numerical tests, we calculate four geoid models using the free air anomaly grid datasets as the input data and the determined computation parameters in Section 3.3 using least squares modification of stokes integral with additive corrections geoid determination method. These geoid models are generated in grid form with a 1 arc-minute spatial resolution. Figure 19 shows the map of each geoid model. The geoid heights provided by these models vary between 45.2 m and 55.4 m, with a 49.6 m mean and 2.0 m standard deviation. These statistics are intentionally given in decimeter precision because the given statistics for the calculated models (see in Table 14) differ in the centimeter digit, which is deemed significant from a practical applications point of view.

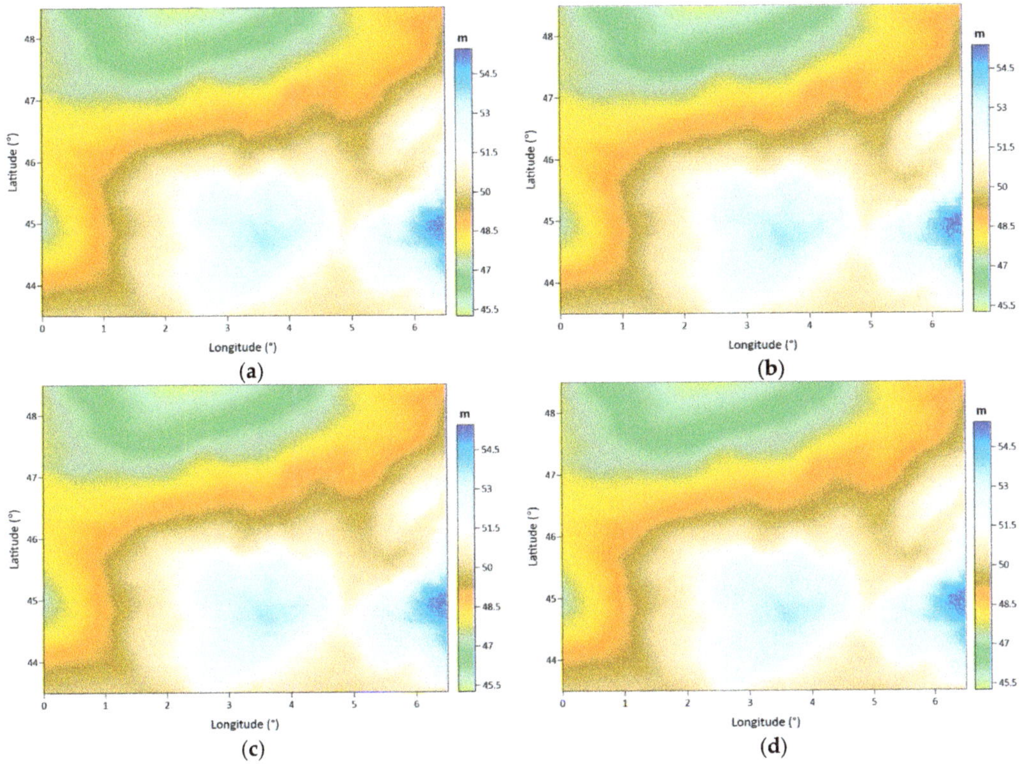

Figure 19. Geoid model maps calculated using free air anomaly grids from: Kriging (**a**), the IDP (**b**), nearest neighbor (**c**), and the ANN (**d**) interpolation methods using 1 arc-minutes spacing.

Table 14. Statistics of geoid models calculated using free air anomaly grids from the Kriging, the IDP, nearest neighbor, and ANN interpolation methods.

Unit: m	Max	Min	Mean	Std	RMSE
Kriging	55.470	45.267	49.616	2.021	49.657
IDP	55.450	45.266	49.616	2.021	49.657
Nearest neighbor	55.462	45.267	49.616	2.021	49.657
ANN	55.451	45.264	49.616	2.021	49.657

Figure 20 shows the difference maps between the calculated geoid models. In the previous steps of the numerical tests, the Kriging interpolation method is the reference method, and the derived values from each interpolation technique are compared with this reference method. Based on a similar opinion, the calculated geoid models with each free air anomaly dataset are compared to the geoid model calculated with the Kriging based free air anomaly grid.

Considering the difference map given in Figure 20a and its corresponding statistics in Table 15, a highly improved fit between the geoid models calculated with free air anomaly grids using Kriging and IDP methods is recognized. These two models fit with a 0.4 cm standard deviation of geoid height differences an almost 0.0 cm mean. The geoid height differences between them vary from −8.4 cm to 5.8 cm, which is quite reasonable.

Figure 20b depicts the geoid height differences between the geoid models based on the Kriging and nearest neighbor free air grids. Both the Figure 20b and the given statistics in Table 15 show that these two geoid models fit with a 0.4 cm standard deviation of geoid height differences, which vary between −10.1 cm and 14.2 cm. In these results, while the IDP and nearest neighbor geoid models fit the Kriging geoid surface with a standard deviation of less than 1 cm, the ANN-based geoid model does not have a similar compatibility with the Kriging geoid. Their geoid height differences vary between −11.9 cm and 7.8 cm, with a 1.58 cm standard deviation. Figure 20c shows the distribution pattern of the geoid height differences based on the ANN and Kriging grids.

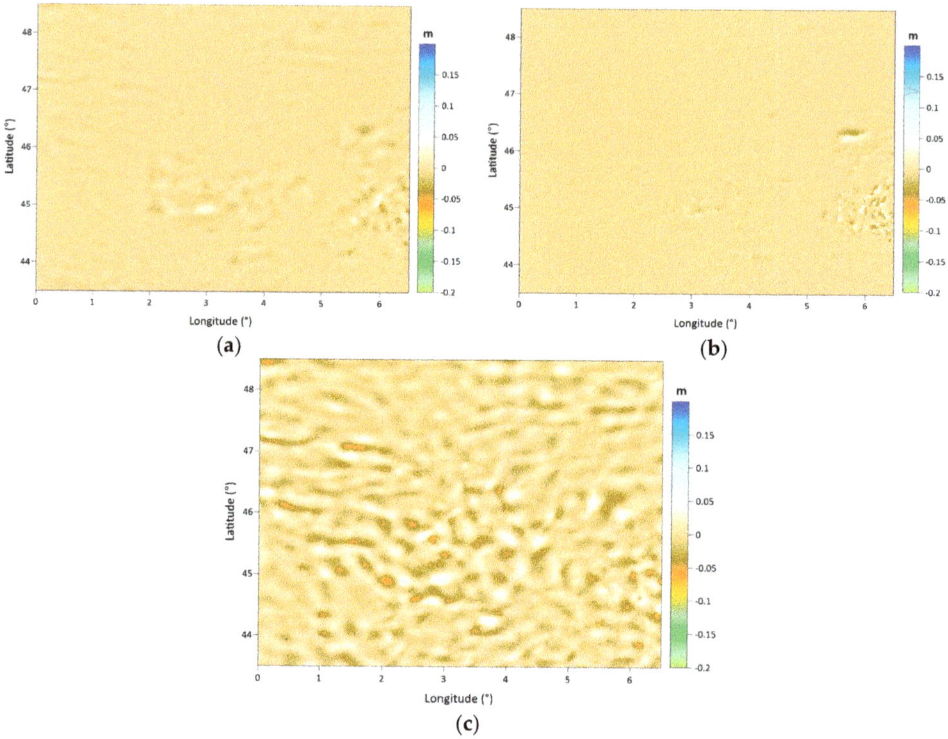

Figure 20. Geoid height difference maps between the Kriging-based geoid model and: the IDP-based geoid model (**a**), nearest neighbor-based geoid model (**b**), and the ANN-based geoid model (**c**) with 1 arc-minute spacing.

Table 15. Geoid height difference statistics between the Kriging-based geoid model and the IDP-based, nearest neighbor-based, and ANN-based geoid models.

	Unit: cm	Max	Min	Mean	Std	RMSE
KRIGING vs.	IDP	5.810	−8.420	−0.006	0.384	0.384
	Nearest neighbor	14.240	−10.050	−0.001	0.408	0.408
	ANN	7.830	−11.940	−0.001	1.580	1.580

The area-based comparisons of the geoid models explain the spatial variations in the model differences and provide information regarding the areal consistency between the compared models. It provides a relative measure of the quality of a calculated geoid model with respect to another geoid model assumed as a reference. In addition to these area-based relative assessments, we also validated the geoid models using independent 75 GPS/leveling benchmarks (see Figure 4a) in the area. Table 16 compares the validation statistics of the models. According to the standard deviations of the geoid height differences at the GPS/leveling benchmarks, the accuracy of the calculated geoid models is 4.1 cm (it is 4.5 cm for geoid model including the ANN-based free air anomalies). The 1.059 m mean value of the geoid height differences corresponds to the datum difference between the GPS/leveling surface and the geoid model surface. The most essential handicap of the GPS/leveling control dataset is that the benchmarks cover a very limited part of the study area, and their distribution is insufficient to represent the rough topographical region where the geoid models are expected to deteriorate in accuracy. Since the only available independent control dataset is this 75-point set, the evaluation of the geoid models could be carried out to a limited extent.

Table 16. Validation statistics of the performed geoids.

Unit: m	Max	Min	Mean	Std	RMSE
Kriging	1.162	0.972	1.059	0.041	1.060
IDP	1.164	0.974	1.059	0.041	1.060
Nearest neighbor	1.163	0.972	1.059	0.041	1.059
ANN	1.165	0.967	1.060	0.045	1.061

4. Conclusions

Gravity data gridding is a crucial process in gravity field mapping and geoid determination applications. Terrestrial gravimetry provides precise gravity observations on the topography, which are gradients of the Earth's gravity potential and essential input data in geoid determination. The gravity observations on the Earth's surface have a high-frequency character, since they contain information regarding the gravitational effects of topographical masses underneath the Earth's surface. In order to use these observations in gravity field mapping and geoid modeling, they need to be reduced to the geoid surface as the geopotential surface to be modelled. In addition to fulfilling the requirements for geoid determination, we reduce the gravity data to the geoid for data interpolation and gridding. A number of reduction formulas exist in the literature, and Bouguer gravity anomalies are employed in gravity interpolation and gridding.

This article is dedicated to clarifying certain issues regarding gravity gridding in gravity field mapping and geoid determination purposes. First of all, we inspected the effects of simple and complete Bouguer anomalies in gravity gridding. The test area has a miscellaneous topography, and it is seen that using simple or complete Bouguer anomalies in data gridding in the plain topographical part does not produce significant differences in the grid values. However, differences up to 55 mGal occur in grid values in the southeastern part of the area, where topography rises to ~4750 m. The grid differences between the simple and complete Bouguer anomalies result in up to 27.3 cm geoid height differences between the models calculated using the LSMSA method.

The impact of the interpolation method on gravity gridding is the second topic investigated in this study. In this regard, the complete Bouguer anomalies are gridded using the Kriging, inverse distance to a power, nearest neighbor, and artificial neural network methods. The first three methods have widespread use in spatial data interpolation applications. They estimate the interpolation point's value, employing the data points with their estimated weights. On the other hand, artificial neural networks are a last-generation method that employs a function whose parameters are determined by a training process. For the training of the system, the dataset is divided into training and test data. The iteratively trained system reveals the improved functional parameters, which are employed in the determination of the values at the grid nodes. Instead of directly employing the data points, it runs the determined function in interpolating the values at the grid nodes, resulting in a smooth appearance in the generated grid. Additionally, each run of the computation algorithm adopts a randomly selected training dataset and weights, outputting a different grid dataset each time. In this algorithm, the number of neurons and iterations are two critical parameters that should be decided carefully based on a trial-and-error procedure.

Based on the test results, it is concluded that the used interpolation method produces differences in the magnitudes and distribution patterns of the grid values. In order to estimate the maximum differences in the grid values depending on the used interpolation algorithm, we compared the generated Bouguer anomaly grid datasets with respect to the one generated using the Kriging method. The Bouguer anomaly differences reach up to 40 mGal on land between the grids derived from the ANN and Kriging methods. The standard deviation of these differences at grid nodes is 5.3 mGal. The differences between the geoid models calculated from these two datasets are 11.9 cm at most, with a 1.6 cm standard deviation. The validation results of these two geoid models at 75 GPS/leveling control benchmarks reveal a 4.1 cm accuracy for the Kriging method and a 4.6 cm accuracy for the ANN. In conclusion, the choice of interpolation algorithm significantly impacts both the gridding of gravity data and the accuracy of the geoid model created using the LSMSA method with the interpolated grid. Progress in data acquisition techniques and sensor technologies in recent decades provides opportunities for determining the geoid models with a sub-centimeter accuracy. The remarkable progress in data precision has rendered every stage crucial in modeling approaches that will impact the computation result. Therefore, using complete Bouguer anomalies and an appropriate interpolation algorithm in gravity gridding for precise determination of the Earth's gravity field and geoid is recommended.

Author Contributions: Conceptualization, O.K., B.E. and S.E.; methodology, O.K. and B.E.; software, O.K., B.E. and S.E.; validation, O.K., B.E. and S.E.; formal analysis, O.K.; investigation, B.E.; resources, O.K. and B.E.; data curation, O.K.; writing—original draft preparation, O.K.; writing—review and editing, B.E. and S.E.; visualization, O.K.; supervision, B.E. and S.E.; project administration, O.K. and B.E.; funding acquisition, O.K. and B.E. All authors have read and agreed to the published version of the manuscript.

Funding: This research was funded by The Scientific and Technological Research Council of Turkey (TUBITAK), application no: 1059B141900691, and the German Academic Exchange Service (DAAD), personal reference no: 91774457. In this context, this study has been realized in TU Munich from 01.10.2020 to 29.09.2021 under the supervision of Prof. Dr. Roland Pail.

Data Availability Statement: The data used in the study can be obtained upon request to the addresses specified in this article.

Acknowledgments: The first author expresses his gratitude to Roland Pail and Thomas Gruber, who supervised his research studies at the Technical University of Munich in Germany. TUBITAK and DAAD provided grant support for the first author's research period during his Ph.D. studies at the Technical University of Munich. The research presented in this article constitutes a part of the first author's Ph.D. thesis study at the Graduate School of Istanbul Technical University (ITU). All the authors thank the two reviewers and the editor of this manuscript for their valuable contributions and critiques.

Conflicts of Interest: The authors declare no conflicts of interest. The outcomes of this study are available to the funders.

References

1. Erol, S.; Erol, B. A Comparative Assessment of Different Interpolation Algorithms for Prediction of GNSS/levelling Geoid Surface Using Scattered Control Data. *Measurement* **2021**, *173*, 108623. [CrossRef]
2. Işık, M.S.; Erol, S.; Erol, B. Investigation of the Geoid Model Accuracy Improvement in Turkey. *J. Surv. Eng.* **2022**, *148*, 05022001. [CrossRef]
3. Sjöberg, L.E.; Gidudu, A.; Ssengendo, R. The Uganda Gravimetric Geoid Model 2014 Computed by The KTH Method. *J. Geod. Sci.* **2015**, *5*, 35–46. [CrossRef]
4. Featherstone, W.E.; Kuhn, M. Height systems and vertical datums: A review in the Australian context. *J. Spat. Sci.* **2009**, *51*, 21–41. [CrossRef]
5. Hofmann-Wellenhof, B.; Moritz, H. *Physical Geodesy*, 1st ed.; Springer: Berlin/Heidelberg, Germany, 2005.
6. Jekeli, C. *Heights, the Geopotential, and Vertical Datums*; 459, Ohio State University Reports, Geodetic Science and Surveying; Department of Civil and Environmental Engineering and Geodetic Science: Columbus, OH, USA, 2000.
7. Ellmann, A.; Märdla, S.; Oja, T. The 5 mm geoid model for Estonia computed by the least squares modified Stokes's formula. *Surv. Rev.* **2019**, *52*, 352–372. [CrossRef]
8. Hackney, R.I.; Featherstone, W.E. Geodetic versus geophysical perspectives of the gravity anomaly. *Geophys. J. Int.* **2003**, *154*, 35–43. [CrossRef]

9. Kiamehr, R. A new height datum for Iran based on the combination of gravimetric and geometric geoid models. *Acta Geod. Geoph.* **2007**, *42*, 69–81. [CrossRef]
10. Heiskanen, W.A.; Moritz, H. *Physical Geodesy*; Institude of Physical Geodesy, Technical University: Graz, Austria, 1967.
11. Abbak, R.A.; Ustun, A. A software package for computing a regional gravimetric geoid model by the KTH method. *Earth Sci. Inform.* **2015**, *8*, 255–265. [CrossRef]
12. Sakil, F.F.; Erol, S.; Ellmann, A.; Erol, B. Geoid modeling by the least squares modification of Hotine's and Stokes' formulae using non-gridded gravity data. *Comput. Geosci.* **2021**, *156*, 104909. [CrossRef]
13. Abbak, R.A.; Sjöberg, L.E.; Ellmann, A.; Ustun, A. A precise gravimetric geoid model in a mountainous area with scarce gravity data: A case study in central Turkey. *Stud. Geophys. Geod.* **2012**, *56*, 909–927. [CrossRef]
14. Sjöberg, L.E. A general model for modifying Stokes' formula and its least-squares solution. *J. Geod.* **2003**, *77*, 459–464. [CrossRef]
15. Sjöberg, L.E. Least squares combination of satellite and terrestrial data in physical geodesy. *Ann. Geophys.* **1981**, *37*, 25–30.
16. Sjöberg, L.E. *Least-Squares Modification of Stokes and Venning-Meinesz Formulas by Accounting for Errors of Truncation, Potential Coefficients and Gravity Data*; Technical Report; Department of Geodesy, Institute of Geophysics, University of Uppsala: Uppsala, Sweden, 1984.
17. Sjöberg, L.E. Refined least-squares modification of Stokes formula. *Manuscr. Geod.* **1991**, *16*, 367–375.
18. Sjöberg, L.E. A solution to the downward continuation effect on the geoid determined by Stokes' formula. *J. Geod.* **2003**, *77*, 94–100. [CrossRef]
19. Sjöberg, L.E. A computational scheme to model the geoid by the modified Stokes formula without gravity reductions. *J. Geod.* **2003**, *77*, 423–432. [CrossRef]
20. Sjöberg, L.E. A Local Least-Squares Modification of Stokes' Formula. *Stud. Geophys. Geod.* **2005**, *49*, 23–30. [CrossRef]
21. Sjöberg, L.E. Topographic Effects in Geoid Determinations. *Geosciences* **2018**, *8*, 143. [CrossRef]
22. Abbak, R.A.; Erol, B.; Ustun, A. Comparison of the KTH and remove-compute-restore techniques to geoid modelling in a mountainous area. *Comput. Geosci.* **2012**, *48*, 31–40. [CrossRef]
23. Abdalla, A. Determination of a Gravimetric Geoid Model of Sudan Using the KTH Method. Master's Thesis, Royal Institute of Technology (KTH), Stockholm, Sweden, January 2009.
24. Abdalla, A.; Mogren, S. Implementation of a rigorous least-squares modification of Stokes' formula to compute a gravimetric geoid model over Saudi Arabia (SAGEO13). *Can. J. Earth Sci.* **2015**, *52*, 823–832. [CrossRef]
25. Agren, J. Regional Geoid Determination Methods for the Era of Satellite Gravimetry: Numerical Investigations Using Synthetic Earth Gravity Models. Ph.D. Thesis, Royal Institute of Technology (KTH), Stockholm, Sweden, October 2004.
26. Agren, J.; Sjöberg, L.E.; Kiamehr, R. Computation of a new gravimetric model over Sweden using the KTH method. In Proceedings of the Integrating Generations, FIG Working Week, Stockholm, Sweden, 14–19 June 2008.
27. Ellmann, A. The Geoid for the Baltic Countries Determined by the Least Squares Modification of Stokes' Formula. Ph.D. Thesis, Royal Institute of Technology (KTH), Stockholm, Sweden, 2004.
28. Yildiz, H.; Forsberg, R.; Agren, J.; Tscherning, C.; Sjöberg, L. Comparison of remove-compute-restore and least squares modification of Stokes' formula techniques to quasi-geoid determination over the Auvergne test area. *J. Geod. Sci.* **2012**, *2*, 53–64. [CrossRef]
29. Abbak, R.A.; Ellmann, A.; Ustun, A. A practical software package for computing gravimetric geoid by the least squares modification of Hotine's formula. *Earth Sci. Inform.* **2022**, *15*, 713–724. [CrossRef]
30. Işık, M.S.; Erol, B.; Erol, S.; Sakil, F.F. High-resolution geoid modeling using least squares modification of Stokes and Hotine formulas in Colorado. *J. Geod.* **2021**, *95*, 49. [CrossRef]
31. Märdla, S.; Ellmann, A.; Agren, J.; Sjöberg, L.E. Regional geoid computation by least squares modified Hotine's formula with additive corrections. *J. Geod.* **2018**, *92*, 253–270. [CrossRef]
32. Ellmann, A.; Vanicek, P. UNB application of Stokes–Helmert's approach to geoid computation. *J. Geodyn.* **2007**, *43*, 200–213. [CrossRef]
33. Vanicek, P.; Kingdon, R.; Kuhn, M.; Ellmann, A.; Featherstone, W.E.; Santos, M.C.; Martinec, Z.; Hirt, C.; Avalos-Naranjo, D. Testing Stokes–Helmert geoid model computation on a synthetic gravity field: Experiences and short-comings. *Stud. Geophys. Geod.* **2013**, *57*, 369–400. [CrossRef]
34. Kaas, E.; Sørensen, B.; Tscherning, C.C.; Veicherts, M. Multi-processing least squares collocation: Applications to gravity field analysis. *J. Geod. Sci.* **2013**, *3*, 219–223. [CrossRef]
35. Tscherning, C.C. Geoid Determination by 3D Least-Squares Collocation. In *Geoid Determination*; Lecture Notes in Earth System Sciences; Sansò, F., Sideris, M., Eds.; Springer: Berlin/Heidelberg, Germany, 2013; Volume 110. [CrossRef]
36. Tscherning, C.C. Least-Squares Collocation. In *Encyclopedia of Geodesy*; Grafarend, E., Ed.; Springer: Cham, Switzerland, 2015. [CrossRef]
37. Barzaghi, R. The Remove-Restore Method. In *Encyclopedia of Geodesy*; Grafarend, E., Ed.; Springer: Cham, Switzerland, 2016. [CrossRef]
38. Abbak, R.A.; Ustun, A.; Ellmann, A. Comparison between simple and complete Bouguer approaches in interpolation of mean gravity anomalies. *J. Geod. Geoinf.* **2012**, *1*, 45–52. [CrossRef]
39. Torge, W. *Geodesy*; 3rd Revised and Extended Edition; Walter de Gruyter GmbH & Co.: Berlin, Germany, 2001.
40. Bajracharya, S. Terrain Effects on Geoid Determination. Master's Thesis, University of Calgary, Calgary, AB, Canada, 2003.

41. Tziavos, I.N.; Sideris, M.G. Topographic Reductions in Gravity and Geoid Modeling. In *Geoid Determination*; Sansò, F., Sideris, M., Eds.; Lecture Notes in Earth System Sciences; Springer: Berlin/Heidelberg, Germany, 2013; Volume 110. [CrossRef]
42. Goos, J.M.; Featherstone, W.E. Experiments with two different approaches to gridding terrestrial gravity anomalies and their effect on regional geoid computation. *Surv. Rev.* **2003**, *37*, 92–112. [CrossRef]
43. Kuhn, M.; Featherstone, W.E.; Kirby, J.F. Complete spherical Bouguer gravity anomalies over Australia. *Aust. J. Earth Sci.* **2009**, *56*, 213–223. [CrossRef]
44. Kiamehr, R.; Sjöberg, L.E. Effect of the SRTM global DEM on the determination of a high-resolution geoid model: A case study in Iran. *J. Geod.* **2005**, *79*, 540–551. [CrossRef]
45. Torge, W.; Müller, J. *Geodesy*, 4th ed.; Walter de Gruyter: Berlin, Germany, 2012. [CrossRef]
46. Janak, J.; Vanicek, P.; Foroughi, I.; Kingdon, R.; Sheng, M.; Santos, M. Computation of precise geoid model of Auvergne using current UNB Stokes-Helmert's approach. *Contrib. Geophys. Geod.* **2017**, *47*, 201–229. [CrossRef]
47. De Gaetani, C.I.; Marotta, A.M.; Barzaghi, R.; Reguzzoni, M.; Rossi, L. *The Gravity Effect of Topography: A Comparison among Three Different Methods*; IntechOpen: London, UK, 2021. [CrossRef]
48. Moritz, H. Geodetic reference system 1980. *Bull. Géod.* **1980**, *54*, 395–405. [CrossRef]
49. Forsberg, R. *A Study of Terrain Reductions, Density Anomalies and Geophysical Inversion Methods in Gravity Field Modelling*; Rep. No. 355; Ohio State Univ.: Columbus, OH, USA, 1984.
50. Forsberg, R.; Tscherning, C.C. *An Overview Manual for the GRAVSOFT Geodetic Gravity Field Modelling Programs*, 3rd ed.; National Space Institute, (DTU-Space): Copenhagen, Denmark, 2014; p. 68.
51. Tscherning, C.C.; Forsberg, R.; Knudsen, P. The GRAVSOFT package for geoid determination. In Proceedings of the 1st Continental Workshop on the geoid in Europe, Research Institute of Geodesy, Topography and Cartography, Prague, Czech Republic, 11–14 May 1992; pp. 327–334.
52. Vanicek, P.; Novak, P.; Martinec, Z. Geoid, topography, and the Bouguer plate or shell. *J. Geod.* **2001**, *75*, 210–215.
53. Vanicek, P.; Tenzer, R.; Sjoberg, L.E.; Martinec, Z.; Featherstone, W.E. New views of the spherical Bouguer gravity anomaly. *Geophys. J. Int.* **2004**, *159*, 460–472. [CrossRef]
54. Erol, B.; Çevikalp, M.R.; Erol, S. Accuracy assessment of the SRTM2gravity high-resolution topographic gravity model in geoid computation. *Surv. Rev.* **2023**, *55*, 546–556. [CrossRef]
55. Goyal, R.; Featherstone, W.; Claessens, S.; Dikshit, O.; Balasubramanian, N. An experimental Indian gravimetric geoid model using Curtin University's approach. *Terr. Atmos. Ocean. Sci.* **2021**, *32*, 813–827. [CrossRef]
56. Varga, M.; Grgic, M.; Bjelotomic Orsulic, O.; Basic, T. Influence of digital elevation model resolution on gravimetric terrain correction over a study-area of Croatia. *Geofizika* **2019**, *36*, 17–32. [CrossRef]
57. Jarvis, A.; Reuter, H.I.; Nelson, A.; Guevara, E. Hole-Filled Seamless SRTM Data V3. International Centre for Tropical Agriculture (CIAT). 2006. Available online: http://srtm.csi.cgiar.org (accessed on 14 March 2024).
58. U.S. Geological Survey. Earth Resources Observation and Science (EROS) Center Archive—Digital Elevation—Shuttle Radar Topography Mission Void Filled. Available online: https://www.usgs.gov/ (accessed on 14 March 2024).
59. Mukul, M.S.; Srivastava, V.; Jade, S.; Mukul, M. Uncertainties in the Shuttle Radar Topography Mission (SRTM) Heights: Insight from the Indian Himalaya and Peninsula. *Sci. Rep.* **2017**, *7*, 41672. [CrossRef]
60. Varga, M.; Bašič, T. Accuracy Validation and Comparison of Global Digital Elevation Models over Croatia. *Int. J. Remote Sens.* **2015**, *36*, 170–189. [CrossRef]
61. Erol, B.; Işık, M.S.; Erol, S. An Investigation on Accuracy Analysis of Global and Regional (High Resolution) Digital Elevation Models. *Afyon Kocatepe Üniversitesi Fen Ve Mühendislik Bilimleri Dergisi* **2020**, *20*, 598–612. [CrossRef]
62. Robeson, S.M. Spherical Methods for Spatial Interpolation: Review and Evaluation. *Cartogr. Geogr. Inf. Syst.* **1997**, *24*, 3–20. [CrossRef]
63. Li, J.; Heap, A.D. A Review of Spatial Interpolation Methods for Environmental Scientists. *Geosci. Aust. Rec.* **2008**, *23*, 137.
64. Golden Software. *Surfer*; Powerful Contouring; Gridding & 3D Surface Mapping; Golden Software, LLC: Golden, CO, USA, 2023.
65. Karaca, O. Assessments on Surface Interpolation Methods for Local Geoid Modelling. Master's Thesis, Istanbul Technical University, Istanbul, Turkey, June 2016.
66. Golden Software. *Surfer User's Guide*; Golden Software, LLC: Golden, CO, USA, 2022; Available online: https://www.goldensoftware.com (accessed on 14 March 2024).
67. Knotters, M.; Heuvelink, G.B.M.; Hoogland, T.; Walvoort, D.J.J. *A Disposition of Interpolation Techniques*; Work Document 190; Wageningen University and Research Centre, Statutory Research Tasks Unit for Nature and the Environment: Wageningen, The Netherlands, 2010.
68. Yang, C.S.; Kao, S.P.; Lee, F.B.; Hung, P.S. Twelve different interpolation methods: A case study of Surfer 8.0. In Proceedings of the XXth ISPRS Congress, Istanbul, Turkey, 12–23 July 2004; pp. 778–785.
69. Babak, O.; Deutsch, C.V. Statistical approach to inverse distance interpolation. *Stoch. Environ. Res. Risk Assess.* **2009**, *23*, 543–553. [CrossRef]
70. Yilmaz, I. A Research on the Accuracy of Landform Volumes Determined Using Different Interpolation Methods. *Sci. Res. Essay* **2009**, *4*, 1248–1259.
71. Erol, B.; Çelik, R.N. Modelling local GPS/levelling geoid with the assessment of inverse distance weighting and geostatistical kriging methods. In Proceedings of the XXXVth ISPRS Congress, Technical Commission IV, Istanbul, Turkey, 12–23 July 2004.

72. Isaaks, E.H.; Srivastava, R.M. *Applied Geostatistics*; Oxford University Press: New York, NY, USA, 1989.
73. Heuvelink, G.B.M. Incorporating process knowledge in spatial interpolation of environmental variables. In Proceedings of the 7th International Symposium on Spatial Accuracy Assessment in Natural Resources and Environmental Sciences, Lisboa, Portugal, 5–7 July 2006; pp. 32–47.
74. Yaprak, S.; Arslan, E. Searching the use of Kriging method on geoid surface modeling. *J. ITU* **2008**, *7*, 51–62. (In Turkish)
75. Demuth, H.; Beale, M.; Hagan, M. *MATLAB Neural Network Toolbox™ 6*; User's Guide; The MathWorks, Inc.: Southbridge, MA, USA, 2018.
76. Erol, S. Time-Frequency Analyses of Tide-Gauge Sensor Data. *Sensors* **2011**, *11*, 3939–3961. [CrossRef]
77. Kavzoglu, T.; Saka, M.H. Modelling local gps/levelling geoid undulations using artificial neural networks. *J. Geod.* **2005**, *78*, 520–527. [CrossRef]
78. Schuh, H.; Ulrich, M.; Egger, D.; Müller, J.; Schwegmann, W. Prediction of Earth orientation parameters by artificial neural networks. *J. Geod.* **2022**, *76*, 247–258. [CrossRef]
79. Seager, J.; Collier, P.; Kirby, J. Modelling Geoid Undulations with an Artificial Neural Network, IJCNN'99. In Proceedings of the International Joint Conference on Neural Networks, Washington, DC, USA, 10–16 July 1999; Volume 5, pp. 3332–3335. [CrossRef]
80. Stopar, B.; Ambrožič, T.; Kuhar, M.; Turk, G. GPS-derived geoid using artificial neural network and least squares collocation. *Surv. Rev.* **2006**, *38*, 513–524. [CrossRef]
81. MathWorks Inc. *MATLAB, Version 9.10.0.1739362 (R2021a) Update 5*; The MathWorks, Inc.: Southbridge, MA, USA, 2021. Available online: https://www.mathworks.com (accessed on 14 March 2024).
82. Albayrak, M.; Özlüdemir, M.T.; Aref, M.M.; Halicioglu, K. Determination of Istanbul geoid using GNSS/levelling and valley cross levelling data. *Geod. Geodyn.* **2020**, *11*, 163–173. [CrossRef]
83. Yilmaz, M.; Arslan, A. Effect of increasing number of neurons using artificial neural network to estimate geoid heights. *Int. J. Phys. Sci.* **2011**, *6*, 529–533. [CrossRef]
84. Erol, B.; Erol, S. Learning-based computing techniques in geoid modeling for precise height transformation. *Comput. Geosci.* **2013**, *52*, 95–107. [CrossRef]
85. Abdalla, A.; Tenzer, R. The evaluation of the New Zealand's geoid model using the KTH method. *Geod. Cartogr.* **2011**, *37*, 5–14. [CrossRef]
86. Sjöberg, L.E. The topographic bias by analytical continuation in physical geodesy. *J. Geod.* **2007**, *81*, 345–350. [CrossRef]
87. Sjöberg, L.E.; Nahavandchi, H. The atmospheric geoid effects in Stokes' formula. *Geophys. J. Int.* **2000**, *140*, 95–100. [CrossRef]
88. Ellmann, A.; Sjöberg, L.E. Ellipsoidal correction for the modified Stokes formula. *Boll. Geod. Sci. Affin.* **2004**, *63*, 153–172.
89. Kiamehr, R. A Hybrid Precise Gravimetric Geoid Model for Iran Based on Recent GRACE and SRTM Data and the Least-Squares Modification of Stokes' Formula. *J. Earth Space Phys.* **2006**, *32*, 7–23.
90. ICGEM. International Centre for Global Earth Models. Available online: http://icgem.gfz-potsdam.de/ (accessed on 27 September 2023).
91. Tocho, C.N.; Antokoletz, E.D.; Gómez, A.R.; Guagni, H.; Piñon, D.A. Analysis of high-resolution global gravity field models for the estimation of International Height Reference System (IHRS) coordinates in Argentina. *J. Geod. Sci.* **2022**, *12*, 131–140. [CrossRef]
92. Zingerle, P.; Pail, R.; Gruber, T.; Oikonomidou, X. The combined global gravity field model XGM2019e. *J. Geod.* **2020**, *94*, 66. [CrossRef]
93. Pail, R.; Fecher, T.; Barnes, D.; Factor, J.F.; Holmes, S.A.; Gruber, T.; Zingerle, P. Short note: The experimental geopotential model XGM2016. *J. Geod.* **2018**, *92*, 443–451. [CrossRef]
94. Duquenne, H. A data set to test geoid computation methods. In Proceedings of the 1st International Symposium of the International Gravity Field Service Meeting, Istanbul, Turkey, 28 August–1 September 2006; Harita Dergisi. pp. 61–65.
95. Valty, P.; Duquenne, H.; Panet, I. Auvergne dataset: Testing several geoid computation methods. In *Geodesy for Planet Earth*; Kenyon, S., Pacino, M.C., Marti, U., Eds.; Springer: Berlin/Heidelberg, Germany, 2012; pp. 465–472. [CrossRef]
96. Reguzzoni, M.; Carrion, D.; De Gaetani, C.I.; Albertella, A.; Rossi, L.; Sona, G.; Batsukh, K.; Toro Herrera, J.F.; Elger, K.; Barzaghi, R.; et al. Open access to regional geoid models: The International Service for the Geoid. *Earth Syst. Sci. Data* **2021**, *13*, 1653–1666. [CrossRef]

geosciences

MDPI

Article

Annual Coastal Boulder Mobility Detected in 2017–2021 Remote Sensing Imagery and Its Relation to Marine Storms (Gulf of Taranto, Mediterranean Sea)

Marco Delle Rose

National Research Council of Italy, Institute of Atmospheric Sciences and Climate, 73100 Lecce, Italy; marco.dellerose@cnr.it

Abstract: Landward displacements of coastal boulders are geomorphological signatures of sea flooding and erosion processes. In this study, using open-access resources that do not require the integration of specialist software, the 2017 to 2021 annual mobility of medium, coarse, and very coarse boulders spread over about 100 km of the eastern coast of the Gulf of Taranto (Italy, Mediterranean Sea) was explored. The boulder displacement data obtained from remote sensing imagery were verified and refined by means of geomorphological field investigation. The main results are the following: (1) A large interannual variability in the boulder mobility was found; (2) storm Detlef, which crossed over the Mediterranean during 11–13 November 2019, was recognized as the cause of a massive displacement phenomenon; and (3) the marine weather conditions driving the investigated morphodynamic process were inferred.

Keywords: open access resources; comparative visual analysis; geomorphological investigation; overtopping flow; Ionian Sea; Apulia region; storm Detlef

1. Introduction

In several coastal areas, different-sized boulders are transported and emplaced above the sea level by waves that surge over the coastline like land-crossing bores. Massive displacements of large clasts are geomorphological signatures of flooding events due to severe storms and tsunamis [1–4]. Since about a decade, specific bands available with commercial Earth observation satellites such as WorldView-2 have allowed the definition of geomorphological and morphometric features of coastal deposits in detail [5,6]. The open access to high-resolution remote sensing imagery via the web-interface provided by Google Earth (GE) has greatly increased the possibilities of understanding boulder formation and mobility [7–9]. The dynamics and morphometry of displaced clasts can help to assess the wave energy on the coast, an important focus of research in physical geography and in applied sciences such as engineering and coastal management [1,2,9].

Boulder mobility accounts for the number of boulders displaced over a stretch of coast by water flows during a given time interval. It has attracted the attention of several scholars in recent years and has been used for various purposes, for example, to test hydrodynamics equations that govern the boulder displacement [10], to explore the ability of contemporary storms to cause geomorphological changes on rocky coasts [11], and to investigate the relationship among transport distance and storm frequency and intensity [12]. Moreover, the measure of the mobility of coastal boulders over pluriannual periods has been addressed in several studies to explore sedimentological features (see, e.g., [13,14]).

By means of geomorphological survey and comparing two sets of very-high-resolution (VHR) imagery provided by GE, Delle Rose et al. [9] recognized a massive displacement phenomenon (involving many dozens of large boulders) that occurred between July 2018 and June 2020 along the 100 km eastern coast of the Gulf of Taranto (southern Apulia region, Italy, central Mediterranean; see Figure 1). These authors considered two storms (Vaia, which occurred on 28–29 October 2018, and Detlef, which occurred on 12–13 November 2019, respectively; see references [15,16] for their synoptic framework) as possible causes for such an impressive displacement of boulders.

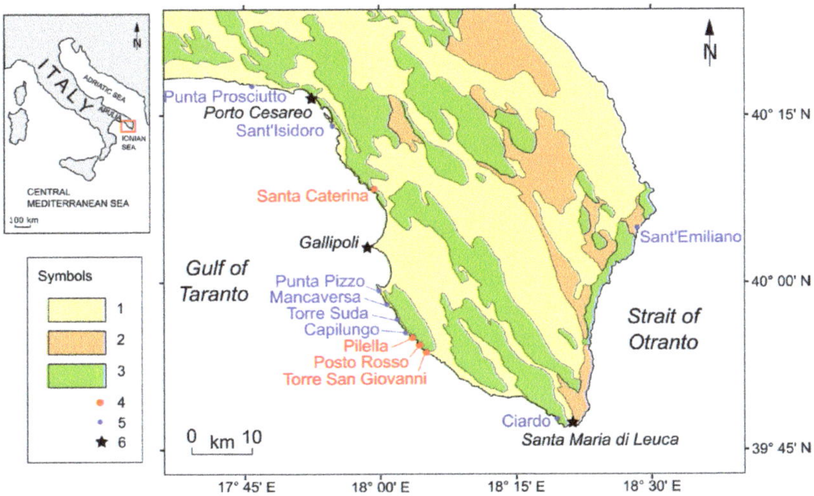

Figure 1. Geological map of the study area (after [9], modified). Symbols: 1, Quaternary rocks (carbonates, bioclastic deposits, marls); 2, Tertiary rocks (carbonates, marls); 3, Cretaceous rocks (limestones, dolostones); 4, site of boulder displacement detected in this study (see text); 5, site of boulder displacement detected by [9]; 6, wind gauge station.

The present work is a continuation of previous studies [4,9,17]. Its aims were threefold: (1) determine the annual mobility of large boulders along the coast in Figure 1 over 4 years (from 2017 fall to 2021 summer), (2) establish whether one of the two mentioned storms can be ruled out as cause of the massive displacement phenomenon, and (3) infer the marine weather conditions driving the investigated morphodynamic process.

2. Materials and Methods

2.1. Study Area

The Gulf of Taranto faces the north Ionian Sea for more than 330 km (Figure 1) and is characterized by a microtidal regime with an astronomical tides of 0.3 m. For such a semi-enclosed basin, maximum significant waves of 6.3 m for a 50-year return period and of 8.2 m for a 100-year return period were calculated, respectively [18,19]. Along the low-lying eastern side of the gulf, jointed carbonate rocks extensively outcrop. These rocks are shaped by typical coastal karst forms including solution pans and pinnacles. The coast is prone to boulder formation and displacement, and because of such characteristics, two selected sites (namely Sant'Isidoro and Torre Suda; see Figure 1) have been periodically surveyed with direct observation since 2017 [17,20,21]. Careful geomorphological surveys are performed especially after the more severe storms.

The two 6-month climatic seasons of October–March (cold–wet, CW) and April–September (warm–dry, WD) characterize the Mediterranean climate over the Gulf of Taranto [22,23]. Indeed, the southwards migration of the Atlantic storm track and the descending mid-atmospheric throughs from Northern Europe are typical of the CW season, while good weather with scarce precipitation prevails from April to September. The Apulia region is characterized by the predominance of northerly winds, although winds from the south are significant for frequency and strength [24]. The Gulf of Taranto is exposed to strong winds generated by the seasonal prevalence of the anticyclones. Typical local storms in the central Mediterranean Sea are usually driven by middle-atmosphere low-pressure centers (troughs in the 500 hPa GPH weather maps) generally following a NW-to-SE path [23]. Their income is characterized by southerly warm currents over the south Mediterranean Sea that can cause strong surface winds, according with the position of the surface pressure minima, especially in fall season. These meteorological conditions can cause, in turn, quite high waves above the southern Apulia coasts because of their potentially long geometrical fetch southwards to the Africa coasts and in accordance with their duration [17,25].

2.2. Remote Sensing Data and Geomorphological Survey

A challenge for this work was to explore the potential of physical geography investigations based on publicly available remote sensing sources and simple and inexpensive methods. Open-access resources that do not require the integration of specialist software such as GE have proven particularly effective in geomorphological research [8,26,27]. Since the first availability of very-high-resolution (sub meter) satellite images (produced by Airbus Defence and Space, Maxar Technologies, and other space technology companies), the online measuring tools of GE have been used to measure the horizontal dimensions of the boulders as well as their transport distance [28–30]. On the other hand, the horizontal accuracy of the images made available in the last few years has been tested on several geographic regions (see, e.g., [31,32]). For the Italian territory, the use of VHR GE images in geomorphology and other disciplines focused on the physical sciences has also been successfully explored (see, e.g., [33–35]).

To detect the morphometric features of the boulders and their mobility at annual scale, the eastern coast of the Gulf of Taranto was visually examined using the four last sets of VHR remote sensing images available on GE (dated July 2017, July 2018, June 2020, and September 2021, respectively) and a set of aerial images (taken during July 2019 at an average flight altitude of 180 m) that was recently made available by the governmental Authority of Apulia at the site "Territorial Planning Service" [36] (*Servizio Assetto del Territorio*, in Italian). The last set allowed us to fill the time gap (i.e., the 2019 summer) of the GE sets series. The visual analysis was performed according to the rudiments of geographic object-based image analysis (GEOBIA) [37]. The comparison between GE and regional aerial images [36] was carried out by means of QGIS software. It must be noted that the use of GE tools and QGIS functions to map large boulders and detect their mobility has been discussed and tested by several authors (see, e.g., [38–40]). Once determined from GE the geographical coordinates, the studied objects were easily found in the field by means of global positioning system (GPS) devices. The on-site geomorphological survey was carried out for the following aims: to verify and refine the morphometric data obtained from the remote sensing imagery; to determine the lithology of the boulders; and to infer, where possible, pre-transport setting and movement type (Appendix A).

2.3. Methodologial Descriptions

Boulders usually are complex solids [41–43]. However, in the first stage of the geomorphological investigation, a boulder is considered an idealized-shaped rectangular cuboid, with the size described by the dimension of its axes: a (major), b (middle), and c (minor). The axes dimensions can be used to establish the flatness index (FI) = $(a + b)/2c$ and the shape (Sh) of the boulders [44,45]. According to the Udden–Wentworth grain-size scale, boulders range from the class of fine boulders, with a middle axis > 0.25 m, to the class of medium blocks, with a middle axis of up to 16.4 m [45]. The middle axis ranges from 0.5 to 1 m for medium boulders, from 1 to 2 m for coarse boulders, and from 2 to 4.1 m for very coarse boulders. Based on the resolution of the used remote sensing images, attention was focused on medium to very coarse boulders. The boulder demarcation was manually performed, taking care to consider the shadowing conditions since the shape of such an object may seemingly change depending on the shadows cast at different times of the day [46].

Following the approach of previous studies [8,30,47], visual analysis was performed at an eye elevation of 50 to 200 m. Using GE, the two axes of the boulder presented as upward-facing in the images can be measured by the ruler tool. They are designated as the apparent major (a_r) and medium axes (b_r), with the minor axis being usually perpendicular to the image [28–30]. Actually, since the boulders may be inclined rather than horizontal, a_r and b_r are equal or smaller than the major and middle axis, respectively. Dealing with boulder displacement, other geometrical features that can be obtained using a computer workstation are the initial (x_i) and final distances (x_f) of the boulders from the coastline and the related transport distance (TD). The geographical coordinates of x_i and x_f were taken by estimating the positions of the geometric center point in the VHR GE images. Then, the TD (i.e., the joining line between the two positions) was measured using the ruler tool.

In the previous work [9], the difference between the measurements of displacement taken with GE's ruler tool and the ones carried out during a field survey with a measuring tape rarely exceeded 5%. Given this accuracy, in this study, the TDs were obtained, in a practical way, by measuring the distances between x_i and x_f in GE images. Several measures of TD were taken also in the field (see examples in Appendix A), with the aim of further testing the difference between the two types of measurements. The quality of the aerial images available at the site [36], as visualized in QGIS, was adequate to compare shape and position of the displaced boulders with those shown in the GE images. It must be noted that similar techniques have already been used successfully in previous studies (see, e.g., [47,48]).

The use of GPS devices allowed us to quickly identify the boulders at the respective x_f and thus to detail the morphometric features inferred from the remote sensing analysis. The prints left on the platform at the x_i by the displaced boulders were recognized in the same easy manner. Where the boulder was detached from the parent rock, the trace (called the socket; see ref. [49]) consists of a fresh and un-weathered rock surface not yet covered by lichen. The print left by a boulder that has already been uprooted from its geological substrate and has been lying on the ground for some time (i.e., the sub-aerial pre-transport setting; see Appendix A) can be detectable for some years under favorable sunlight conditions before being covered by lichen [9].

Surface karst forms such as solution pans and pinnacles have been used to establish the pre-transport setting and movement type for several boulders [9,17]. These inferences are based on the fact that fresh and un-weathered faces of boulders are indicative of the absence of karst processes before displacement (see examples in Appendix A). Also, ephemeral tracks (they are usually visible for only a few years) recognizable during the field geomorphological investigation, like the drag and impact marks [2,9,50], can give hints about the type of movement to which the boulders have been subjected. Finally, organic remains of marine plants covering the faces of displaced boulders suggest their initial submerged position.

To obtain an overview of the marine conditions in the eastern Gulf of Taranto during 2017–2021, publicly available datasets were examined. Special attention was given to the storminess conditions, which more than likely caused the landward boulder displacements. A marine storm is "*a succession of sea states in which the significant wave height exceeds a fixed threshold for a duration of at least equal to 12 h*" [51]. Moreover, to be relevant by the geomorphological point of view, a storm must have "*the potential to significantly alter the underlying morphology and expose the backshore to waves, currents and/or inundation*" [52].

The web archives of the GLOBO-BOLAM-MOLOCH model cascade were used to make a screening of the central Mediterranean 2017–2021 storms [53]. To identify the middle-atmosphere pressure conditions that allowed us to check the presence of stormy conditions, the 500 hPa geopotential height maps of the GLOBO model archive were carefully screened. Then, in order to obtain information about the offshore wind field (fetch length and direction, average wind speed at 10 m height, and storm duration) the archives of the mesoscale models BOLAM (10 km horizontal resolution) and MOLOCH (3 km horizontal resolution) were used. The storms were selected considering a minimum duration of 12 h together with a minimum fetch usually greater than 500 km, alongside a minimum wind speed of 12 m/s throughout the considered time. These are typical conditions achieved by the storms affecting the Gulf of Taranto and correspond to significant wave heights generally above 3 m (sea state $\geq$ 5 of the Douglas sea scale). As performed in the previous work [9], using windstorm characteristics extracted by the BOLAM-MOLOCH model, the characteristic wave height H_0 in the marine storms was calculated (see Appendix B).

The nearshore wind conditions during the storms, useful for completing the description of the weather features, were extracted from the datasets of the wind gauge stations placed along the studied stretch of coast (Figure 1). Finally, the minimum values of wave height theoretically capable of displacing the boulders were inferred from commonly used hydrodynamic equations [1,54–56] to relate the results of the present and previous works [9,17] (see Appendix B for details).

3. Results

The annual boulder mobility from fall 2017 to summer 2021 for the eastern coast of the Gulf of Taranto and the related marine conditions are reported below. The newly recognized displacements of medium to very coarse boulders are described in Section 3.1, while the multi-temporal check performed on the displacements recognized by [9] is shown in Section 3.2. The marine weather conditions and the main features of the storms for the above time interval are given in Section 3.3.

3.1. New Detection of Boulder Displacement

Multi-temporal analysis allowed the recognition of 38 boulder displacements that occurred during the investigated time period, spread over four previously undetected sites (in red in Figure 1). In Table 1, these sites are reported from north to south. For 25 out of 38 boulders, both the pre- and post-transport positions were recognized; thus, the TD was established. To expose the physical geography results, selected images downloaded from both GE and the site [36] were elaborated in the Inkscape program (an open-source vector graphics editor).

Table 1. Acronym code (ID Code), number of identified Boulders, and number of established transport distances (No. of TD).

Site Name	ID Code	Boulders	No. of TD
Santa Caterina	SC	2	1
Pilella	PIL	14	11
Posto Rosso	RO	16	10
Torre San Giovanni	SG	6	3

Numerical data and complementary field measures and observations are given in Appendix A. The initial and final coordinates of the displaced boulders are in Tables A1, A3, A5 and A7. The lengths of the boulder axes measured on VHR GE imagery and on-site, respectively, together with other features (inclination on the ground; initial and final distances from the coastline; transport distance; lithology; shape; flatness index; pre-transport setting; inferred type of movement) of the displaced boulders are in Tables A2, A4, A6 and A8.

The northernmost site, Santa Caterina, is quite far from the other three sites (about 25 km), while these latter are close together (a few kilometers from each other; Figure 1). The stretch of the coast of Santa Caterina is characterized by a vertical cliff up to 5 m high; differently, at the other new sites, the height above the sea level of the cliff edge (H_c) rarely exceeds 1.5 m.

The two boulders displaced at Santa Caterina (Tables 1, A1 and A2) experienced different movements (Figure 2). SCa was positioned at the cliff edge at least up to July 2019. In its place, after June 2020, a socket is present (Figure A1 in Appendix A), while the detached boulder is not detectable in the images. Likely, it fell back onto the sea bottom after the impact of a high-energy run-up wave. The socket is hardly visible in the September 2021 image, probably because of an unfavorable sunlight condition. The boulder SCb is clearly distinguishable, with respect to the surroundings, both in the July 2017 and July 2018 images. SCb was still in the same position in July 2019 (Figure 2a–c), while it showed a new position in both June 2020 and September 2021 (Figure 2d,e). Currently, it lies in a shrub vegetation area (Figure A2 in Appendix A). The displacement of SCb is quantifiable in a TD of about 15 m (Table A1) from SW to NE (Figure 2f). Both the SCa and SCb boulders were displaced between July 2019 and June 2020 (Figure 2c,d).

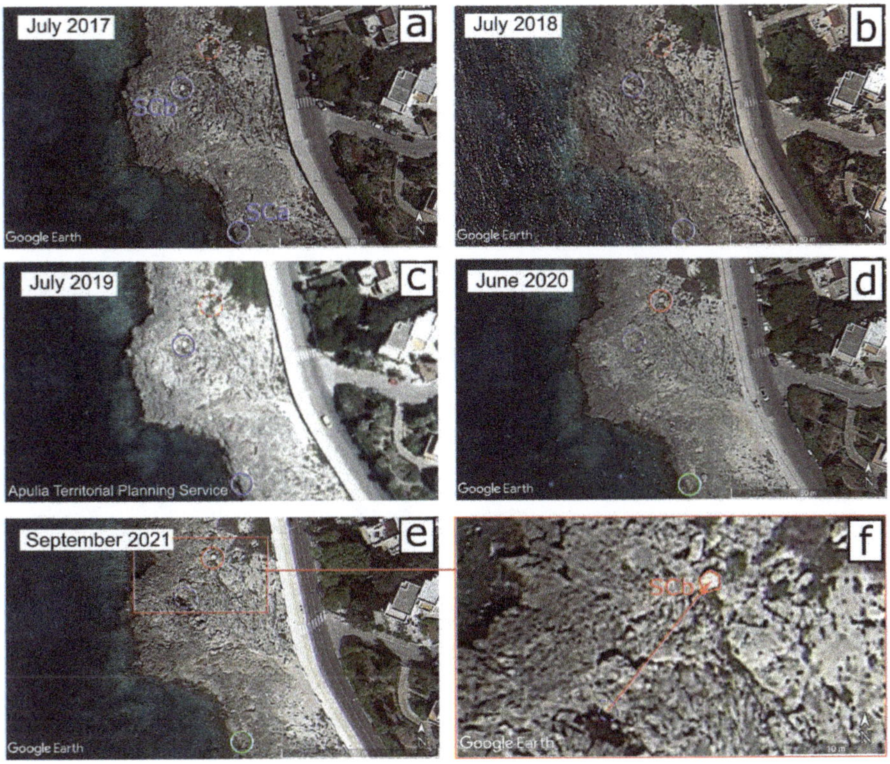

Figure 2. Santa Caterina coast (40°08′16.93″–40°08′19.47″ N, 17°59′16.11″–17°59′21.97″ E) detection of boulder displacements: (**a**,**b**,**d**,**e**) GE images, eye elevation of 127 m; (**c**) regional aerial image; (**f**) transport vector of SCb boulder, eye elevation of 43 m. Blue symbols, pre-displacement position; red symbols, post-displacement position; green symbols, socket of boulder disappeared into the sea. Dotted lines help the comparison.

Along the stretch of the coast of Pilella (Figure 1 for location), 14 displaced boulders were identified by multi-temporal analysis. For 11 of these boulders, both the initial and final positions were detected and, thus, the TD calculated (Table 1). The middle axis ranges from 1 to 1.5 m, except for one case (PILj), which is nearly 2 m (Table A4); TD ranges from 1.5 to 5 m (Table A3), with vectors oriented about SSW–NNE. The initial positions of these 14 boulders do not change in the July 2017, July 2018, and July 2019 images; likewise, the final positions do not change in the June 2020 and September 2021 images. Two boulders of this site are inclined of 55–60° with respect to the ground (Table A4). Such a geomorphological setting can make axis measurement from remote sensing images rather unreliable. As an example, for the PILn boulder (see Figure A3 for its arrangement on the coast), b_r was found to be one-third smaller than b_f (Table A4).

The northernmost boulders of the Pilella site are shown in Figure 3. Comparing the boulder shadow casts of the July 2017 to July 2019 images, detailed planar shapes of PILl, PILm, and PILn can be inferred (Figure 3a–c). In the June 2020 and September 2021 images, PILl, PILm, and PILn are overturned in comparison with the positions of the previous images (Figure 3d–f). Their exposed surfaces, not covered by lichen, are whitish and easily distinguishable from the surrounding gray texture. Consequently, they were overturned by water flow, roughly oriented SSW–NNE, between July 2019 and June 2020.

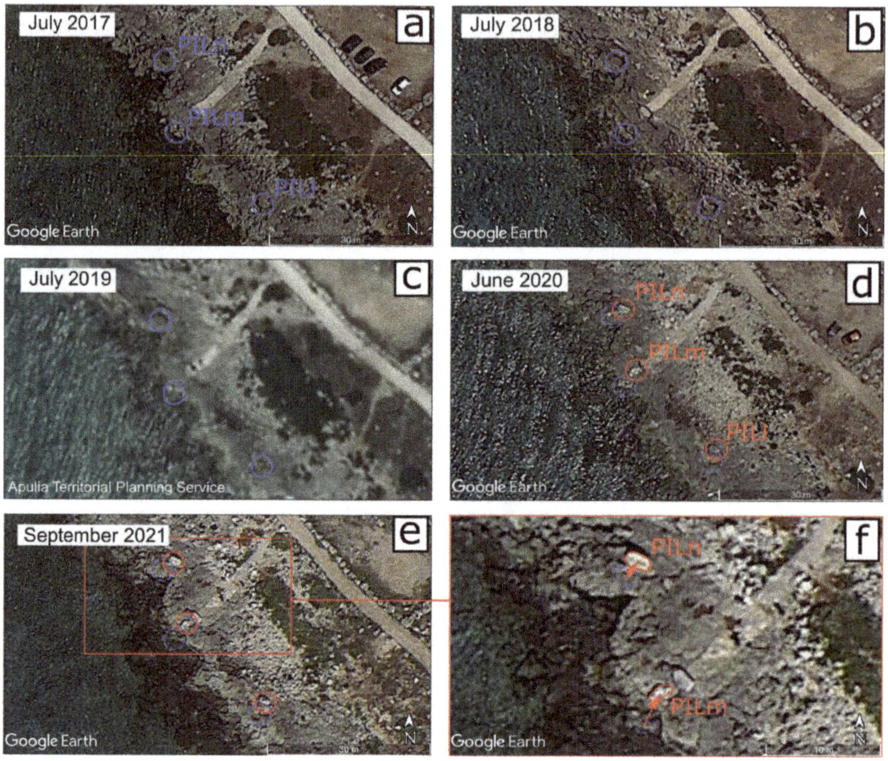

Figure 3. Pilella coast (39°55′26.18″–39°55′27.61″ N, 18°03′25.23″–18°03′28.54″ E) detection of boulder displacements: (**a**,**b**,**d**,**e**) GE images, eye elevation of 73 m; (**c**) regional aerial image; (**f**) transport vector of PILm and PILn boulders, eye elevation of 38 m. Blue symbols, pre-displacement position; red symbols, post-displacement position. Dotted lines help the comparison.

The site of Posto Rosso, where 16 displaced boulders were identified (Table 1), is located about 1.5 km southeast of Pilella (Table A5). For 6 of these boulders, the initial position was not established, while the final position was established for all the 16 boulders. Their middle axis ranged from about 1 to 2 m and the transport distance from 2 to 17 m (Table A6); vectors are oriented about SSW–NNE. In Figure 4, the analyzed time series remote sensing images of the central sector of the stretch of the Posto Rosso coast are reported. Due to the different sun–boulder-sensor geometries of the July 2017 to July 2019 images, the planar shape of the boulders can be defined in detail (Figure 4a–c). The ROe, ROg, and ROh, boulders appear in the series of images taken since June 2020; instead, the ROl boulder appears after September 2021 (Figure 4d,e; see also Figure A5b in Appendix A for the geomorphological setting). The faces of these four boulders observed in the field were fresh and did not show solution microforms, while traces of organic remains of marine plants were found. It follows that they were on the sea bottom before being displaced by waves (see Appendix A for submerged pre-transport setting).

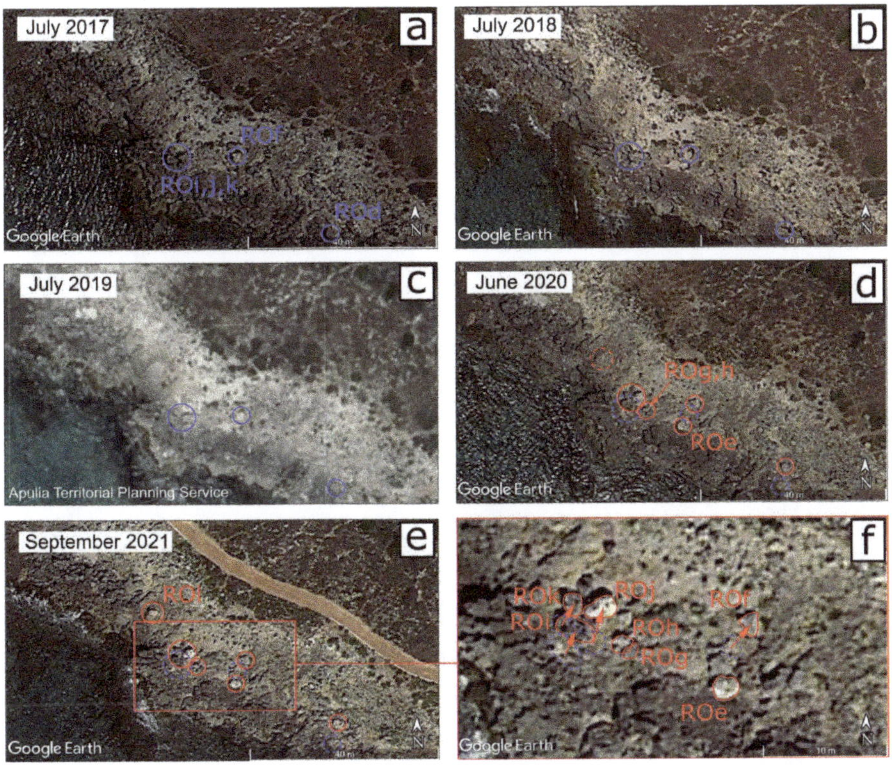

Figure 4. Posto Rosso coast (39°54′47.70″–39°54′49.57″ N, 18°04′15.79″–18°04′20.12″ E) detection of boulder displacements: (**a**,**b**,**d**,**e**) GE images, eye elevation of 93 m; (**c**) regional aerial image; (**f**) transport vectors of ROf, RIi, ROj, and ROk boulders, eye elevation of 35 m. Blue symbols, pre-displacement position; red symbols, post-displacement position. Dotted lines help the comparison.

Except for the ROl boulder, the changes in position of the boulder displaced at the Posto Rosso site occurred between July 2019 and June 2020 (Figure 4c,d), while no changes apparently occurred between July 2017 and July 2018 (Figure 4a,b) and July 2018 and July 2019 (Figure 4b,c). Given the above, it can be inferred that the ROl boulder was emplaced above the sea level between June 2020 and September 2021. As a whole, among the four new sites, Posto Rosso is the one with the greatest number of displaced boulders. These calcarenite boulders were detached from a surface that corresponds to the stratigraphic contact with the limestone (see Figure A5a,c,d in Appendix A). Apparently, the stratigraphy here is crucial for the boulder formation.

The southernmost site identified by multi-temporal analysis is Torre San Giovanni (Figure 1 for location). For three out the six boulders displaced, the pre-displacement position and TD were determined (Table A7). Their middle axis ranges from about 1 to more than 1.5 m (Table A8). The transport vectors are up to almost 10 m in length and approximately S–N oriented, as shown in Figure 5f. All the boulders of this site experienced position changes between July 2019 and June 2020 (Figure 5a–e). The SGc and SGd boulders have apparently completed their movement because they were stopped by other boulders lying on the coastal platform (Figure A6 in Appendix A).

On the displaced boulders whose initial position was not detected in the remote sensing images (Table A7), decaying organic matter (likely remains of marine plants) was observed. This suggests that they were in submerged conditions before being moved.

As a whole, visual analysis of the 2017–2021 time series remote sensing images allowed to establish that 37 out of 38 boulders of the newly detected sites were displaced between July 2019 and June 2020. One boulder (ROl; see above) changed position between June 2020 and September 2021. For 25 boulders, the TD was determined (Table A9).

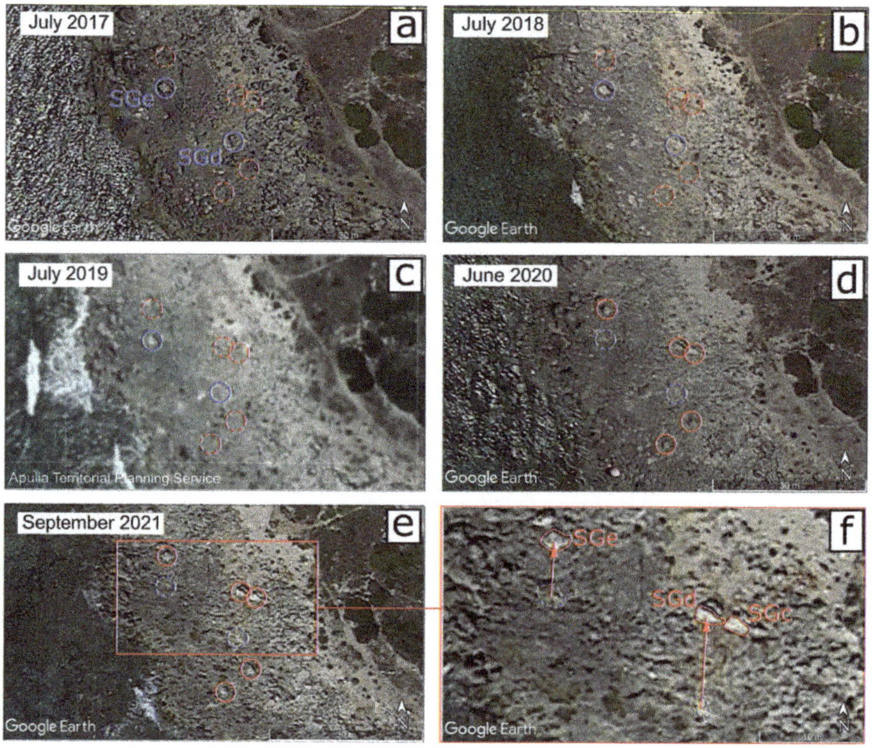

Figure 5. Torre San Giovanni coast (39°54′03.59″–39°54′05.11″ N, 18°05′06.89″–18°05′10.90″ E) detection of boulder displacements: (**a**,**b**,**d**,**e**) GE images, eye elevation of 85 m; (**c**) regional aerial image; (**f**) transport vectors of SCd and SCe boulders, eye elevation of 41 m. Blue symbols, pre-displacement position; red symbols, post-displacement position. Dotted lines help the comparison.

3.2. Checking Previously Detected Displacement

The multi-temporal investigation was performed on the 81 displaced boulders previously detected by bi-temporal images analysis [9] with the aim to detect their annual mobility for the 4 years considered herein (see Sections 1 and 2). In what follows, some features of the boulder displacement are given with reference to the sites of Punta Prosciutto, Torre Suda, Mancaversa, and Punta Pizzo (see Figure 1).

Eighteen boulders that changed position between July 2018 and June 2020 were previously recognized along the Punta Prosciutto headland [9]. By means of the new analysis, it was found that all 18 boulders changed position between July 2019 and June 2020. For 13 boulders, both the initial and final positions were established; thus, the TD was determined (Table A10). Several changes in positions are shown in Figure 6. The case of the cluster composed of the PRi, PRj, PRk, and PRl boulders is particularly enlightening. Before the displacement, they were some meters to the southwest and differently arranged in comparison with the post-displacement arrangement (Figure 6f). One SSW–NNE transport vector of about 5 m is drawn in the figure; actually, each of the boulders was transported from about 4 to 6 m. The boulder PRk is now vertically positioned; thus, it shows middle and minor axes in remote sensing images. The other three boulders of the cluster lie inclined on the coastal platform.

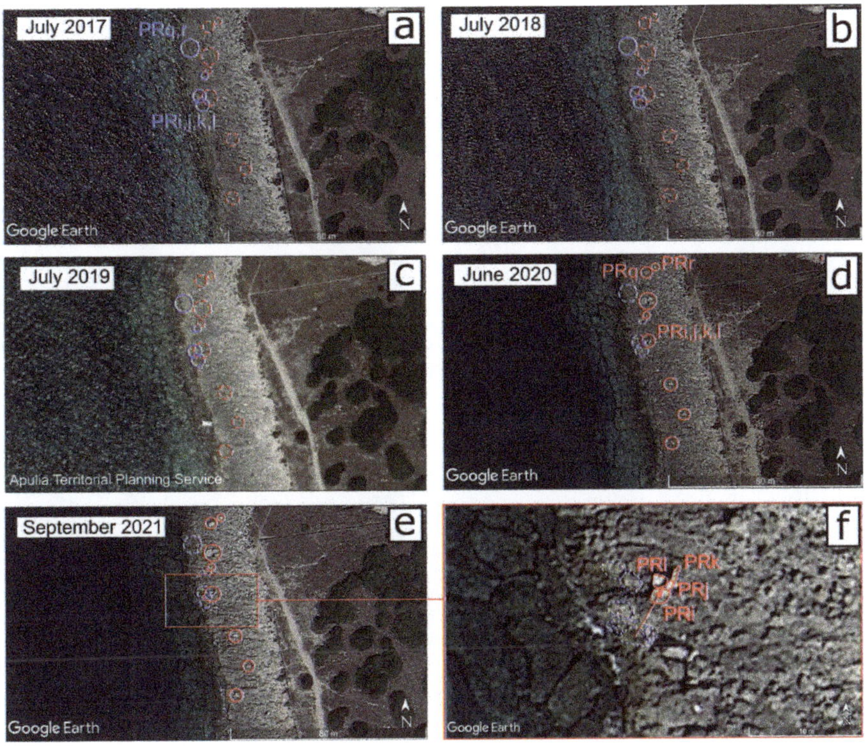

Figure 6. Punta Prosciutto headland (40°17′32.93″–40°17′35.96″ N, 17°45′41.75″–17°45′49.46″ E) detection of boulder displacements: (**a**,**b**,**d**,**e**) GE images, eye elevation of 167 m; (**c**) regional aerial image; (**f**) transport vector of the PRi, j, k, l cluster, eye elevation of 35 m. Blue symbols, pre-displacement position; red symbols, post-displacement position. Dotted lines help the comparison.

The boulders PRq and PRr appear to be the result of the breaking of a singular boulder that likely occurred during transport (Figure 6a–c). Again, this geomorphological process took place between July 2019 and July 2020 (Figure 6c,d). Moreover, PRr was overturned between July 2020 and September 2021, thus changing position again (Figure 6d,e).

The largest displaced boulder detected by Delle Rose et al. [9] (named SUa) was found at Torre Suda (Figure 1). The size of this boulder is 5.4 × 4.6 × 1.9 m; it was displaced by about 9 m from SSE to NNW (Figure 7). Due to its middle axis dimension, SUa belongs to the class of fine blocks. It apparently slid and rotated about 180° over a submarine terrace plain just below the sea level, constrained in moving from the low cliff from which it was detached (Figure 7f). This constraint could have caused such an anomalous displacement (see also Figure A7 in Appendix A). SUa is placed in the initial position in the July 2017 to July 2019 images (Figure 7a–c), while its socket is not distinguishable both in the June 2020 and September 2021 images because it is close to the sea level.

By multi-temporal analysis, it was found that the 14 displacements detected at Torre Suda [9] evidently occurred between July 2019 and June 2020, although one of the displaced boulders (the SUi boulder; see Appendix A) had already undergone a first position change between July 2018 and July 2019.

Apart from SUa, another nine boulders detected by [9] have middle axes that exceed 2 m in length (very coarse boulders), including the MAa boulder (that was detected at Mancaversa site; Figure 1). Due to an unfavorable sun–object-sensor geometry, MAa is barely detectable in the July 2018 image. (Its presence can be deduced from the narrow shadow cast that marks its NE side; see Figure 8b.) Differently, in both the July 2017 and July 2019 images, the planar shape of MAa is well defined by topographic shadow. This boulder was moved between July 2019 and June 2020 (Figure 8c,d). It was detached from the parent rock and lifted and rotated due to the impact of an overtopping wave (Appendix A; Figures 8f and A7).

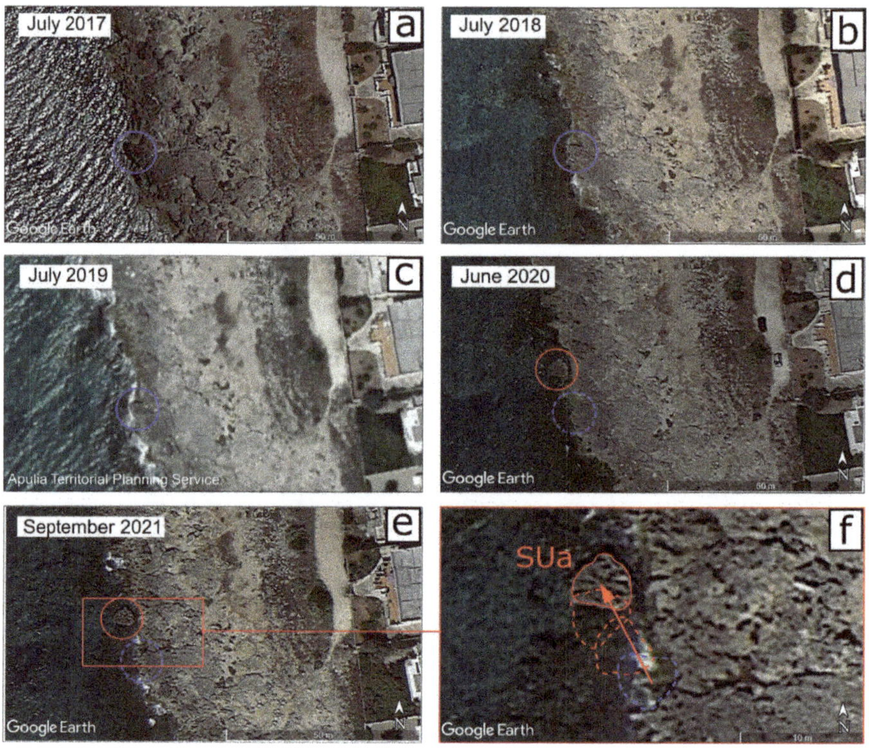

Figure 7. Torre Suda coast (39°57′05.72″–39°57′07.66″ N, 18°01′44.75″–18°01′49.31″ E) detection of boulder displacements: (**a**,**b**,**d**,**e**) GE images, eye elevation of 99 m; (**c**) regional aerial image; (**f**) transport vector of SUa boulder, eye elevation of 35 m. Blue symbols, pre-displacement position; red symbols, post-displacement position. Dotted lines help the comparison.

Different from the cluster composed of the PRi, PRj, PRk, and PRl boulders (see above), other clusters were found with few changes in mutual spatial arrangement of the boulders after displacement. The case of the cluster composed of the PIa, PIb, PIc, PId, and PIe boulders is considered below (Figure 9). This cluster appears evenly moved by about 5 m from SSW to NNE (Figure 9f). Apparently, no collision between the boulders occurred. The 18 displacements found at Punta Pizzo [9] occurred between July 2019 and June 2020, as detected in the multi-temporal images.

In conclusion, by checking the displacements detected by [9], it was found that all the 81 boulders were moved between July 2019 and June 2020. Moreover, 1 of these 81 boulders had already undergone a first position change between July 2018 and July 2019, while another changed position yet again between June 2020 and September 2021.

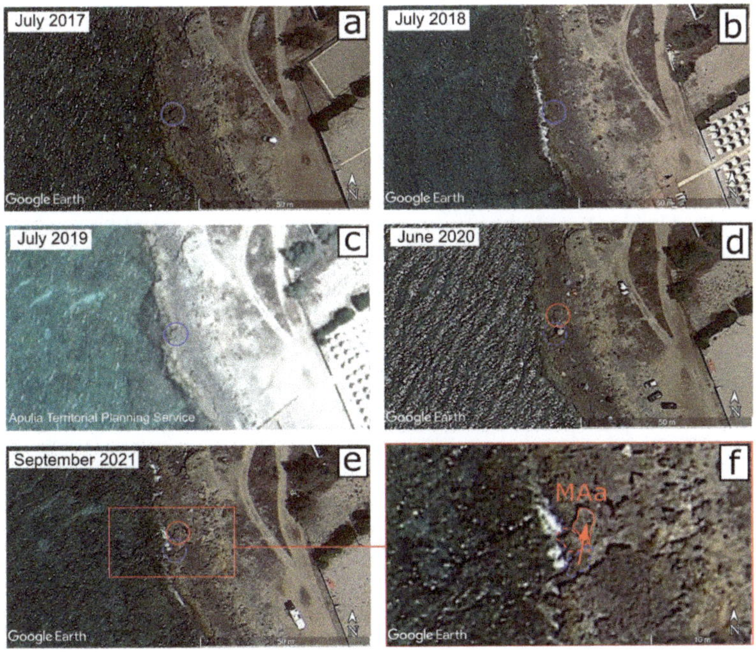

Figure 8. Mancaversa coast (39°58′19.56″–39°58′21.50″ N, 18°00′41.53″–18°00′46.09″ E) detection of boulder displacements: (**a**,**b**,**d**,**e**) GE images, eye elevation of 99 m; (**c**) regional aerial image; (**f**) transport vector of MAa boulder, eye elevation of 37 m. Blue symbols, pre-displacement position; red symbols, post-displacement position. Dotted lines help the comparison.

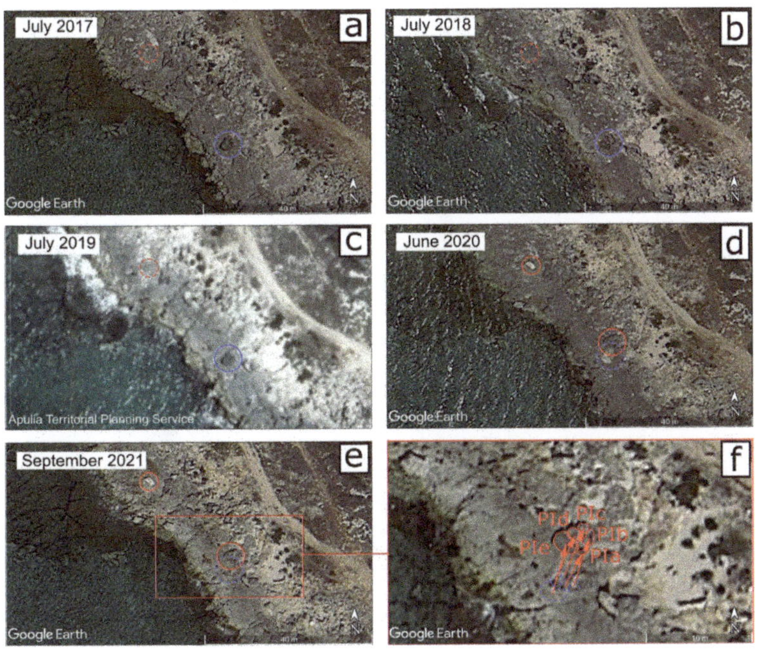

Figure 9. Punta Pizzo coast (39°59′33.92″–39°59′35.48″ N, 17°59′45.88″–17°59′49.52″ E) detection of

boulder displacements: (**a,b,d,e**) GE images, eye elevation of 80 m; (**c**) regional aerial image; (**f**) transport vectors of PIa, b, c, d, e boulders, eye elevation of 35 m. Blue symbols, pre-displacement position; red symbols, post-displacement position. Dotted lines help the comparison.

3.3. Marine Weather Conditions and Severe Storms

The main features of the storms that could have caused the detected boulder displacements are reported in Table 2. All the identified storms (see Section 2.3 and Appendix B for the methodological approach) occurred during the CW seasons. Marine conditions with sea state 5 occurred in 4 days of the 2017–2018 CW season. The 2018–2019 CW season has been characterized by only two severe storms (including Vaia), both with sea state 6. Four storms with sea state from 5 to 7 (the highest value belonging to Detlef) hit the study area during the 2019–2020 CW season, which therefore was the stormiest period of the considered 4 years. Finally, two storms, both with sea state 6, occurred during the 2020–2021 CW season.

Table 2. Main features of the storms inferred using the GLOBO-BOLAM-MOLOCH model cascade. *R*, duration; *F*, fetch length; *U*, wind speed; H_0, characteristic wave height. Sea state (degree, description) according to the Douglas sea scale.

Days	R (h)	F (km)	*U* (m/s)	H_0 (m)	Sea State
6 November 2017	12	600	14–16	3.4	5, rough
14 November 2017	12	800	16–18	4.0	5, rough
29 November 2017	15	900	14–16	3.9	5, rough
17 March 2018	12	700	16–18	4.0	5, rough
28–29 October 2018 [1]	48	700	12–14	5.7	6, very rough
2–3 February 2019	12	800	18–20	4.6	6, very rough
12–13 November 2019 [2]	24	900	20–22	8.8	7, high
24 November 2019	12	800	18–20	4.6	6, very rough
22 December 2019	24	400	14–16	4.7	6, very rough
2 March 2020	12	600	12–14	3.4	5, rough
3 December 2020	18	600	16–18	5.4	6, very rough
28–29 December 2020	12	600	18–20	4.6	6, very rough

[1] storm Vaia; [2] storm Detlef.

The H_0 associated with sea state 7 (storm Detlef) is 8.8 m. It is significantly larger than those associated with lower sea states, including the one of storm Vaia (5.7 m). Except for these two meteorological events, no other storms are associated with H_0 greater than 4.7 m (Table 2).

The nearshore wind conditions during the storms that occurred in the 2019–2020 CW season can be inferred from two out of the three wind gauge stations placed along the eastern coast of the Gulf of Taranto (Figure 1). Unfortunately, the Gallipoli station was not operational during this period (see Appendix B). The data extracted from the datasets of the Santa Maria di Leuca and Porto Cesareo stations are reported in Appendix C. They confirm that stronger wind conditions occurred during storm Detlef in comparison with the ones that occurred during the other storms of the 2019–2020 CW season. During 12–13 November 2019, the wind speed at 30 min averages measured at the Santa Maria di Leuca station ranged between 15 and 18 m/s for about 24 h, with a peak value of 24 m/s. The wind speed and duration during other storms were lower (Table A11). The measurements of the Porto Cesareo station confirm the above. The wind speed was around 20 m/s for almost the first half of November 13, while only during December 22 did it exceed 15 m/s for about 4 h (Figures A10–A13 in Appendix C).

4. Discussion

The following issues are discussed below: the annual mobility of the large clasts (Section 4.1), the cause of the massive displacement phenomenon (Section 4.2), and the marine weather conditions (Section 4.3). Complementary arguments are in Appendices C and D.

4.1. Annual Mobility of the Coastal Boulders

Along the eastern coast of the Gulf of Taranto, one hundred and nineteen boulders that changed position between July 2017 and September 2021 were detected by remote sensing image analysis (Sections 3.1 and 3.2). Moreover, two boulders (SUi and PRr) were apparently moved twice during this lapse of time. No boulder was displaced between July 2017 and July 2018. One boulder (SUi) was certainly displaced between July 2018 and July 2019. One hundred and eighteen displacements (including the second transport of the SUi boulder) occurred between July 2019 and June 2020 (Table 3). The PRr boulder experienced its second displacement between June 2020 and September 2021, which is the same interval of the change in position of ROl boulder.

Table 3. Annual intervals in which the detected displacements occurred.

		CW Season	No. of Displacements
1	July 2017–July 2018	2017–2018	0
2	July 2018–July 2019	2018–2019	1
3	July 2019–June 2020	2019–2020	118
4	June 2020–September 2021	2020–2021	2

The above result allows us to better define the period in which the massive displacement phenomenon recognized in the previous work [9] occurred, which is now limited to the interval 3 of Table 3. Moreover, storm Vaia must be ruled out as a possible causal storm.

Although some coastal boulders can be moved under rough sea conditions due to favorable combinations of hydrodynamic processes [57,58], the landward displacement of a high number of large boulders is typical of high-energy processes like tsunamis or exceptional storms [7,54,56]. During July 2017–September 2021, a western Mediterranean earthquake caused very small waves in the Gulf of Taranto unable to determine boulder displacements [9,59]. Consequently, the cause of the massive boulder displacement that occurred during the interval 3 (Table 3) must be found among the four storms that occurred during the 2019–2020 CW season (Table 2).

4.2. Event Responsible for the Massive Displacement

To relate the findings of the present study to those of ref. [9], the values of calculated wave height H at the initial position x_i (obtained from H_0 considering the inland decay in height [60,61]; see Appendix B) were compared to the minimum wave height H_m required to move the SA boulders (Figure 10). This graph has been used by some authors to identify the causal events of different boulder deposits [2,50]. The hydrodynamic equations to calculate H and H_m are reported in Appendix B.

A value of H_s = 1.5 m was taken as representative of the average rising of the sea level (storm surge + tide) during storm Detlef; for the weaker storms that occurred during the 2019–2020 CW season (Tables 2 and 3), a value of H_s = 0.5 m was considered (see ref. [9] for details). In Figure 10, both the circles (boulders investigated in the present study) and squares (boulders investigated by [9]) form a trend that agrees with the line shape. H_m exceeds the height of the bore only for a few boulders. This graph suggests, again, that both storm Detlef was the cause of the massive displacement, and some boulders could be moved even during the weaker storms of the interval 3 (see Appendix D for more inferences). Most of the obtained TDs are few meters, while only a small percentage of boulders were moved beyond 10 m (Appendix A, Figure A7).

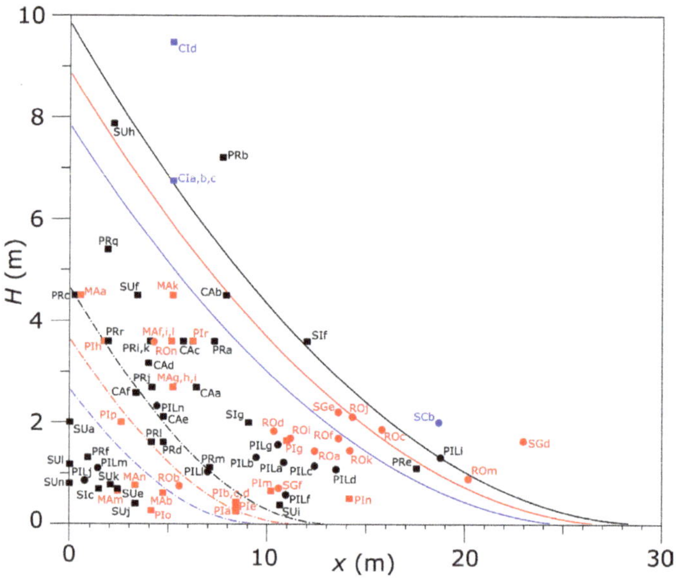

Figure 10. Comparison between wave height required to move the boulders H_m (circles and squares) and the height of the land-crossing bore H (lines) as a function of the distance x from the cliff edge and the coast height H_c (colors) for different characteristic wave height H_0. Continuous line: wave height for storm Detlef H_0 = 8.8 m, and a total sea level increase (storm surge + tide) H_s = 1.5 m (see text). Dot-dashed line: wave height for H_0 = 4.7 m and H_s = 0.5 m. Black, H_c = 0.5 m; red, H_c = 1.5 m; blue, H_c = 2.5 m; circles: boulders investigated in the present study; squares: boulders investigated in the previous study [9].

The next research target is to compute the wave characteristics approaching the nearshore, for the case herein investigated, using the open-source SWAN (Simulating WAves Nearshore) code. However, the data collected and processed so far support that exceptional nearshore wave conditions drove the massive displacement phenomenon caused by storm Detlef along the eastern side of the Gulf of Taranto (see also Section 4.3). On the other hand, by on-site investigations performed after storm Detlef, 6 and 11 displaced boulders were found at Torre Suda (Sui second displacement, SUj, SUk, SUl, SUm, and SUn) and Sant'Isidoro (SIc, SIf, SIg, and eight additional boulders that were not detected by remote sensing imagery analysis because of their small size or short displacement), respectively [4,17,20,21] (Figure 1 for sites' location). After storm Vaia, instead, only one case of boulder displacement at the Torre Suda site was detected (SUi, first displacement; see Figure A8a,b in Appendix A), while no displacements were found at the Sant'Isidoro site. Other evidence supporting the recognition of storm Detlef as the cause of the massive displacement can be found in Appendix D.

4.3. Marine Weather Conditions

The 12–13 November storm attracted the interest of several weather bureaus. It was named "Detlef" by the Meteorological Institute of Berlin Free University [62], and the name was adopted in some studies [4,63,64] (in [17], storm Detlef is provisionally named "Ionics19"). It was also named "Trudy" and "Bernardo" by the U.S. National Aeronautics and Space Administration and the Spanish Meteorological Agency, respectively [65–67]. On 10 November, a wide trough elongated from Northern Europe toward the western Mediterranean, creating a center of low atmosphere pressure. The cyclone transformed into a subtropical storm by 0600 UTC on 11 November [68] and then moved eastward over the central Mediterranean. Later, it reached the northern Adriatic Sea, causing an exceptional sea level height in the Venice Lagoon [16,69,70]. Regarding the effects of storm Detlef on the study area, it must be noted that a strong wind over the eastern side of the Gulf of Taranto started soon after midnight on 12 November from the southeast and persisted for 24 h from almost the same direction, with an offshore fetch of about 900 km and an average wind speed of 20–22 m/s. The calculated characteristic wave height is 8.8 m (Table 2), which is greater than the maximum significant wave for a 100-year return period (Section 2).

A similar synoptic condition caused the 24 November storm with a weaker depression in the middle atmosphere. Again, the strongest winds blew from the south-eastern direction over the southern Apulia coast, with an intensity of 18–20 m/s (Table 2). In this case, the fetch was of about 800 km, while the duration of these conditions of about 12 h. Both the 22 December 2019 and the 2 March 2020 storms were caused by two middle-atmosphere lows coming from the north-eastern direction with respect to the Italian peninsula [9]. This difference caused a different impact over the study area, with more variability of the wind speed and direction. In the case of December 2019, the prevailing direction was from the western direction over the southern Apulia coast, with a shorter fetch because of the geographic shape of the Gulf of Taranto (Figure 1). The BOLAM model shows a wind of about 14–16 m/s at 10 m over the sea, associated with a duration of 24 h and a fetch of about 400 km. In the case of March 2020, the strongest wind blew from the south, with a lower intensity of about 12–14 m/s, a duration of 12 h, and a fetch of about 600 km. The calculated characteristic wave height for the storms that occurred between 24 November 2019 and 2 March 2020 ranges from 3.4 to 4.7 m (Table 2), values lower than that of the maximum significant wave for a 50-year return period (Section 2).

5. Conclusions

Exploring the 2017–2021 boulder mobility along the eastern coast of the Gulf of Taranto, a large interannual variability was found. A massive displacement occurred between 2019 fall and 2020 summer (involving at least 118 large clasts), while only 0–2 displacements occurred during each of the other explored years.

The cause of such a phenomenon must be found among the storms that occurred during the 2019–2020 CW season. Thus, differently from the previous statement [9], storm Vaia (28–29 October 2018) must be rejected as a possible causative storm. Considering several geomorphological, meteorological, and hydrodynamic evidences, storm Detlef (12–13 November 2019) is recognized as the causal event. In any case, it cannot be ruled out that the other storms of the 2019–2020 CW season could have moved some boulders.

For storm Detlef, the calculated characteristic wave height was just under 9 m, greater than the maximum significant wave for a 100-year return period. Differently, the characteristic wave heights for the other storms of the 2019–2020 CW season ranged from 3.4 to 4.7 m, values lower than the maximum significant wave for a 50-year return period. It must be highlighted that, along the studied coast, storm Vaia caused the displacement of only one boulder among those detected by the remote sensing analysis and the geomorphological survey.

This study expands the geomorphological dataset and deepens the marine weather knowledge on the massive displacement phenomenon identified by [9]. Such a phenomenon is the largest (for the number of boulders involved and length of the affected coast) among those that have occurred (and documented) in the Mediterranean Sea (see ref. [4]). This result confirms that overtopping flows due to storm waves can leave a significant geomorphological signature in coastal boulder deposits.

Funding: This research received no external funding.

Data Availability Statement: Data supporting the reported results can be found at the following links: https://protezionecivile.puglia.it/annali-e-dati-idrologici-elaborati (accessed on 10 September 2023), for the hydrological annals; http://www.isac.cnr.it/dinamica/projects/forecasts/index.html (accessed on 10 September 2023), for wind and wave data; http://93.51.158.171/web/simop/home (accessed on 10 September 2023), for wind data. Google Earth accessed on 28 January 2024.

Acknowledgments: The author thanks Gianluca Pappaccogli for the assistance in the processing of the remote sensing images; Paolo Martano for the calculation of the characteristic wave height and the description of the marine conditions; Luca Marzo, Luca Orlanducci, and Anna Lisa Signore for the assistance during the geomorphological surveys; and two anonymous reviewers for constructive comments on an earlier version of this work. The elaboration of remote sensing images (Figures 2–9) was carried out in the free open-source Inkscape editor.

Conflicts of Interest: The author declares no conflicts of interest.

Appendix A

The data of the detected displaced boulders retrieved from remote sensing resources (see Sections 2 and 3) together with morphometric features determined by on-site investigation are reported below (Tables A1–A8). Geographical coordinates and TD were taken from the September 2021 GE image. In the course of the field survey, the inclination to the horizontal position of the boulders I and the dimensions of their axes (a_f, major; b_f, middle; c_f, minor) were measured and the lithology (Li) of the boulders (which, in turn, allows us to determine ρ_s) established.

The PTS refers to some physical conditions (scenario) of the boulder before the displacement. The boulder can be isolated (sub-aerial, SA, or submerged, SB), laterally limited by joints (joint-bounded, JB), or located at the cliff edge (CE) [1,71]. These scenarios determine the possible three MTs (saltation/lifting, sliding, and overturning/rolling) during boulder displacement. SA and SB boulders can be lifted, slid, or rolled by the water flow; due to the movement constraints, JB boulders can only be lifted, while CE boulders can be lifted or rolled [54,56].

Figure A1 shows the socket left on the parent rock by the SCa boulder (Section 3.1). The socket was ephemeral, so that a few years after its formation, it has become almost indistinguishable from its rock parent (Figure A1a,b). The pre-transport setting of this boulder is "cliff edge" (CE) (see ref. [56]). All the other boulder detected in the present work are "sub-aerial" (SA) or "submerged" (SB) (see Tables A2, A4, A6 and A8).

Figure A1. Santa Caterina site; the socket left by the SCa boulder (2.5 m over the sea level); (**a**) after about 1 year from the detachment (the image was taken 1 November 2020), the rock surface is fresh and not weathered; (**b**) after about 4 years from the detachment (the image was taken 22 September 2023, with 1 m long tape measure for scale), the rock surface is almost indistinguishable from the parent rock (see Figure 2d,e).

The post-transport position of SCa in unknown (Table A1). It was not recognized from GE images or in the field, and probably, SCa fell into the sea after detaching from its initial position.

Table A1. Initial (pre-displacement) and final (post-displacement) geographical coordinates of the storm-displaced boulders at Santa Caterina coast; ind., indeterminable; A.I., annual interval in which the displacement occurred (see Table 3).

	Initial Position		Final Position		
ID	Latitude	Longitude	Latitude	Longitude	A.I.
SCa	40°08′17.09″ N	17°59′19.21″ E	ind.	ind.	3
SCb	40°08′18.36″ N	17°59′18.47″ E	40°08′19.03″ N	17°59′18.85″ E	3

Differently from the socket of SCa, due to the different distance to the coastal cliff (Table A2), the surface of the SCb boulder is still fresh and unaltered (Figure A2). Moreover, since SCb lies in a shrub vegetation area, it is easily recognizable in remote sensing images (Figure 2d,e).

Figure A2. Santa Caterina site; the SCb boulder after the displacement occurred between July 2019 and September 2020 (the image was taken 22 September 2023, with 1 m long tape measure for scale). The whitish surface of SCb is uncovered by lichen and therefore still easily detectable in the field as well as by remote sensing images (see Section 3.1).

Table A2. Main features of the storm-displaced boulders at Santa Caterina coast; major and middle axes measured on GE remote sensing imagery: a_r and b_r (in m); dimensions of the axes taken in the field: a_f, b_f, and c_f (in m); inclination on the ground I; initial distances from the coastline x_i; final distances from the coastline x_f (in m). TD, transport distance (in m); Li, lithology: C, calcarenite, L, limestone; Sh, shape: O, oblate ($b/a > 0.6$, $c/b < 0.6$); E, equant ($b/a > 0.6$, $c/b > 0.6$); B, bladed ($b/a < 0.6$, $c/b < 0.6$); P, prolate ($b/a < 0.6$, $c/b > 0.6$). FI, flatness index; PTS, pre-transport setting: JB, joint-bounded; SA, sub-aerial; SB, submerged; CE, cliff edge. MT, movement type: ST, saltation; SL, sliding; OV, overturning. ind., indeterminable. Note: Abbreviations are the same for Tables A4, A6 and A8.

ID	a_r	b_r	a_f	b_f	c_f	I	x_i	x_f	TD	Li	Sh	FI	PTS	MT
SCa	3.3	1.2	-	-	-	-	0.8	ind.	-	C	-	-	CE	-
SCb	1.8	1.5	1.8	1.6	0.7	5°	18.8	21.6	15.4	C	O	2.5	SA	OV

Table A3. Initial (pre-displacement) and final (post-displacement) geographical coordinates of the storm-displaced boulders at the Pilella coast. Note: See caption of Table A1 for abbreviations.

	Initial Position		Final Position		
ID	Latitude	Longitude	Latitude	Longitude	A.I.
PILa	39°55′22.66″ N	18°03′35.76″ E	39°55′22.72″ N	18°03′35.80″ E	3
PILb	39°55′22.60″ N	18°03′35.42″ E	39°55′22.68″ N	18°03′35.51″ E	3
PILc	39°55′22.27″ N	18°03′34.62″ E	39°55′22.35″ N	18°03′34.66″ E	3
PILd	39°55′22.24″ N	18°03′34.49″ E	39°55′22.33″ N	18°03′34.50″ E	3
PILe	ind.	ind.	39°55′23.54″ N	18°03′32.43″ E	3
PILf	39°55′23.83″ N	18°03′31.60″ E	39°55′23.88″ N	18°03′31.63″ E	3
PILg	39°55′23.97″ N	18°03′31.61″ E	39°55′24.11″ N	18°03′31.72″ E	3
PILh	ind.	ind.	39°55′24.08″ N	18°03′30.58″ E	3
PILi	39°55′24.39″ N	18°03′31.08″ E	39°55′24.49″ N	18°03′31.12″ E	3
PILj	39°55′25.09″ N	18°03′27.70″ E	39°55′25.10″ N	18°03′27.78″ E	3
PILk	39°55′25.38″ N	18°03′27.86″ E	ind.	ind.	3
PILl	39°55′26.41″ N	18°03′27.20″ E	39°55′26.49″ N	18°03′27.26″ E	3
PILm	39°55′26.86″ N	18°03′26.56″ E	39°55′26.96″ N	18°03′26.64″ E	3
PILn	39°55′27.29″ N	18°03′26.47″ E	39°55′27.32″ N	18°03′26.54″ E	3

Several displaced boulders show a high value of *I*. As an example, the PILn boulder (Figure A3) is inclined of about 55° with respect to the ground (Table A4). Such an arrangement can cause the underestimation of the horizontal dimensions of the boulders from remote sensing imagery (Section 3).

Figure A3. Pilella site; the PILn boulder lies on a smaller boulder and forms an angle of about 55° with respect to the ground. Note the difference in length between b_r and b_f (see the caption of Table A2 for symbols).

As mentioned in Section 2.3, after estimating the center point of the boulders at the initial and final positions, several TDs were verified onsite by means of a measuring tape. At the Pilella coast, the measurements taken almost never differed more than 5% from those detected in the remote sensing images (see also Appendix B in ref. [9]). As an example, for the PILm boulder, 3.4 m was measured in the field against 3.7 m in the image.

Table A4. Main features of the storm-displaced boulders at the Pilella coast. Note: See caption of Table A2 for abbreviations.

ID	a_r	b_r	a_f	b_f	c_f	I	x_i	x_f	TD	Li	Sh	FI	PTS	MT
PILa	1.6	0.9	1.6	1.0	0.5	0–5°	10.8	12.4	3.6	C	O	2.6	SA	OV
PILb	2.0	1.0	1.8	1.0	0.4	0–5°	9.4	10.5	4.1	C	B	3.5	SA	OV
PILc	1.5	1.0	1.5	0.9	0.4	~20°	12.4	14.7	2.5	C	B	3	SA	OV
PILd	2.0	1.2	1.9	1.0	0.6	0–5°	13.6	15.6	2.6	C	B	2.4	SA	OV
PILe	2.3	1.6	2.5	1.5	0.3	~15°	ind.	13.2	-	C	O	6.7	SB	ST,OV
PILf	2.2	1.5	2.4	1.2	0.5	~60°	10.9	12.5	1.5	C	B	3.6	SA	SL
PILg	1.7	1.4	1.5	1.3	0.6	~10°	10.6	15.5	4.8	C	O	2.3	SA	OV
PILh	2.2	1.0	1.9	1.0	0.5	5°	ind.	3.9	-	L	B	2.9	SB	ind.
PILi	1.9	1.1	2.2	1.1	0.4	5°	18.8	20.4	2.6	C	B	4.1	SA	ST,OV
PILj	3.1	1.9	3.0	1.8	0.6	5°	0.7	1.9	1.6	C	O	4	SB	SL
PILk	2.0	1.4	-	-	-	-	6.9	ind.	-	-	-	-	SA	-
PILl	2.0	1.1	2.1	1.1	0.8	0–5°	7.0	8.9	3.3	L	P	2	SA	OV
PILm	2.1	1.1	2.0	1.0	0.4	0–5°	1.1	3.7	3.2	C	B	3.7	SA	OV
PILn	2.2	1.2	2.1	1.8	0.4	~55°	4.4	6.2	1.9	C	O	4.9	SA	OV

Most of the boulders examined at Pilella clearly show fresh and not weathered surfaces, while the surfaces covered by lichen and with karst dissolution features are in contact with the ground. They must be considered as SA overturned boulders.

The initial position of both the PILe and PILh boulders was not detected in the time series remote sensing images (Table A4). At its final position, the upper surface of PILe is whitish and is covered by decaying marine plants (Figure A4a). This suggests that it was below sea level before the transport. PILh is currently just over the edge of the cliff, and it presents some solution pans over its upper surface (Figure A4b). Before the displacement, it was roughly at the sea level. For the PILf and PILj boulders, some tracks referable as drag marks were observed along their trajectory, and consequently, their type of movement is considered SL (Table A4). For the remaining boulders, no elements useful to define with good confidence the MT were identified.

Figure A4. Pilella site; (**a**) the PILe boulder (green arrow indicates decaying marine plants); (**b**) the PILh boulder (blue arrows indicate solution pans).

Table A5. Initial (pre-displacement) and final (post-displacement) geographical coordinates of the storm-displaced boulders at the Posto Rosso coast. Note: See caption of Table A1 for abbreviations.

	Initial Position		Final Position		
ID	**Latitude**	**Longitude**	**Latitude**	**Longitude**	**A.I.**
ROa	30°54′45.53″ N	18°04′20.88″ E	30°54′47.70″ N	18°04′20.99″ E	3
ROb	30°54′46.19″ N	18°04′20.51″ E	30°54′46.21″ N	18°04′20.55″ E	3
ROc	30°54′46.87″ N	18°04′20.58″ E	30°54′46.90″ N	18°04′20.69″ E	3
ROd	30°54′47.88″ N	18°04′18.76″ E	30°54′48.03″ N	18°04′18.79″ E	3
ROe	ind.	ind.	30°54′48.30″ N	18°04′17.88″ E	3
ROf	30°54′48.40″ N	18°04′17.90″ E	30°54′48.45″ N	18°04′17.96″ E	3
ROg	ind.	ind.	30°54′48.39″ N	18°04′17.57″ E	3
ROh	ind.	ind.	30°54′48.43″ N	18°04′17.54″ E	3
ROi	30°54′48.38″ N	18°04′17.34″ E	30°54′48.44″ N	18°04′17.36″ E	3
ROj	30°54′48.40″ N	18°04′17.44″ E	30°54′48.51″ N	18°04′17.47″ E	3
ROk	30°54′48.45″ N	18°04′17.59″ E	30°54′48.52″ N	18°04′17.37″ E	3
ROl	ind.	ind.	30°54′48.79″ N	18°04′17.15″ E	4
ROm	30°54′49.29″ N	18°04′16.85″ E	30°54′49.40″ N	18°04′16.95″ E	3
ROn	30°54′49.02″ N	18°04′15.97″ E	30°54′48.48″ N	18°04′16.41″ E	3
ROo	ind.	ind.	30°54′49.68″ N	18°04′16.38″ E	3
ROp	ind.	ind.	30°54′49.71″ N	18°04′16.43″ E	3

Many coarse boulders were detected at the Posto Rosso site (Section 3.1, Table A6). The ROa boulder was detached and overturned (Figure A5a). The boulder ROl also appears to have been overturned (Figure A5b). The ROn boulder, despite its large size, was transported for 17.5 m due to the water flow (Figure A5c,d). This measure, taken using a measuring tape, differs from the one detected in the remote sensing image by less than 0.5 m (Table A6). In Figure A5c, one phase of estimating the position of the geometric center of the boulder before the movement can be observed. Again, ROn is not overturned, and no drag marks are observed over the platform surface. Thus, it appears to have undergone a saltation movement.

Figure A5. Posto Rosso site; (**a**) the overturned ROa boulder and (in the foreground) its socket (with 1 m long tape measure for scale); (**b**) the geomorphological arrangement of ROl boulder that was displaced during the 2020–2021 CW season (see Section 3.1); (**c**) the socket of the ROn boulder; (**d**) the ROn boulder was detached from its initial position and displaced 17–17.5 m inland (see text).

Table A6. Main features of the storm-displaced boulders at the Posto Rosso coast. Note: See caption of Table A2 for abbreviations.

ID	a_r	b_r	a_f	b_f	c_f	I	x_i	x_f	TD	Li	Sh	FI	PTS	MT
ROa	2.7	1.4	2.7	1.3	0.8	0–5°	12.3	18.5	5.8	C	P	2.5	SA	OV
ROb	2.8	1.3	2.9	1.4	0.6	0–5°	5.7	6.7	1.9	C	B	3.6	SA	SL
ROc	1.6	1.3	1.7	1.4	0.5	0–5°	15.9	18.2	2.5	C	O	3.1	SA	ST,OV
ROd	2.5	1.9	2.4	1.5	0.7	~20°	10.4	14.7	4.5	C	O	2.8	SA	OV
ROe	2.1	1.6	2.3	1.6	0.5	0–5°	ind.	11.2	-	C	O	3.9	SB	ind.
ROf	1.8	1.4	1.7	1.4	0.7	0–5°	13.6	15.8	2.1	C	O	2.2	SA	ST,OV
ROg	1.5	0.9	1.3	0.9	0.4	~25°	ind.	14.6	-	C	B	2.7	SB	ST
ROh	1.6	0.9	1.7	0.9	0.7	0–5°	ind.	14.4	-	C	P	1.8	SB	ST,OV
ROi	2.8	1.3	2.7	1.4	0.7	~10°	11.1	12.6	1.9	C	B	2.9	SA	ST,OV
ROj	2.6	1.5	2.4	1.7	0.8	~20°	14.4	16.7	3.1	C	B	2.6	SA	OV
ROk	1.7	1.2	1.7	1.3	0.7	~15°	14.2	16.1	2.1	C	O	2.1	SA	OV
ROl	1.8	1.3	1.9	1.2	0.3	~20°	ind.	9.3	-	C	O	5.2	SB	OV
ROm	1.5	0.8	1.4	0.8	0.5	0–5°	20.3	25.2	4.8	C	P	2.2	SA	ST,OV
ROn	3.1	1.8	2.8	2.1	0.5	~20°	4.2	20.7	17.1	C	O	4.9	SA	ST
ROo	1.3	0.9	1.4	0.9	0.5	~10°	ind.	25.3	-	C	B	2.3	SB	ST,OV
ROp	1.5	0.8	1.7	1.1	0.5	~25°	ind.	26.4	-	C	O	2.8	SB	ST,OV

Table A7. Initial (pre-displacement) and final (post-displacement) geographical coordinates of the storm-displaced boulders at the Torre San Giovanni coast. Note: See caption of Table A1 for abbreviations.

	Initial Position		Final Position		
ID	Latitude	Longitude	Latitude	Longitude	A.I.
SGa	ind.	ind.	39°54′03.86″ N	18°05′09.25″ E	3
SGb	ind.	ind.	39°54′03.99″ N	18°05′09.47″ E	3
SGc	ind.	ind.	39°54′04.46″ N	18°05′09.50″ E	3
SGd	39°54′04.18″ N	18°05′09.34″ E	39°54′04.49″ N	18°05′09.38″ E	3
SGe	39°54′04.52″ N	18°05′08.72″ E	39°54′04.73″ N	18°05′08.76″ E	3
SGf	39°54′05.57″ N	18°05′08.45″ E	39°54′05.61″ N	18°05′08.53″ E	3

Several of the displaced boulders are rather flat (e.g., PILe and SGd; see Tables A4 and A8, respectively). The flat SGd boulder (FI = 6; Figure A6) was transported almost 10 m from an initial position more than 20 m away from the coastline.

For the SGd, SGe, and SGf boulders, the TD was verified onsite. Again, these measures do not differ by more than 5% from those detected in the remote sensing images.

Figure A6. Torre San Giovanni site; SGc (in the foreground) and SGd boulders. As a result of the water flow transport, they formed two imbricated clusters together with other small boulders.

Table A8. Main features of the storm-displaced boulders at the Torre San Giovanni coast. Note: See caption of Table A2 for abbreviations.

ID	a_r	b_r	a_f	b_f	c_f	I	x_i	x_f	TD	Li	Sh	FI	PTS	MT
SGa	1.5	0.9	1.4	1.1	0.4	0–5°	ind.	14.9	-	C	O	3.1	SB	ST,OV
SGb	1.8	1.1	1.8	1.0	0.3	0–5°	ind.	22.3	-	C	B	4.7	SB	ST,OV
SGc	1.9	1.1	1.8	1.2	0.3	0–5°	ind.	31.4	-	C	O	5	SB	ST,OV
SGd	2.1	0.9	2.3	1.3	0.3	~45°	22.8	28.6	9.5	C	B	6	SA	ST,OV
SGe	2.4	1.7	2.2	1.7	0.7	0–5°	13.8	18.2	6.5	C	O	2.8	SA	ST,OV
SGf	1.6	0.8	1.5	1.2	0.6	~30°	10.4	11.7	2.4	C	O	2.2	SA	OV

When both the initial and final positions of the detected boulders were established, the transport distance was calculated (Sections 4.1 and 4.2). In Tables A9 and A10, a list of these 84 boulders is given.

Table A9. Displaced boulders detected in this study for which TD was calculated and whose position changed between July 2019 and June 2020.

Site	Boulder ID [1]
Santa Caterina	SCb
Pilella	PILa, PILb, PILc, PILd, PILf, PILg, PILi, PILj, PILl, PILm, PILn
Posto Rosso	ROa, ROb, ROc, ROd, ROf, ROi, ROj, ROk, ROm, ROn
Torre San Giovanni	SGd, SGe, SGf

[1] For geographical coordinates, transport distance, dimensions of the axes, pre- and post-displacement distances from the coastline, lithology, shape, flatness index, pre-displacement setting, and movement type, see Tables A1–A8.

Table A10. Displaced boulders detected by [9] whose position results changed between July 2019 and June 2020 and for which TD was calculated.

Site	Boulder ID [1]
Punta Prosciutto	PRa, PRb, PRc, PRd, PRe, PRf, PRi, PRj, PRk, PRl, PRm, PRq, PRr
Sant'Isidoro	SIf, SIg
Punta Pizzo	PIa, PIb, PIc, PId, PIe, PIg, PIh, PIj, PIm, PIn, PIo, PIp, PIr
Mancaversa	MAa, MAb, MAc, MAf, MAg, MAh, MAi, MAj, MAk, MAl, MAm, MAn
Torre Suda	SUa, SUe, SUf, SUh, SUi, SUj, SUk, SUl, SUn
Capilungo	CAa, CAb, CAc, CAd, CAe, CAf
Ciardo	CIa, CIb, CIc, CId

[1] For geographical coordinates, transport distance, dimensions of the axes, pre- and post-displacement distances from the coastline, lithology, shape, flatness index, pre-displacement setting, and movement type, see Appendix A in ref. [9].

Based on the above sets of data, it can be deduced that more than 65% of medium-to-coarse boulders were moved less than 5 m, while less than 15% were moved more than 10 m, including the SCb and ROn boulders (Figure A7; Tables A1, A2, A5 and A6 for morphometric data). This result agrees with those of other studies performed with different methods on Mediterranean coasts [72,73].

As reported in Section 3, two boulders underwent two displacements during 2017–2021. In the case of the SUi boulder (Section 3.2), this singular mobility can be due to a slipway ramp that channels the water flows (Figure A8).

The PTS of the MAa boulder was considered as joint-bounded (JB) in the previous work [9]. It was detached from the parent rock along a vertical joint system, leaving a small step on the ground surface (Figure A9). In any case, MAa did not jump over the step, as it was moved in opposite direction. The boulder could therefore be considered SA. Since the slope angle (θ) is zero, the minimum wave height calculated by the Equations (A4) and (A7) (Appendix B) is the same (H_m = 4.5 m; see Figure 10).

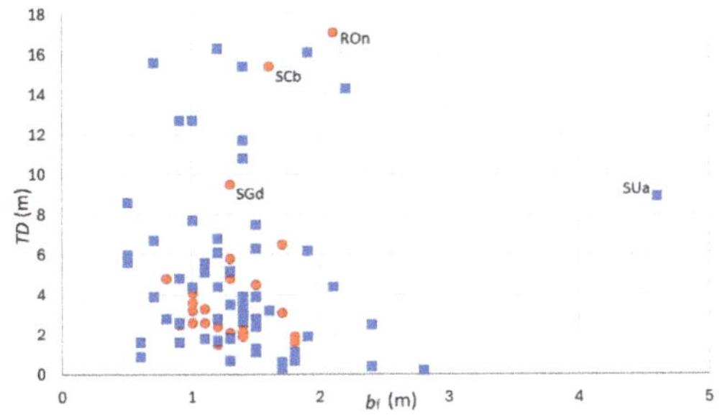

Figure A7. Transport distance (TD) vs. middle axis measured in the field (b_f). Red circles, boulders detected in the present study; blue squares, boulders detected in the previous study [9].

Figure A8. Torre Suda site; (**a**) the socket of the SUi boulder (image taken the 23 February 2019); (**b**) SUi boulder lying on a slipway ramp after the first movement (image taken 23 February 2019); (**c**,**d**) SUi boulder displaced from the ramp due to the second movement (images taken 17 October 2020).

Figure A9. Mancaversa site; the MAa boulder lying about 4.5 m from its initial position (in the foreground); rock joints are indicated by green arrows (see text).

For the MAa boulder, the TD was verified onsite. Again, this measure does not differ by more than 5% from that detected in the remote sensing image.

Appendix B

As mentioned in Section 2.3, the windstorm characteristics (duration R, length of the fetch over the sea surface F, fetch direction D, and wind speed U) determined by the forecast map archive of the BOLAM-MOLOCH model [53] were used to calculate the spectral peak wave height from the similarity relation equations [74,75]. They relate T (wave period), H_0, and R to the wind speed at 10 m height offshore U and to the fetch over the sea F:

$$gT/U = 0.286(gF/U^2)^{1/3}, \tag{A1}$$

$$gH_0/U^2 = 0.0016(gF/U^2)^{1/2}, \tag{A2}$$

$$gR/U = 68.8(gF/U^2)^{2/3}, \tag{A3}$$

where g is the gravity acceleration. This procedure has already been used in previous works [9].

The duration R of stability of wind conditions, estimated from the wind maps, was used as a limiting factor to calculate an effective fetch F for Equations (A1) and (A2) whenever the actual geometric fetch appears to be longer. It must be noted that H_0 is nearly equivalent to the significant wave height [76,77].

Data of wind gauge stations placed along the coast (Figure 1) have been used to infer the nearshore wind conditions during storms presumed to be responsible for changes in boulder position. Due to the incompleteness and inhomogeneity of the datasets, these conditions can be only partially defined. The Santa Maria di Leuca and Gallipoli stations are managed by the Apulia Civil Protection that publishes comprehensive annual reports (named *Annali Idrologici*, Hydrological Annals) [78–82]. These report the daily average and maximum together with four 30 min averages at four selected hours of the day (00, 06, 12, and 18). However, the Gallipoli station has been in operation only since November 2020. The Porto Cesareo station belongs to the Apulia Meteomarine Network (*SIMOP*, *Sistema Informativo Meteo Oceanografico delle Coste Pugliesi*, in Italian); its data are publicly available online and consist of 15 min measures of wind speed and direction [83]. The Porto Cesareo station has been in operation since February 2019.

The displacement of boulders due to wave impact on the coast starts when the hydrodynamics forces exceed the resistance forces of net friction [54,55]. Nandasena et al. [1,71] and Nandasena [56] elaborated a set of equations to calculate the minimum wave height (H_m) required to set in motion and cause onshore transport of the boulders, depending on PTS and MT. For SA and SB boulders moved by saltation/lifting, sliding, and rolling, H_m is as follows, respectively:

$$H_m = \frac{2c(\rho_s/\rho_w - 1)\cos\theta}{C_L}, \tag{A4}$$

$$H_m = \frac{2c(\rho_s/\rho_w - 1)(\mu\cos\theta + \sin\theta)}{C_D(c/b) + \mu C_L}, \tag{A5}$$

$$H_m = \frac{2c(\rho_s/\rho_w - 1)(\cos\theta + (c/b)\sin\theta)}{C_D(c^2/b^2) + C_L}, \tag{A6}$$

For JB boulders moved by saltation/lifting, H_m is as follows:

$$H_m = \frac{2c(\rho_s/\rho_w - 1)(\cos\theta + \mu\sin\theta)}{C_L}, \tag{A7}$$

For CE boulders moved by saltation/lifting and rolling, H_m is as follows, respectively:

$$H_m = \frac{2c(\rho_s/\rho_w - 1)}{C_L - \mu C_D(c/b)}, \tag{A8}$$

$$H_{\rm m} = \frac{2c(\rho_s/\rho_w - 1)}{C_L - C_D(c^2/b^2)}, \quad \text{(A9)}$$

In the Equations (A4)–(A9): b = b-axis length (m); c = c-axis length (m); ρ_s = density of boulder (kg m^{-3}); ρ_w = density of seawater (kg m^{-3}); θ = bed slope angle (degrees); μ = coefficient of static friction along the ground surface; C_L = lift coefficient; C_D = drag coefficient. As is known, the values of the coefficients C_D and C_L are crucial in determining the reliability of the results [61,84,85]. Their correct determination would require considering the influence of environmental factors like the substrate roughness and flow turbulence and performing hydrodynamic site-specific measurements. Considering that, in the present work, the use of the above equations is simply aimed at a comparison with the results of the previous work [9], the values of 1.95 for C_D, 0.178 for C_L, and 0.7 for μ were used (see, e.g., refs. [1,9,48,86,87]). Again, based on the local geology, the following values of density (ρ_s) were used: 1.8 kg m^{-3} for calcarenite; 2.2 kg m^{-3} for limestone.

To estimate the decrease in wave height over the shore, say the effective wave height H impacting the boulder at the distance x from the shoreline, the equation of Cox and Machemehl [60] was used:

$$H = \left[(R + H_s - H_c)^{1/2} - 5x/(Tg^{1/2})\right]^2, \quad \text{(A10)}$$

In Equation (A10): x = distance from the coastline (m); g = gravity acceleration; T was estimated by Equation (A1); the breaking wave height R was directly estimated as its minimum value H_0 from Equation (A2); the average coastline height above the mean sea level H_c was corrected by the total sea level increase (storm surge + tide) H_s.

Appendix C

The 2019–2020 data extracted from the datasets of the Santa Maria di Leuca (Table A11) and Porto Cesareo stations (Figures A10–A13) are reported below (see Section 3.3).

Table A11. Wind speed daily average (Ua), wind speed daily maximum (Um), and 30 min average wind speed (U_{30}) and direction (D) at four selected hours. Data recorded by the Santa Maria di Leuca wind gauge (data extracted from [80,81]).

			00:00		06:00		12:00		18:00	
Storm	**Ua**	**Um**	U_{30}	**D**	U_{30}	**D**	U_{30}	**D**	U_{30}	**D**
12 November 2019 [1]	8.5	24.0	9.8	106°	18.1	107°	13.4	145°	11.7	164°
13 November 2019 [1]	6.0	19.8	15.4	169°	12.7	167°	12.1	173°	4.3	172°
24 November 2019	5.5	19.3	4.9	145°	7.4	129°	14.5	107°	16.3	112°
22 December 2019	4.9	18.1	8.2	191°	9.4	179°	14.0	222°	7.8	259°
2 March 2020	3.7	12.7	7.4	153°	3.0	53°	6.1	139°	8.3	146°

[1] storm Detlef.

During a marine storm, wind speed is significantly higher over the sea than land, while wind direction is affected by more or less apparent changes in crossing the coastline. Land station measurements of wind speed can be halved in comparison with offshore values [88,89]. However, it must be noted that the Santa Maria di Leuca station has an elevation of 26 m a.s.l. and is placed about 100 m inland, while the Porto Cesareo station has an elevation of 12 m a.s.l. and is placed few tens of meters inland. Since the two stations are close to the coastline, the anemometric values reported above can be considered as indicative for the nearshore conditions. The consistency with the wind speed values extracted from the GLOBO-BOLAM-MOLOCH model cascade (Table 2) confirms the above.

Figure A10. Wind speed and direction (dashed line) recorded by the Porto Cesareo wind gauge during storm Detlef; image downloaded 13 November 2023 from the site of the Apulian Meteomarine Network [83].

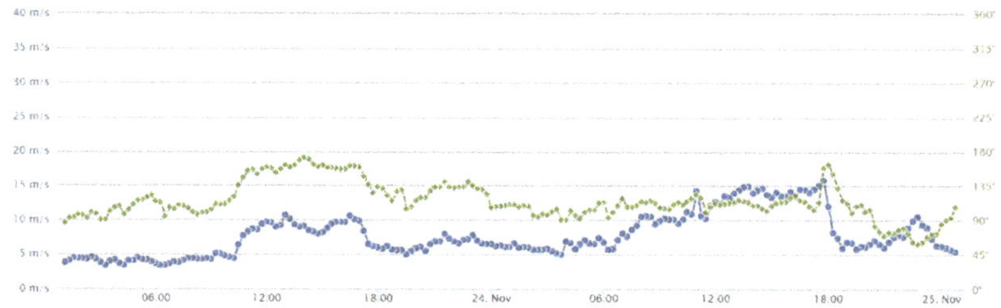

Figure A11. Wind speed and direction (dashed line) recorded by the Porto Cesareo wind gauge during the 24 November 2019 storm; image downloaded 13 November 2023 from the site of the Apulian Meteomarine Network [83].

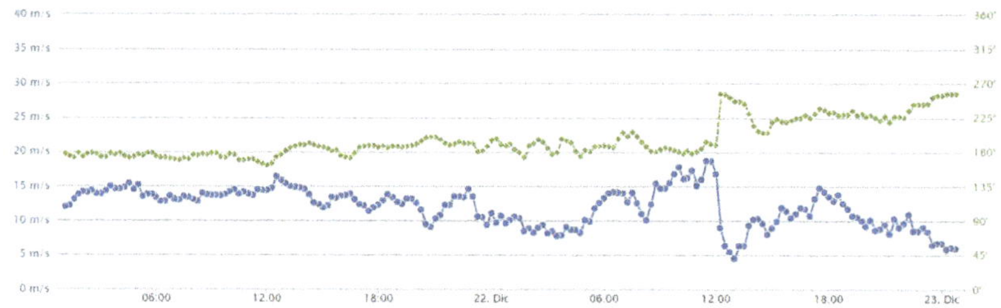

Figure A12. Wind speed and direction (dashed line) recorded by the Porto Cesareo wind gauge during the 22 December 2019 storm; image downloaded 13 November 2023 from the site of the Apulian Meteomarine Network [83].

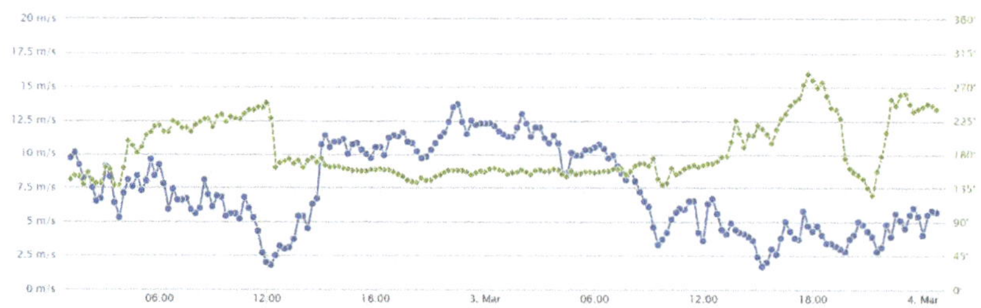

Figure A13. Wind speed and direction (dashed line) recorded by the Porto Cesareo wind gauge during the 2 March 2020 storm; image downloaded 13 November 2023 from the site of the Apulian Meteomarine Network [83].

Appendix D

By using the storm power index (SPI) introduced by Dolan and Davis [90], the intensity of the storms reported in Table 2 can be calculated. SPI is defined as the "*storm's duration times the square of maximum significant wave height*" [90]. A number of studies have shown a strong correlation between SPI and the wave impact on the coasts (see, e.g., [91–93]). Considering that H_0 is nearly equivalent to the significant wave height [76,77], the storm intensities can be assessed as reported in Figure A14.

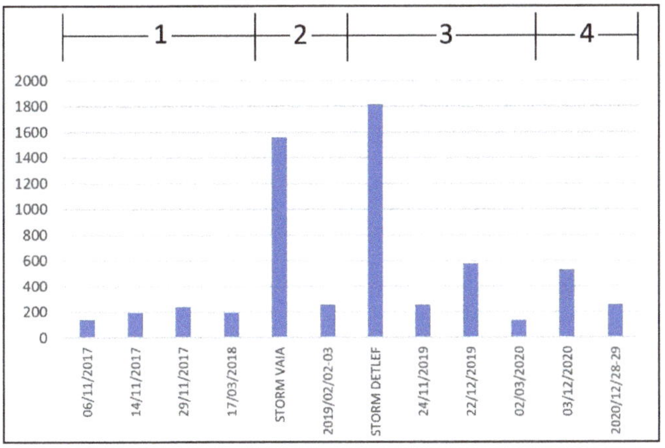

Figure A14. Storm power index (in m^2 h) of the storms in Table 2. Annual intervals of Table 3 are reported in the upper side.

Considering the obtained values of SPI, the quite different ability of the two most severe storms to determine geomorphological signatures in the study area is surprising. It can be guessed that the interplay between wind speed and length and duration in time of the fetch involved can determine a sort of threshold-like behavior between storms that do not cause massive changes (including Vaia) and the most intense one (Detlef). Some insights into the reasons for the different effects of the two storms on the studied coast were discussed in Delle Rose and Martano [4] and Delle Rose et al. [9].

As stated above (Section 4.3), some boulders displaced during the interval 3 (Table 3) could be caused by the storms that occurred between 24 November 2019 and 2 March 2020 (Figure 10). Considering Figure A14, the storm that occurred on 22 December 2019 (SPI about 580) seems to be the main suspect. However, considering that storm Vaia (SPI about 1570) was able to move only one of the detected boulders (Section 3.2), this probability is rather low. On the other hand, the 3 December 2020 storm can be considered the most probable cause for the two displacements that occurred during the interval 4 (Table 3).

The strong power of storm Detlef on the eastern coast of the Gulf of Taranto is supported also by the resulting damage [17,21]. Promenade walls were destroyed and breakwater blocks displaced toward the coastal roads. The most affected places were Porto Cesareo, Santa Caterina, and Gallipoli (Figure 1). In the last town, several boats sunk, and numerous infrastructures along the seafront were destroyed. Differently, no damage was reported for the other considered storms.

A quantitative comparison with other phenomena of boulder displacements can help to better understand the work caused by storm Detlef on the studied coast. Several authors used the minimum flow velocity V to set the boulders in motion [1,71] as a measure of the nearshore wave energy (see, e.g., [48,86,94]). For the biggest displaced boulders herein considered, V ranges from 2.7 to 8.1 m/s (Table A12).

Table A12. Minimum flow velocity (V) (in m/s) required to set the boulders in motion; a_f, b_f, c_f, x_i, and TD (in m). V was computed with the equations of Nandasena et al. [1,71]. Note: See caption of Table A2 for abbreviations.

Boulder ID	Size ($a_f \times b_f \times c_f$)	x_i	TD	PTS	MT	V
PRq [1]	2.7 × 1.4 × 0.4	1.9	11.7	SA	ST	5.9
SIg [1]	1.7 × 1.5 × 0.5	9	2.4	SA	OV	4.4
SCb	1.8 × 1.6 × 0.7	18.8	15.4	SA	OV	5.4
PIh [1]	3.1 × 2.2 × 0.4	1.7	14.3	SA	ST	5.9
MAa [1]	3.4 × 2.1 × 0.5	0.6	4.4	SA	SL	2.8
SUa [1]	5.4 × 4.6 × 1.9	0	8.9	SB	SL	4.4
CAd [1]	2.6 × 2.4 × 0.9	3.9	2.5	SA	OV	5.6
PILj	3.0 × 1.8 × 0.6	0.7	1.6	SB	SL	2.7
ROd	2.4 × 1.5 × 0.7	10.4	4.5	SA	OV	4.3
ROn	2.8 × 2.1 × 0.5	4.2	17.1	SA	ST	6.6
SGe	2.2 × 1.7 × 0.7	13.8	6.5	SA	ST	7.8
CIc [1]	2.2 × 1.4 × 0.5	5.2	3.3	SA	ST	8.1

[1] Boulder detected by [9]; the MAa boulder, considered JB by [9], is here considered SA (see Appendix A).

Values of V up to 14 m/s were calculated to explain boulder displacements due to inundation events driven by typhoons in the Asia-Pacific region [39,79,94–96]. Instead, for the Mediterranean region, V values up to 12 m/s are reported in the literature to displace SA or SB boulders by severe storms [1,48]. Therefore, the values in Table A12 appear consistent (in terms of order of magnitude) with those found in the literature.

References

1. Nandasena, N.A.K.; Paris, R.; Tanaka, N. Reassessment of hydrodynamic equations to initiate boulder transport by high energy events (storms, tsunamis). *Mar. Geol.* **2011**, *281*, 70–84. [CrossRef]
2. Engel, M.; May, S.M. Bonaire's boulder field revisited: Evidence for Holocene tsunami impact on the Leeward Antilles. *Quat. Sci. Rev.* **2012**, *54*, 126–141. [CrossRef]
3. Kennedy, A.B.; Cox, R.; Dias, F. Storm waves may be the source of some "tsunami" coastal boulder deposits. *Geophys. Res. Lett.* **2021**, *48*, e2020GL090775. [CrossRef] [PubMed]
4. Delle Rose, M.; Martano, P. The Imprint of Recent Meteorological Events on Boulder Deposits along the Mediterranean Rocky Coasts. *Climate* **2022**, *10*, 94. [CrossRef]
5. Collin, A.; Etienne, S.; Planes, S. High-energy events, boulder deposits and the use of very high resolution remote sensing in coral reef environments. *J. Coast. Res. Spec. Issue* **2013**, *65*, 690–695. [CrossRef]
6. Collin, A.; Etienne, S.; Jeanson, M. Three-dimensional structure of coral reef boulders transported by stormy waves using the very high resolution WorldView-2 satellite. *J. Coast. Res. Spec. Issue* **2016**, *75*, 572–576. [CrossRef]
7. Cox, R. Megagravel deposits on the west coast of Ireland show the impacts of severe storms. *Weather* **2020**, *75*, 72–77. [CrossRef]
8. Ruban, D.A. Finding Coastal Megaclast Deposits: A Virtual Perspective. *J. Mar. Sci. Eng.* **2020**, *8*, 164. [CrossRef]
9. Delle Rose, M.; Martano, P.; Orlanducci, L. Coastal Boulder Dynamics Inferred from Multi-Temporal Satellite Imagery, Geological and Meteorological Investigations in Southern Apulia, Italy. *Water* **2021**, *13*, 2426. [CrossRef]
10. Kennedy, D.M.; Woods, J.L.D.; Naylor, L.A.; Hansom, J.D.; Rosser, N.J. Intertidal boulder-based wave hindcasting can underestimate wave size: Evidence from Yorkshire, UK. *Mar. Geol.* **2019**, *411*, 98–106. [CrossRef]
11. Hastewell, L.J.; Schaefer, M.; Bray, M.; Inkpen, R. Intertidal boulder transport: A proposed methodology adopting Radio Frequency Identification (RFID) technology to quantify storm induced boulder mobility. *Earth Surf. Process. Landf.* **2019**, *44*, 681–698. [CrossRef]
12. Gomez-Pazo, A.; Perez-Alberti, A.; Trenhaile, A. Tracking clast mobility using RFID sensors on a boulder beach in Galicia, NW Spain. *Geomorphology* **2021**, *373*, 107514. [CrossRef]
13. Autret, R.; Dodet, G.; Suanez, S.; Roudaut, G.; Fichaut, B. Long–term variability of supratidal coastal boulder activation in Brittany (France). *Geomorphology* **2018**, *304*, 184–200. [CrossRef]
14. Hoffmeister, D.; Curdt, C.; Bareth, G. Monitoring the sedimentary budget and dislocated boulders in western Greece—Results since 2008. *Sedimentology* **2020**, *67*, 1411–1430. [CrossRef]
15. Ferrarin, C.; Valentini, A.; Vodopivec, M.; Klaric, D.; Massaro, G.; Bajo, M.; De Pascalis, F.; Fadini, A.; Ghezzo, M.; Menegon, S.; et al. Integrated sea storm management strategy: The 29 October 2018 event in the Adriatic Sea. *Nat. Hazards Earth Syst. Sci.* **2020**, *20*, 73–93. [CrossRef]

16. Ferrarin, C.; Bajo, M.; Benetazzo, A.; Cavaleri, L.; Chiggiato, J.; Barbariol, F.; Bastianini, M.; Bertotti, L.; Davolio, S.; Magnusson, L.; et al. Local and large-scale controls of the exceptional Venice floods of November 2019. *Prog. Oceanogr.* **2021**, *197*, 102628. [CrossRef]
17. Delle Rose, M.; Fidelibus, C.; Martano, P.; Orlanducci, L. Storm-induced boulder displacements: Inferences from field surveys and hydrodynamic equations. *Geosciences* **2020**, *10*, 374. [CrossRef]
18. Morucci, S.; Picone, M.; Nardone, G.; Arena, G. Tides and waves in the central Mediterranean Sea. *J. Oper. Oceanogr.* **2016**, *9*, s10–s17. [CrossRef]
19. Dentale, F.; Furcolo, P.; Pugliese Carratelli, E.; Reale, F.; Contestabile, P.; Tomasicchio, G.R. Extreme wave analysis by integrating model and wave buoy data. *Water* **2018**, *10*, 373. [CrossRef]
20. Ciricugno, L.; Delle Rose, M.; Fidelibus, C.; Orlanducci, L.; Mangia, M. Sullo spostamento di massi costieri causato da onde "estreme" (costa ionica salentina). *Geol. Territ.* **2019**, *16*, 15–23. (In Italian)
21. Delle Rose, M.; Ciricugno, L.; Fidelibus, C.; Martano, P.; Marzo, L.; Orlanducci, L. Considerazioni geologiche su processi morfodinamici causati sulla costa ionica salentina da recenti tempeste. *Geol. Territ.* **2020**, *18*, 5–15. (In Italian)
22. Spano, D.; Snyder, R.L.; Cesaraccio, C. Mediterranean Climates. In *Phenology: An Integrative Environmental Science*; Schwartz, M.D., Ed.; Springer: Dordrecht, The Netherlands, 2003; pp. 139–156.
23. Lionello, P.; Malanotte-Rizzoli, P.; Boscolo, R.; Alpert, P.; Artale, V.; Li, L.; Luterbacher, J.; May, W.; Trigo, R.; Tsimplis, M.; et al. The Mediterranean Climate: An Overview of the Main Characteristics and Issues. *Dev. Earth Environ. Sci.* **2006**, *4*, 1–26.
24. Zito, G.; Ruggiero, L.; Zuanni, F. Aspetti meteorologici e climatici della Puglia. In Proceedings of the First Workshop on "Clima, Ambiente e Territorio nel Mezzogiorno", Taormina, Italy, 11–12 December 1989; CNR: Roma, Italy, 1991; pp. 43–73. (In Italian).
25. Mastronuzzi, G.; Pignatelli, C. The boulder berm of Punta Saguerra (Taranto, Italy): A morphological imprint of the Rossano Calabro tsunami of April 24, 1836? *Earth Planets Space* **2012**, *64*, 829–842. [CrossRef]
26. Zhao, Q.; Yu, L.; Li, X.; Peng, D.; Zhang, Y.; Gong, P. Progress and Trends in the Application of Google Earth and Google Earth Engine. *Remote Sens.* **2021**, *13*, 3778. [CrossRef]
27. Goudie, A. Coasts. In *Landscapes of the Anthropocene with Google Earth*; Goudie, A., Ed.; Springer Nature: Cham, Switzerland, 2023; pp. 129–162.
28. Kennedy, A.B.; Mori, N.; Yasuda, T.; Shimozono, T.; Tomiczek, T.; Donahue, A.; Shimura, T.; Imai, Y. Extreme block and boulder transport along a cliffed coastline (Calicoan Island, Philippines) during Super Typhoon Haiyan. *Mar. Geol.* **2017**, *383*, 65–77. [CrossRef]
29. Medina, F.; Mhammdi, N.; Emran, A.; Hakdaoui, S. A case of uplift and transport of a large boulder by the recent winter storms at Dahomey beach (Morocco). In Proceedings of the IX Symposium on the Iberian Atlantic Margin, Coimbra, Portugal, 4–7 September 2018; p. 2.
30. Haslett, S.K.; Wong, B.R. Reconnaissance survey of coastal boulders in the Moro Gulf (Philippines) using Google Earth imagery: Initial insights into Celebes Sea tsunami events. *Bull. Geol. Soc. Malays.* **2019**, *68*, 37–44. [CrossRef]
31. Khuwaja, Z.; Arain, J.; Ali, R.; Meghwar, S.L.; Jatoi, M.A.; Shaikh, F.A. Accuracy Measurement of Google Earth Using GPS and Manual Calculations. In Proceedings of the International Conference Sustainable Development in Civil Engineering, Jamshoro, Pakistan, 23–25 November 2017; Mehran University: Jamshoro, Pakistan, 2018; p. 9.
32. Guo, J.; Tu, H.J.; Li, H.; Zhao, Y.; Zhou, J. Horizontal Accuracy Assessment of Google Earth Data Overtypical Regions of Australia Using Worldview. *Int. Arch. Photogramm. Remote Sens. Spatial Inf. Sci.* **2021**, *XLIII-B3-2021*, 763–768.
33. Pulighe, G.; Baiocchi, V.; Lupia, F. Horizontal accuracy assessment of very high resolution Google Earth images in the city of Rome, Italy. *Int. J. Digit. Earth* **2016**, *9*, 1753–8947. [CrossRef]
34. Guerriero, L.; Di Martire, D.; Calcaterra, D.; Francioni, M. Digital Image Correlation of Google Earth Images for Earth's Surface Displacement Estimation. *Remote Sens.* **2020**, *12*, 3518. [CrossRef]
35. Delle Rose, M.; Martano, P. Wind–Wave Conditions and Change in Coastal Landforms at the Beach–Dune Barrier of Cesine Lagoon (South Italy). *Climate* **2023**, *11*, 128. [CrossRef]
36. Regione Puglia—Area Politiche per la Mobilita e Qualita Urbana—Servizio Assetto del Territorio. Available online: http://webapps.sit.puglia.it/freewebapps/Idrogeomorfologia/index.html (accessed on 16 June 2023). (In Italian).
37. Blaschke, T.; Hay, G.J.; Kelly, M.; Lang, S.; Hofmann, P.; Addink, E.; Queiroz Feitosa, R.; van der Merr, F.; van der Werff, H.; van Coillie, F.; et al. Geographic Object-Based Image Analysis—Towards a new paradigm. *J. Photogramm. Remote Sens.* **2014**, *87*, 180–191. [CrossRef] [PubMed]
38. Hoffmeister, D. Mapping of subaerial coarse clasts. In *Geological Records of Tsunamis and Other Extreme Waves*; Engel, M., Pilarczyk, J., May, S.M., Brill, D., Garrett, E., Eds.; Elsevier: Amsterdam, The Netherlands, 2020; pp. 169–184.
39. Terry, J.P.; Lau, A.Y.A.; Nguyen, K.A.; Liou, Y.-A.; Switzer, A.D. Clustered, Stacked and Imbricated Large Coastal Rock Clasts on Ludao Island, Southeast Taiwan, and Their Application to Palaeotyphoon Intensity Assessment. *Front. Earth Sci.* **2021**, *9*, 792369. [CrossRef]
40. Vaccher, V.; Hastewell, L.; Devoto, S.; Corradetti, A.; Mantovani, M.; Korbar, T.; Furlani, S. The application of UAV-derived SfM-MVS photogrammetry for the investigation of storm wave boulder deposits on a small rocky island in the semi-enclosed Northern Adriatic Sea. *Geomat. Nat. Hazards Risk* **2024**, *15*, 2295817. [CrossRef]
41. Erdmann, W.; Kelletat, D.; Scheffers, A. Boulder transport by storms—Extreme-waves in the coastal zone of the Irish west coast. *Mar. Geol.* **2018**, *399*, 1–13. [CrossRef]

42. Oetjen, J.; Engel, M.; Pudasaini, S.P.; Schuettrumpf, H. Significance of boulder shape, shoreline configuration and pre-transport setting for the transport of boulders by tsunamis. *Earth Surf. Process. Landf.* **2020**, *45*, 2118–2133. [CrossRef]
43. Nandasena, N.A.K.; Scicchitano, G.; Scardino, G.; Milella, M.; Piscitelli, A.; Mastronuzzi, G. Boulder displacements along rocky coasts: A new deterministic and theoretical approach to improve incipient motion formulas. *Geomorphology* **2022**, *407*, 108217. [CrossRef]
44. Blott, S.G.; Pye, K. Particle shape: A review and new methods of characterization and classification. *Sedimentology* **2008**, *55*, 31–63. [CrossRef]
45. Blair, T.C.; Mcpherson, J.G. Grain-size and textural classification of coarse sedimentary particles. *J. Sed. Res.* **1999**, *69*, 6–19. [CrossRef]
46. Noormets, R.; Felton, E.A.; Crook, K.A.W. Sedimentology of rocky shorelines: 2. Shoreline megaclasts on the north shore of Oahu, Hawaii–origins and history. *Sediment. Geol.* **2002**, *150*, 31–45. [CrossRef]
47. Causon Deguara, J.; Gauci, R. Evidence of extreme wave events from boulder deposits on the south-east coast of Malta (Central Mediterranean). *Nat. Hazards* **2017**, *86*, 543–568. [CrossRef]
48. Mottershead, D.N.; Soar, P.J.; Bray, M.J.; Hastewell, L.J. Reconstructing boulder deposition histories: Extreme wave signatures on a complex rocky shoreline of Malta. *Geosciences* **2020**, *10*, 400. [CrossRef]
49. Knight, J.; Burningham, H. Boulder dynamics on an Atlantic-facing rock coastline, northwest Ireland. *Mar. Geol.* **2011**, *283*, 56–65. [CrossRef]
50. Barbano, M.S.; Pirrotta, C.; Gerardi, F. Large boulders along the south-eastern Ionian coast of Sicily: Storm or tsunami deposit? *Mar. Geol.* **2010**, *275*, 140–154. [CrossRef]
51. Boccotti, P. *Wave Mechanics for Ocean Engineering*; Elsevier Science: Amsterdam, The Netherlands, 2000; pp. 1–496.
52. Harley, M. Coastal Storm Definition. In *Coastal Storms*; Ciavola, P., Coco, G., Eds.; John Wiley & Sons: Hoboken, NJ, USA, 2017; pp. 1–21.
53. GLOBO-BOLAM-MOLOCH Forecasts. Available online: https://www.isac.cnr.it/dinamica/projects/forecasts/bolam/ (accessed on 10 September 2023).
54. Nott, J. Waves, coastal boulder deposits and the importance of the pre-transport setting. *Earth Planet. Sci. Lett.* **2003**, *210*, 269–276. [CrossRef]
55. Noormets, R.; Felton, E.A.; Crook, K.A.W. Sedimentology of rocky shorelines: 3. Hydrodynamics of megaclasts emplacement and transport on a shore platform, Oahu, Hawaii. *Sediment. Geol.* **2004**, *172*, 41–65. [CrossRef]
56. Nandasena, N.A.K. Perspective of incipient motion formulas: Boulder transport by high-energy waves. In *Geological Records of Tsunamis and Other Extreme Waves*; Nandasena, N.A.K., Engel, M., Pilarczyk, J., May, S.M., Brill, D., Garrett, E., Eds.; Elsevier: Amsterdam, The Netherlands, 2020; pp. 641–659.
57. Carbone, F.; Dutykh, D.; Dudley, J.M.; Dias, F. Extreme wave runup on a vertical cliff. *Geophys. Res. Lett.* **2013**, *40*, 3138–3143. [CrossRef]
58. Knight, J.; Burningham, H.; Griffiths, D.; Yao, Y. Coastal boulder movement on a rocky shoreline in northwest Ireland from repeat UAV surveys using Structure from Motion photogrammetry. *Geomorphology* **2023**, *440*, 108883. [CrossRef]
59. Ganas, A.; Briole, P.; Bozionelos, G.; Barberopoulou, A.; Elias, P.; Tsironi, V.; Valkaniotis, S.; Moshou, A.; Mintourakis, I. The 25 October 2018 Mw = 6.7 Zakynthos earthquake (Ionian Sea, Greece): A low-angle fault model based on GNSS data, relocated seismicity, small tsunami and implications for the seismic hazard in the west Hellenic Arc. *J. Geodyn.* **2020**, *137*, 101731. [CrossRef]
60. Cox, J.C.; Machemehl, J. Overland bore propagation due to overtopping wave. *J. Waterw. Port Coast. Ocean. Eng.* **1986**, *112*, 161–163. [CrossRef]
61. Cox, R.; Ardhuin, F.; Dias, F.; Autret, R.; Beisiegel, N.; Earlie, C.S.; Herterich, J.G.; Kennedy, A.; Paris, R.; Raby, A.; et al. Systematic review shows that work done by storm waves can be misinterpreted as tsunami-related because commonly used hydrodynamic equations are flawed. *Front. Mar. Sci.* **2020**, *7*, 4. [CrossRef]
62. 12 Novembre 2019: Scirocco Impetuoso, Marea Eccezionale a Venezia e Alluvioni al sud Italia con la Depressione "Detlef". Available online: http://www.nimbus.it/eventi/2019/191115MareaEccezionaleVenezia.htm (accessed on 13 June 2023). (In Italian).
63. Melet, A.; Buontempo, C.; Mattiuzzi, M.; Salamon, P.; Bahurel, P.; Breyiannis, G.; Burgess, S.; Crosnier, L.; Le Traon, P.Y.; Mentaschi, L.; et al. European Copernicus Services to Inform on Sea-Level Rise Adaptation: Current Status and Perspectives. *Front. Mar. Sci.* **2021**, *8*, 703425. [CrossRef]
64. Irazoqui Apecechea, M.; Melet, A.; Armaroli, C. Towards a pan-European coastal flood awareness system: Skill of extreme sea-level forecasts from the Copernicus Marine Service. *Front. Mar. Sci.* **2023**, *9*, 1091844. [CrossRef]
65. GEOS Captures Tropical Cyclone-like System "Trudy" over the Mediterranean Sea. Available online: https://gmao.gsfc.nasa.gov/research/science_snapshots/2020/medicane_Trudy.php (accessed on 13 June 2023).
66. Segunda Borrasca con Nombre de la Temporada. La borrasca Bernardo Afectara Especialmente al E de las islas Baleares. Available online: https://twitter.com/AEMET_Esp/status/1193226687611383811 (accessed on 13 June 2023). (In Spanish).
67. Medicane "Trudy" (Detlef, Bernardo) Makes landfall in Algeria. Available online: https://watchers.news/2019/11/12/medicane/ (accessed on 13 June 2023).
68. Tropical Cyclone Reports (Mediterranean Sea). Subtropical Storm Detlef. 2019. Available online: http://zivipotty.hu/2019_detlef.pdf (accessed on 21 July 2023).

69. Miglietta, M.M.; Buscemi, F.; Dafis, S.; Papa, A.; Tiesi, A.; Conte, D.; Davolio, S.; Flaounas, E.; Levizzani, V.; Rotunno, R. A high-impact meso-beta vortex in the Adriatic Sea. *Q. J. R. Meteorol.* **2023**, *149*, 637–656. [CrossRef]
70. Cavaleri, L.; Bajo, M.; Barbariol, F.; Bastianini, M.; Benetazzo, A.; Bertotti, L.; Chiggiato, J.; Ferrarin, C.; Trincardi, F.; Umgiesser, G. The 2019 flooding of Venice and its implications for future predictions. *Oceanography* **2020**, *33*, 42–49. [CrossRef]
71. Nandasena, N.A.K.; Tanaka, N.; Sasaki, Y.; Osada, M. Boulder transport by the 2011 Great East Japan tsunami: Comprehensive field observations and whither model predictions? *Mar. Geol.* **2013**, *346*, 292–309. [CrossRef]
72. Biolchi, S.; Denamiel, C.; Devoto, S.; Korbar, T.; Macovaz, V.; Scicchitano, G.; Vilibic, I.; Furlani, S. Impact of the October 2018 Storm Vaia on Coastal Boulders in the Northern Adriatic Sea. *Water* **2019**, *11*, 2229. [CrossRef]
73. Roig-Munar, F.X.; Martin-Prieto, J.A.; Rodriguez-Perea, A.; Gelabert, B. Gloria storm effects on the coastal boulders East of Minorca (Balearic Islands). *Environ Earth Sci* **2023**, *82*, 176. [CrossRef]
74. Hasselmann, K.; Ross, D.B.; Muller, P.; Sell, W. A parametric wave prediction model. *J. Phys. Oceanogr.* **1975**, *6*, 200–228. [CrossRef]
75. Hsu, S.A. (Ed.) *Coastal Meteorology*; Academic Press: Cambridge, MA, USA, 1988; 260p.
76. Burroughs, L. Wave forecasting by manual methods. In *Guide to Wave Analysis and Forecasting*; World Meteorological Organization, Ed.; Secretariat of World Meteorological Organization: Geneva, Switzerland, 1998; pp. 43–56.
77. Barua, D.K. Wave hindcasting. In *Encyclopedia of Coastal Science*; Finkl, C.W., Makowski, C., Eds.; Springer International Publishing: Berlin/Heidelberg, Germany, 2019; pp. 1859–1864.
78. Regione Puglia. Annali Idrologici. 2017. Available online: https://protezionecivile.puglia.it/documents/3171874/3243680/annale2017.pdf/ (accessed on 10 September 2023). (In Italian).
79. Regione Puglia. Annali Idrologici. 2018. Available online: https://protezionecivile.puglia.it/documents/3171874/3243680/annale2018.pdf/ (accessed on 10 September 2023). (In Italian).
80. Regione Puglia. Annali Idrologici. 2019. Available online: https://protezionecivile.puglia.it/documents/3171874/3243680/annale2019.pdf/ (accessed on 10 September 2023). (In Italian).
81. Regione Puglia. Annali Idrologici. 2020. Available online: https://protezionecivile.puglia.it/documents/3171874/3243680/annale2020rev2.pdf/ (accessed on 10 September 2023). (In Italian).
82. Regione Puglia. Annali Idrologici. 2021. Available online: https://protezionecivile.puglia.it/documents/3171874/3243680/annale2021.pdf/ (accessed on 10 September 2023). (In Italian).
83. SIMOP, Sistema Informativo Meteo Oceanografico delle coste Pugliesi, Autorita di Bacino della Puglia. Available online: http://93.51.158.171/web/simop/home (accessed on 10 September 2023).
84. Bathurst, J.C. Field Measurement of Boulder Flow Drag. *J. Hydraul. Eng.* **1996**, *122*, 167–169. [CrossRef]
85. Rovere, A.; Casella, E.; Harris, D.L.; Lorscheid, T.; Nandasena, N.A.K.; Dyer, B.; Sandstrom, M.R.; Stocchi, P.; D'Andrea, W.J.; Raymo, M.E. Giant boulders and Last Interglacial storm intensity in the North Atlantic. *Proc. Natl. Acad. Sci. USA* **2017**, *114*, 12144–12149. [CrossRef]
86. Terry, J.P.; Karoro, R.; Gienko, G.A.; Wieczorek, M.; Lau, A.Y.A. Giant palaeotsunami in Kiribati: Converging evidence from geology and oral history. *Isl. Arc* **2021**, *30*, e12417. [CrossRef]
87. Huang, S.Y.; Yen, J.Y.; Wu, B.L.; Shih, N.W. Field observations of sediment transport across the rocky coast of east Taiwan: Impacts of extreme waves on the coastal morphology by Typhoon Soudelor. *Mar. Geol.* **2020**, *421*, 106088. [CrossRef]
88. Schwing, F.B.; Blanton, J.O. The use of land and sea based wind data in a simple circulation model. *J. Phys. Oceanogr.* **1984**, *14*, 193–197. [CrossRef]
89. Phillips, M.R.; Rees, E.F.; Thomas, T. Winds, sea levels and NAO influences: An evaluation. *Glob. Planet. Change* **2013**, *100*, 145–152. [CrossRef]
90. Dolan, R.; Davis, E. An intensity scale for Atlantic coast northeast storms. *J. Coast. Res.* **1992**, *8*, 840–853.
91. Mendoza, E.T.; Trejo-Rangel, M.A.; Salles, P.; Appendini, C.M.; Lopez-Gonzalez, J.; Torres-Freyermuth, A. Storm characterization and coastal hazards in the Yucatan Peninsula. *J. Coast. Res.* **2013**, *65*, 790–795. [CrossRef]
92. Anfuso, G.; Rangel-Buitrago, N.; Cortes-Useche, C.; Iglesias Castillo, B.; Gracia, F.J. Characterization of storm events along the Gulf of Cadiz (eastern central Atlantic Ocean). *Int. J. Climatol.* **2016**, *36*, 3690–3707. [CrossRef]
93. Martzikos, N.; Afentoulis, V.; Tsoukala, V. Storm clustering and classification for the port of Rethymno in Greece. *Water Util. J.* **2018**, *20*, 67–79.
94. Terry, J.P.; Dunne, K.; Jankaew, K. Prehistorical frequency of high-energy marine inundation events driven by typhoons in the bay of Bangkok (Thailand), interpreted from coastal carbonate boulders. *Earth Surf. Process. Landf.* **2016**, *41*, 553–562. [CrossRef]
95. May, S.M.; Engel, M.; Brill, D.; Cuadra, C.; Lagmay, A.M.F.; Santiago, J.; Suarez, J.K.; Reyes, M.; Bruckner, H. Block and boulder transport in Eastern Samar (Philippines) during Supertyphoon Haiyan. *Earth Surf. Dynamic.* **2015**, *3*, 543–558. [CrossRef]
96. Terry, J.P.; Goff, J. Strongly aligned coastal boulders on Ko Larn Island (Thailand): A proxy for past typhoon-driven high-energy wave events in the Bay of Bangkok. *Geogr. Res.* **2019**, *57*, 344–358. [CrossRef]

geosciences

MDPI

Article

Evaluating the Impact of DEM Spatial Resolution on 3D Rockfall Simulation in GIS Environment

Maria P. Kakavas [1,*], Paolo Frattini [2], Alberto Previati [2] and Konstantinos G. Nikolakopoulos [1]

[1] Department of Geology, Division of Applied Geology and Geophysics, University of Patras, 26504 Patras, Greece; knikolakop@upatras.gr

[2] Department of Earth and Environmental Science, Universita degli Studi di Milano-Bicocca, 20126 Milan, Italy; paolo.frattini@unimib.it (P.F.); alberto.previati@unimib.it (A.P.)

* Correspondence: kakava.maria@upatras.gr

Abstract: Rockfalls are natural geological phenomena characterized by the abrupt detachment and freefall descent of rock fragments from steep slopes. These events exhibit considerable variability in scale, velocity, and trajectory, influenced by the geological composition of the slope, the topography, and other environmental conditions. By employing advanced modeling techniques and terrain analysis, researchers aim to predict and control rockfall hazards to prevent casualties and protect properties in areas at risk. In this study, two rockfall events in the villages of Myloi and Platiana of Ilia prefecture were examined. The research was conducted by means of HY-STONE software, which performs 3D numerical modeling of the motion of non-interacting blocks. To perform this modeling, input files require the processing of base maps and datasets in a GIS environment. Stochastic modeling and 3D descriptions of slope topography, based on Digital Elevation Models (DEMs) without spatial resolution limitations, ensure multiscale analysis capabilities. Considering this capability, seven freely available DEMs, derived from various sources, were applied in HY-STONE with the scope of performing a large number of multiparametric analyses and selecting the most appropriate and efficient DEM for the software requirements. All the necessary data for the multiparametric analyses were generated within a GIS environment, utilizing either the same restitution coefficients and rolling friction coefficient or varying ones. The results indicate that finer-resolution DEMs capture detailed terrain features, enabling the precise identification of rockfall source areas and an accurate depiction of the kinetic energy distribution. Further, the results show that a correct application of the model to different DEMs requires a specific parametrization to account for the different roughness of the models.

Keywords: rockfalls; HY-STONE; DEMs; spatial resolution

1. Introduction

Estimating a rockfall risk, a process that involves risk assessment, susceptibility modeling, and delineating hazard zones, represents a challenging endeavor in rockfall research [1]. A variety of methods have been developed recently, with many focused on measuring the rockfall's intensity, identifying specific areas of concern, and determining the frequency of rockfall occurrences [2]. The complexity of rockfall risk analysis, which necessitates multiscale approaches, has led to the integration of various disciplines and techniques. Remote sensing, in particular, has proven to be a useful tool in rockfall investigations, providing precise models of potential hazard areas and capturing intricate details such as the lithological composition and angles of slopes and source zones [3]. These capabilities are essential, particularly because many hazardous rock slopes are inaccessible, where crucial data can only be obtained via remote sensing techniques. Slope characteristics play a critical role in rockfall hazard assessments, as they significantly influence both the initiation and behavior of rockfalls [4,5]. Digital Elevation Models (DEMs) are utilized to extract slope data and topographical parameters, which are essential for assessing and categorizing these geological features. One of the most beneficial features of DEMs is their capacity to support the generation, analysis, and mapping of slope parameters through Geographic Information System (GIS) tools [5]. The spatial resolutions of DEMs are a crucial aspect in rockfall simulations: lower-resolution DEMs tend to oversimplify the terrain's roughness, thereby reducing the accuracy of the predicted rockfall areas [6]; conversely, higher-resolution DEMs typically provide more detailed results for various rockfall-related factors [7]. Consequently, DEMs of varying resolutions are incorporated into numerous rockfall simulation software to meet the diverse requirements of rockfall analysis [8–19].

A variety of software has been developed to facilitate two-dimensional (2D) and three-dimensional (3D) rockfall trajectory simulations using DEMs. Currently, the accuracy of trajectory calculations has reached a high level due to advancements in numerical simulation techniques [20]. A notable example of such software is RocFall, developed by Stevens, W.D. (1998), which is recognized for its capability to calibrate various slope materials. In this process, DEMs indirectly contribute by providing slope profiles [21,22]. In RocFall, the spatial resolution plays a critical role in the simulations, where smaller cell sizes result in increased bounce heights and kinetic energies [23]. Recent studies involving seven different spatial resolutions ranging from 0.05 m to 90 m have shown that only those with higher pixel sizes can produce realistic trajectories [24]. Rotomap, another model that uses the lumped-mass approach, illustrates how higher-accuracy DEMs more accurately reflect the actual rockfall trajectories [25]. Furthermore, the impact of DEM spatial resolution on rockfall modeling has been explored through the empirical model CONEFALL, which has proven most effective for regional-scale studies [26]. Testing various spatial resolutions within this software has shown that DEMs with 1 m and 5 m resolutions are suitable for small-scale rockfall events, while those with 12.5 m and 25 m are better suited for larger scales. However, DEMs with a 100 m resolution are found to be unsuitable for such studies due to their low accuracy [26].

Conversely, STONE is a lumped-mass 3D rockfall simulation software that directly utilizes DEMs to model topography in raster format [27]. Renowned for its capability to simulate rockfall trajectories, STONE stands out due to its algorithm's ability to operate without the constraints of spatial resolution [28,29]. In conducting hazard assessments for both regional and large-scale events, STONE applies resampling techniques to DEMs, demonstrating that finer resolutions provide more detailed and accurate simulations of rockfall trajectories [28]. Furthermore, the precision of DEMs in STONE significantly influences the lateral dispersion, particularly noting an increase in dispersion with finer resolutions on slopes angled at 45 degrees [30]. Meanwhile, in the RAMMS software, although cell size does not influence the outcomes in flatter terrains, it is crucial in rough terrains where fine-resolution DEMs are recommended for accurately depicting terrain irregularities [31].

HY-STONE is another software designed for the 3D numerical modeling of rockfall processes. It simulates the trajectory of non-interacting blocks through various stages, including free fall, impact, and rolling, with a hybrid approach [30,32,33]. Notable features of HY-STONE include its ability to utilize a 3D vector representation of slope topography derived from raster Digital Elevation Models (DEMs), along with its compatibility with both raster and vector ArcGIS formats [27,34,35]. The software uses three coefficients (two restitution coefficients, normal and tangential, and a rolling friction coefficient), which are spatially described according to the DEM's accuracy by demonstrating that simulated runout distance increases as the DEM's resolution decreases [36,37]. A calibration process of the coefficients is therefore necessary [38].

The main objective of this study is to assess how the DEM's cell size impacts the rockfall data generated by the HY-STONE software, given that the spatial resolution of DEMs significantly influences the rockfall simulation process. Secondly, the paper aims to investigate the effect of the resolution on the calibration of model parameters. This research employs HY-STONE to analyze two rockfall events in Western Greece, specifically in the Myloi and Platiana settlements of the Peloponnese region. The goal is to determine how the accuracy of various DEMs affects the software's output and to validate these results with field observations. To this end, seven freely available DEMs (with spatial resolutions ranging from 5 cm to 90 m) are utilized, including the UAV DEM, DEM from the Greek Cadastral, ALOS AW3D30 DEM, ASTER GDEM, SRTM30 DEM, SRTM90 DEM, and TanDEM_X, to explore their effects on the modeling outcomes.

2. Software and Materials

HY-STONE is a software developed for 3D numerical modeling of rockfall processes, including the simulation of the motion of non-interacting blocks, utilizing a high-resolution 3D description of slope geometry for accurate multiscale stochastic modeling [30,32,33]. It is capable of modeling the free fall, impact, and rolling of blocks, with energy loss simulations upon impact or by rolling based on a hybrid (cinematic and dynamic) approach. It exports the results as raster grids and vector files, detailing the frequency of rockfall, fly height, velocity, kinetic energy, and information on each simulated block's kinematics and dynamics.

The software features a 3D vector description of slope topography by means of a vector triangular network (TRN) obtained from a raster DEM [27,34]. The TRN offers a continuous description of the topography by transforming the DEM's grid points into vertices of a network of contiguous, non-overlapping triangles. In the XYZ global (inertial) Cartesian system, the 3D space is defined by the orthogonal axes X, Y, and Z, where Z represents the elevation. The xp, yp, and zp local (triangle) in the Cartesian system are positioned within this global context, oriented on each triangle's plane, allowing for localized terrain descriptions on the triangle's surface. In the context of rockfall simulation, the software uses the normals to the triangles to compute the angles of impact and to adjust the trajectory of rock blocks upon collision with the ground.

Furthermore, HY-STONE takes into account the slope's lithology, land use, and roughness to inform the rolling friction coefficient ($\tan\Phi_r$, named A_T in this work), as well as the normal (E_N) and tangential (E_T) restitution coefficients that control energy loss upon impact. These coefficients are derived from the literature or experience and can be calibrated against field observations. In particular, E_N and E_T are essential for managing the energy dissipation of blocks, influenced by factors that include the type of surface lithology and the existing vegetation or land use patterns. The coefficients range from 0 (indicating no restitution) to 1 (indicating complete restitution). In the GIS, raster datasets are typically produced by first reclassifying maps of "unique condition units," which are areas defined by specific environmental characteristics such as lithology and vegetation or land use. These maps are then transformed into a raster format. The initial values are typically sourced from scholarly literature and subsequently fine-tuned through multiple simulation iterations to ensure they match well with the field observations, including historical and experimental data. The third coefficient (A_T), which determines the energy dissipated as blocks roll over the slope's surface, depends on various attributes such as surface lithology, land use, and surface roughness. This coefficient is typically a positive floating number between 0 and 1 and is assigned to each model cell. These data are also generated by reclassifying unique condition unit maps and necessitating calibration through iterative simulations to ensure precision and reliability in the modeling outcomes.

For the stochastic simulation, a normal distribution is used to simulate the natural variability of the restitution and friction coefficients within ±30% and 20% of their initial values, respectively. These specific distributions are chosen to realistically represent the variability and uncertainty inherent in natural rockfall events, enhancing the predictive accuracy of the simulations.

In summary, the software requires five spatially distributed input data, provided as ASCII raster files in the ESRI GridAscii format as follows: (i) DEM, (ii) rockfall source locations, (iii) normal (E_N) restitution coefficients, (iv) tangential (E_T) restitution coefficients, and (v) the rolling friction coefficient (A_T).

This study applies HY-STONE to execute a series of analyses, repeated seven times using DEMs of varying accuracies, aiming to conclude the most appropriate one for rockfall simulation. The resolutions explored include a 0.05 m spatial resolution UAV DEM generated from field surveys, a 5 m resolution DEM from the Greek Cadastral, three 30 m resolution DEMs (ALOS AW3D30 DEM, ASTER GDEM, and SRTM30 DEM), and two 90 m resolution DEMs (SRTM90 DEM and TanDEM_X). Two well-known rockfall events in the villages of Myloi and Platiana of Ilia prefecture were examined (Figure 1).

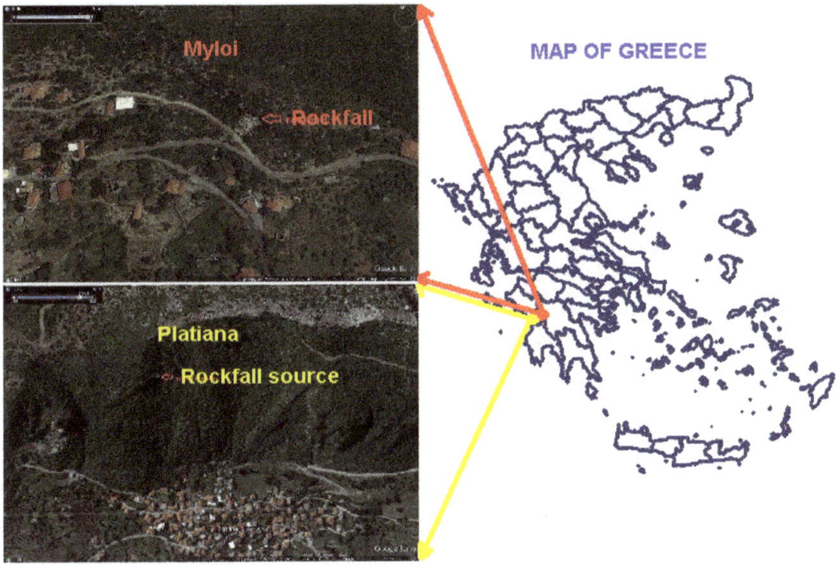

Figure 1. Pictures of the rockfall events in the Myloi and Platiana region in relation to the Hellenic region.

3. Study Areas

3.1. Myloi Settlement

Located in the Ilia prefecture of Western Greece's Peloponnese region, the village of Myloi is geologically situated within the Olonos-Pindos and Gavrovo-Tripolis geotectonic units. According to the engineering geological map of Greece provided by the Hellenic Survey of Geology and Mineral Exploration (HSGME), the area predominantly consists of Upper Cretaceous limestones. The mountainous terrain above the Myloi settlement is characterized by extensive areas of barren limestones and cherts. Research by Depountis et al. (2010) indicates that the predominant fault orientations in the region are east–west (E-W) and northeast–southwest (NE-SW) [39]. Certain weathering processes have significantly weakened the cohesion of the underlying limestone, leading to the detachment of rock blocks and the occurrence of rock sliding. These geological formations are characterized by a slope angle ranging from 45 to 60 degrees [40]. Additionally, the Ilia prefecture experienced devastating wildfires in August 2007, which obliterated the vegetation across approximately 870 km^2 of forested and agricultural land. This extreme event resulted in over sixty fatalities, hundreds of injuries, and extensive damage. The loss of vegetation, compounded by heavy rainfall, reactivated the mechanisms responsible for various scales of rockfall events. Despite the authorities designating the broader Myloi area as reforestable land, a significant rockfall occurred near Andritsaina town on 26 January 2019. This event was triggered by the combined effects of intense rainfall and the complete absence of a root system, which led to weathering and subsequent loss of cohesion. As a result, detached boulders rolled down the slope, causing substantial damage to two houses and blocking the main roads at the village entrance (Figure 2). The slope fragments displaced onto the road and at the end of the slope profile were estimated to have a total volume of 147 m^3.

Figure 2. Pictures of the rockfall in the Myloi region illustrate the final position of rock blocks. Red arrows indicate the final positions of the rock masses at the conclusion of the rockfall event.

3.2. Platiana Settlement

Similar to Myloi, Platiana is another village in the Ilia prefecture of Peloponnese, located approximately 25 km from Myloi (Figure 3). The surrounding area is primarily composed of Upper Cretaceous limestones, as indicated by the engineering geological map of Greece, and is part of the Pindos geotectonic unit according to the Hellenic Survey of Geology and Mineral Exploration (HSGME). This region is a segment of the broader Pindic Unit that has been thrust westward. Near the settlement, there are significant outcrops of limestone with varying thicknesses from thin to medium layers. According to Vagenas (2020) [40], these limestone layers are locally intercalated with chert layers.

Platiana, like Myloi, experienced the devastating wildfires of 2007 and has been declared reforestable land over the past decade. Following the natural reforestation efforts, the vegetation levels have been fully restored (Figure 4). However, from 2008 to 2018, the mountain slopes adjacent to Platiana were completely barren. Similar to other cases, the lack of vegetation exposed the slopes to corrosion factors that compromised their structural integrity and triggered the release of internal stresses. As a result, rocks of various sizes became detached from the main body and collapsed. Notably, on 4 December 2018, after heavy rainfall, a massive boulder, approximately 10 m^3 in volume, slid down the slope. Fortunately, it came to a stop in a churchyard, causing no casualties. During its downward course, the boulder penetrated two dry stone walls that served as field boundaries (Figure 3).

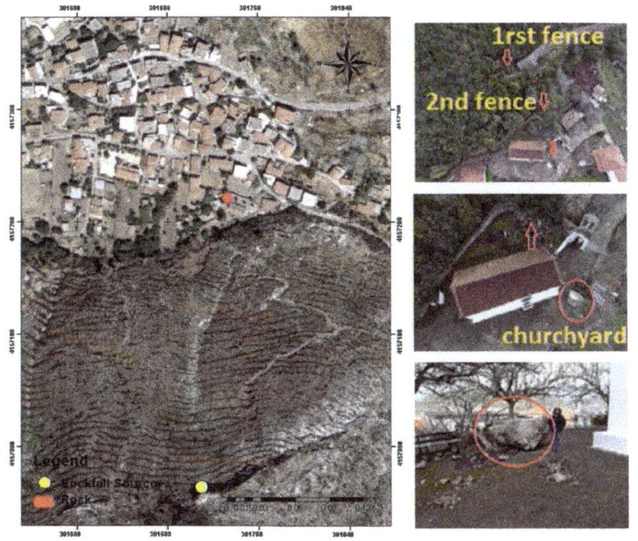

Figure 3. Picture taken from the Platiana region after the rockfall event. The slope picture is taken from the Greek Cadastral (2008).

Figure 4. Reforestation procedures in the slope next to the Platiana village. The slope pictures are taken from the Greek Cadastral.

4. Results

In this section, the results of the rockfall simulations through HY-STONE software are presented. Two different methods were accomplished; the first method is by using the rockfall simulations with DEMs of various resolutions—the same restitution coefficients (E_N and E_T) and rolling friction coefficient (A_T) calibrated for the highest resolution DEM. The second method consisted of adjusting the parameters to calibrate the model to each resolution in order to study the variations of calibrated coefficients as a function of the DEM's resolution.

4.1. Results by Using the Same Restitution Coefficients and Rolling Friction Coefficient

4.1.1. Myloi Rockfall

HY-STONE utilizes spatially distributed data for rockfall modeling, which involves accurately mapping topography, identifying rockfall sources, and evaluating the slope surface characteristics related to energy dissipation from block impacts and rolling. Additionally, detailed information on the geometry of falling blocks, such as their shape and size, along with the initial conditions and algorithm settings, is essential. To achieve this, seven DEMs were used, including the UAV DEM, DEM from the Greek Cadastral, ALOS

AW3D30 DEM, ASTER GDEM, SRTM30 DEM, SRTM90 DEM, and TanDEM_X, spanning resolutions from 0.3 m to 90 m to provide a precise topographical representation.

Given the extensive tree coverage, the UAV's original DSM underwent processing in Cloud Compare Software to manually remove vegetation, obtaining the UAV DEM. This is a necessary step before conducting a HY-STONE simulation because trees may be interpreted as barriers [33]. Specifically, Figure 5 provides the differences in vegetation coverage before (left image) and after (right image) the procedure in the Cloud Compare Software. The left image shows the landscape with full vegetation, where the dense canopy masks the underlying ground features. After processing, the right image shows some residual green areas, which are actually areas where the vegetation density was difficult to resolve completely using the software; however, these might be shadows and low-lying shrubs that remain post-processing. What might seem like gaps in the right image are actually areas where the vegetation has been effectively removed.

Figure 5. Myloi area before (**left** image) and after (**right** image) removing the vegetation through Cloud Compare Software.

The restitution coefficients, E_N and E_T, along with the rolling friction coefficient A_T, were established through back-calibration on the highest resolution DEM, based on the 2018 rockfall event (Table 1).

Table 1. Coefficients of restitution and rolling friction in Myloi rockfall, calibrated on the UAV DEM.

Description	E_N	E_T	A_T
Houses	0.20	0.20	1.00
Road-asphalt	0.55	0.75	0.37
Debris well vegetated	0.35	0.60	0.55
Debris without vegetation	0.40	0.65	0.42

Given that HY-STONE operates on a stochastic model, simulations were performed with 3000 rocks to provide a more accurate representation of the rockfall dynamics. The outcomes of the HY-STONE analysis for the Myloi rockfall are displayed in terms of transit frequency. The following maps (Figure 6) illustrate the cumulative count of rockfall trajectories that pass through each model cell.

Figure 6a, derived from the UAV DEM, illustrates the areas of high transit frequency predominantly in the central sections of the rockfall's left and right flanks. Moving outward from the central rockfall mass, the transit frequency gradually diminishes to zero towards the periphery. This pattern aligns with the actual conditions observed on-site, where the bulk of the rockfall material halted in the high transit zones while only a few scattered blocks reached the outer edges.

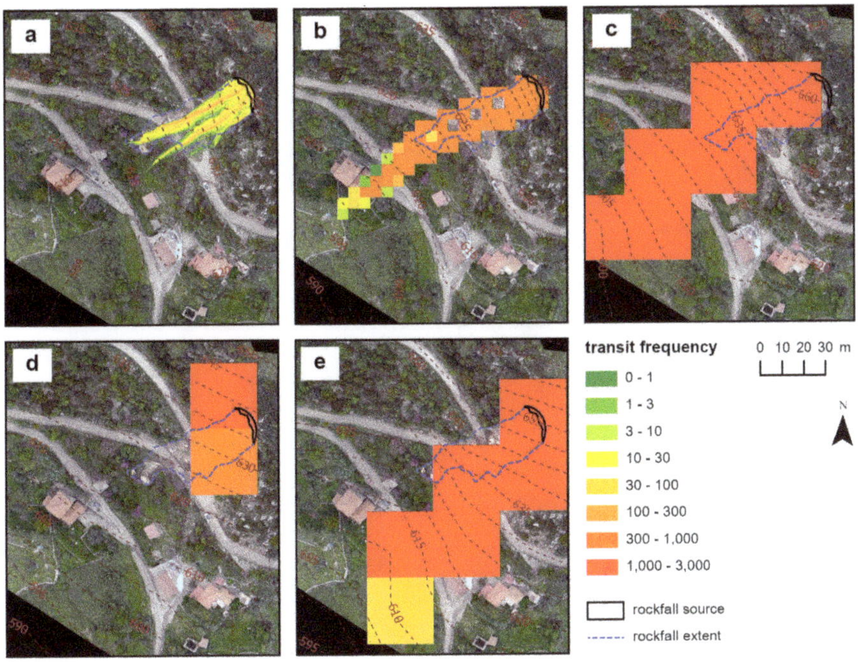

Figure 6. HY-STONE results for the site of Myloi using parameters (E_N, E_T, and A_T) calibrated on the UAV DEM. The transit frequencies of blocks are shown for (**a**) UAV DEM, (**b**) Greek Cadastral DEM, (**c**) ALOS AW3D30 DEM, (**d**) ASTER GDEM, and (**e**) SRTM30 DEM.

The result from using the Greek Cadastral DEM, depicted in Figure 6b, shows that the rock movement ceased around the houses, where the transit frequency values were observed to be medium to low.

The simulations using the ALOS AW3D30 DEM, ASTER GDEM, and SRTM30 DEM (Figure 6c–e) show a distinct distribution of only high or low frequencies without intermediate values. Given the small scale of the Myloi rockfall event (about 30 m of maximum runout), the 30 m resolution DEMs do not yield reliable results. Consequently, the SRTM90 DEM and TanDEM_X, with their coarser resolution, are deemed unsuitable for modeling rockfalls of this particular magnitude, and therefore, simulations using these DEMs were not performed.

The software also produces output files labeled min-ecv.asc, med-ecv.asc, and max-ecv.asc (and optionally, min-ecv.img, med-ecv.img, and max-ecv.img), which represent the minimum, average, and maximum translational kinetic energy of the blocks, respectively, calculated for each model cell in Joules (J). Subsequently, the maximum translational kinetic energies for the Myloi rockfall are presented in kJ in Figure 7a–e.

Figure 7a illustrates the calculated maximum translational kinetic energy using the UAV DEM, which estimates this energy as ranging from 0 to 500 kJ. The highest energy levels were recorded in the central mass of the rockfall, while lower energy levels were observed at the periphery. The energy simulation ceases at the points where the rock movement stops. The Greek Cadastral DEM displays maximum kinetic energies with values of 0 J at the rockfall source and significantly higher values ranging from 200–500 kJ on the right side of the rockfall mass (Figure 7b). Similar kinetic energy ranges are depicted in the simulations using the ALOS AW3D30 DEM and SRTM30 DEM (Figure 7c,e), showing small ranges of variation. Conversely, the ASTER GDEM registered very low values of maximum kinetic energies, not exceeding 10 kJ (Figure 7d).

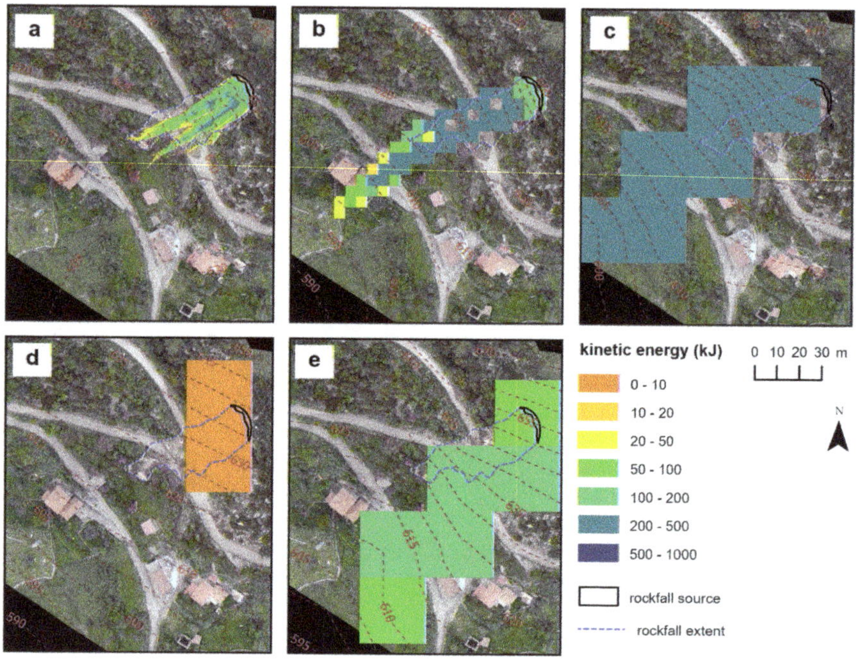

Figure 7. HY-STONE results for the site of Myloi using parameters (E_N, E_T, and A_T) calibrated on the UAV DEM. The maximum translation kinetic energies are shown for (**a**) UAV DEM, (**b**) Greek Cadastral DEM, (**c**) ALOS AW3D30 DEM, (**d**) ASTER GDEM, and (**e**) SRTM30 DEM.

The comparison of the slope profile from the various DEMs reveals that finer resolutions depict a more detailed and precise topography. Figures 8 and 9, from the UAV DEM and Greek Cadastral DEM, respectively, demonstrate a more accurate slope decline compared to the broader and smoother profiles from coarser-resolution DEMs. The UAV DEM, with the finest resolution, shows a detailed descent pattern (Figure 8a), whereas the Greek Cadastral DEM, though slightly coarser (Figure 8b), still maintains a good level of detail. In contrast, DEMs such as the ALOS AW3D30, ASTER GDEM, and SRTM30 DEM (Figure 8c–e) offer progressively less detail as their resolution decreases, which simplifies the terrain's complexities. Additionally, the contour lines from the ASTER GDEM (Figure 7d) show a wider spacing compared to those from the ALOS AW3D30 (Figure 7c) and SRTM30 DEMs (Figure 7e). This broader spacing of contour lines and the less detailed slope profile could account for the notably lower values of maximum translational kinetic energy that were recorded by the ASTER GDEM (Figure 7d).

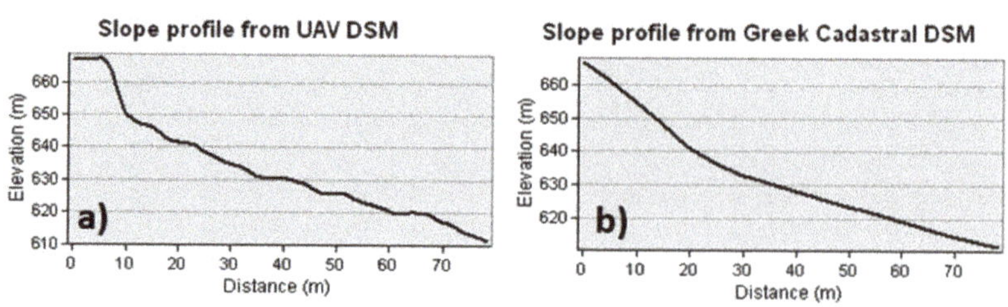

Figure 8. *Cont.*

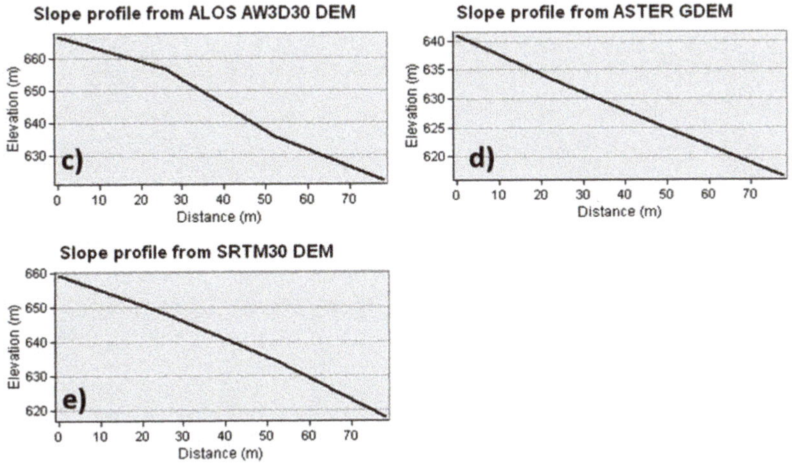

Figure 8. Slope profiles from (**a**) UAV DSE, (**b**) Greek Cadastral DEM, (**c**) ALOS AW3D30 DEM, (**d**) ASTER GDEM, and (**e**) SRTM30 DEM.

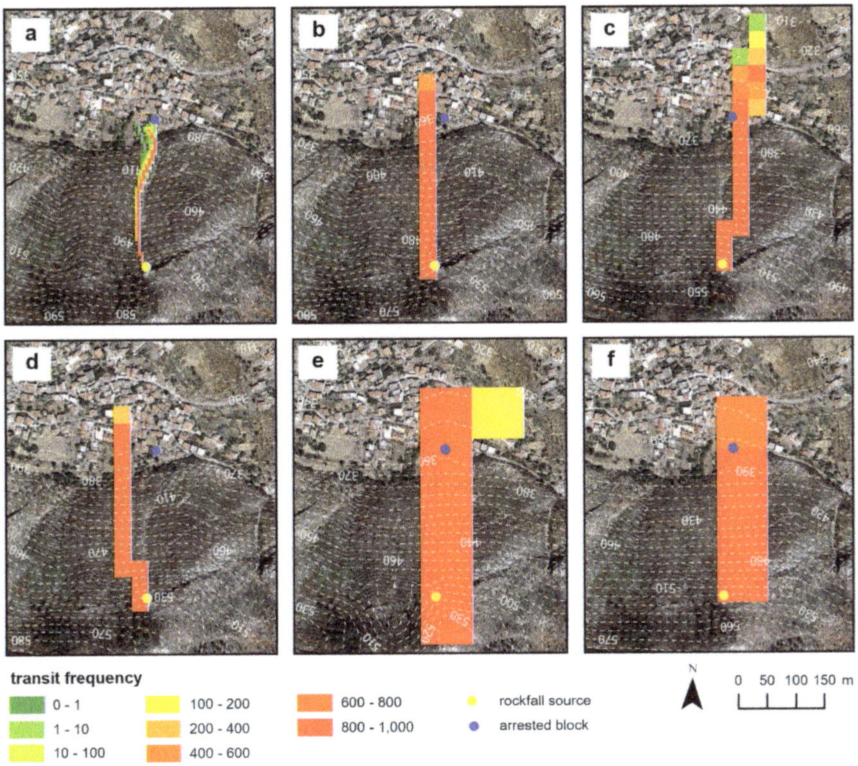

Figure 9. HY-STONE results, in terms of transit frequency for the site of Platiana. (**a**) Greek Cadastral DEM, (**b**) ALOS AW3D30 DEM, (**c**) ASTER GDEM, (**d**) SRTM30 DEM, (**e**) SRTM90 DEM, and (**f**) TanDEM_X.

4.1.2. Platiana Rockfall

For the Platiana rockfall analysis, six DEMs, including the Greek Cadastral DEM, ALOS AW3D30 DEM, ASTER GDEM, SRTM30 DEM, SRTM90 DEM, and TanDEM_X, were used, covering a resolution range from 5 m to 90 m. The coefficients of restitution E_N and E_T and the rolling friction coefficient A_T were back-calibrated on the higher-resolution DEM based on the rockfall event (Table 2). In this scenario, two masonry stone walls are located at the base of the slope and simulated with low values of E_N and E_T and high values of A_T, similar to buildings. Given that HY-STONE operates on a stochastic model, simulations were performed with 1000 rocks to provide a more accurate representation of the rockfall dynamics.

Table 2. Coefficients of restitution and rolling friction in Platiana rockfall, calibrated on the Greek Cadastral DEM.

Description	E_N	E_T	A_T
Houses	0.20	0.20	1.00
Road-asphalt	0.55	0.70	0.25
Debris well vegetated	0.35	0.60	0.42
Debris without vegetation	0.40	0.65	0.35
Stone wall	0.20	0.20	1.00

Figure 9a illustrates the transit frequency of the rocks using the Greek Cadastral DEM. High values are observed throughout the rockfall path, with a decrease in frequency from the midpoint to the endpoint of the rock trajectory. The transit frequency data extends almost up to the point where the rocks stop moving. More accurately, the rockfall trajectory ended just to the left of the actual endpoint.

All other considered DEMs simulated rock transits that extended beyond the actual rockfall path. The ALOS AW3D30 DEM exhibited a trajectory that passed from the left side of the real rock endpoint and captured high values from the rockfall source to the endpoint, with the values diminishing past the point where the rocks stopped (Figure 9b). The ASTER GDEM showed similar results but followed a distinct trajectory that crossed the rock stop point (Figure 9c). Meanwhile, the SRTM30 DEM also recorded high values at the rockfall source, which gradually decreased along a separate pathway, showing low values beyond the rock endpoint (Figure 9d). As is the case with the ALOS AW3D30 DEM, the SRTM30 DEM rockfall simulation also passed from the left side of the real endpoint.

The SRTM90 DEM and TanDEM_X, both with a 90 m spatial resolution, simulated the rock transit frequency in a less realistic manner (Figure 9e,f). Each recorded high values from the rockfall source to the endpoint and then displayed lower values past the endpoint. Specifically, the simulation using the SRTM90 DEM depicted a trajectory that extended beyond the actual endpoint, terminating on its right side. Meanwhile, the simulation with TanDEM_X also surpassed the real endpoint but continued along its initial trajectory.

The following data presents the maximum translational kinetic energies for the Platiana rockfall, expressed in kJ (Figure 10).

The outcomes based on the maximum translational kinetic energy from the Greek Cadastral are displayed in Figure 10a. The map, generated using HY-STONE within a GIS environment, shows high to very high kinetic energies near the rockfall source and in the central area, which then progressively decrease to very low levels towards the end of the slope. The maximum translational kinetic energy values range from 0 to 1500 kJ.

The computation of the maximum translational kinetic energy using the ALOS AW3D30 DEM is depicted in Figure 10b. This DEM outlined a single trajectory where kinetic energies near the rockfall source area are very low, ranging from 0 to 1 kJ, escalating to high energies between 100 to 1500 kJ in the middle of the path and tapering to medium energies from 0.2 to 100 kJ towards the end of the path. Conversely, the ASTER GDEM and SRTM30 DEM delineated multiple paths: low energies in the rockfall source area (0–1 kJ), high energies in the middle part (100–1500 kJ), and medium energies in the residential areas

(0.1–100 kJ), as shown in Figure 10c,d. All the DEMs with a 30 m spatial resolution thus calculated energies extending beyond the actual endpoint of the rockfall.

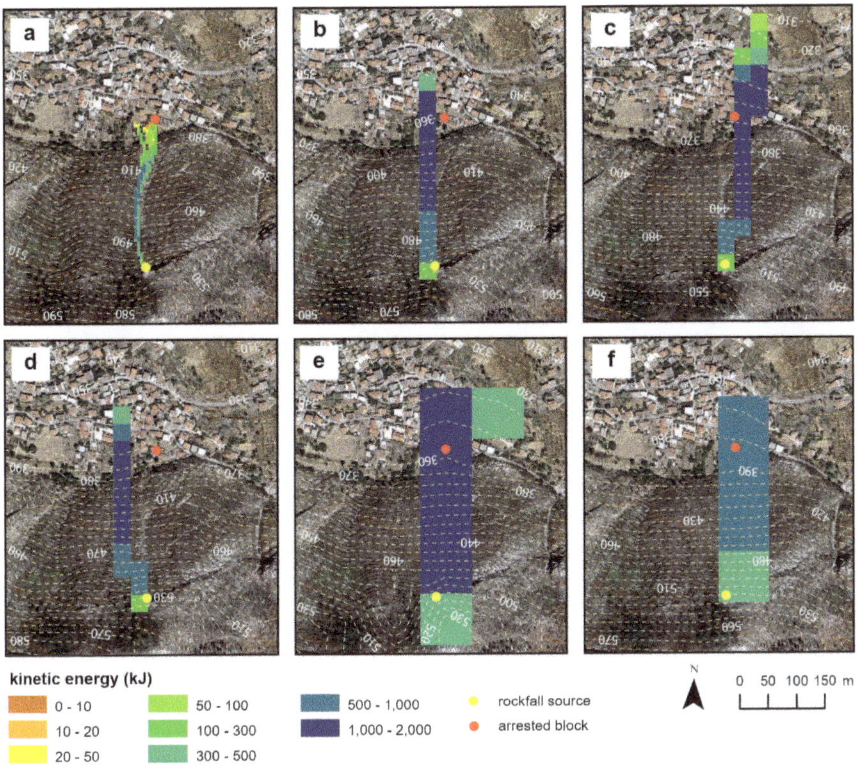

Figure 10. HY-STONE results in terms of the maximum translation kinetic energy for the site of Platiana. (**a**) Greek Cadastral DEM, (**b**) ALOS AW3D30 DEM, (**c**) ASTER GDEM, (**d**) SRTM30 DEM, (**e**) SRTM90 DEM, and (**f**) TanDEM_X.

Figure 10e illustrates the computation using the SRTM90 DEM, which generated two paths that were characterized by very high kinetic energies (500–1500 kJ) at the endpoint of the rockfall and low energies (0.2–1 kJ) near the source area. In contrast, the TanDEM_X mapped a single trajectory (Figure 10f) that displays a progression of medium, high, very high, back to medium, and then low energies, starting from the source area and extending beyond the actual endpoint of the rockfall. Both DEMs calculated kinetic energies that continued past the real endpoint of the rockfall.

The slope profiles (Figure 11) from the DEMs reveal differences in the elevation representation that correlate with the DEMs' spatial resolutions. The Greek Cadastral DEM (Figure 11a) provides a detailed topography due to its higher resolution, with an abrupt elevation drop, suggesting a potentially steeper terrain. In contrast, DEMs with coarser resolutions, such as the ALOS AW3D30 DEM (Figure 11b), ASTER GDEM (Figure 11c), SRTM30 DEM (Figure 11d), SRTM90 DEM (Figure 11e), and TanDEM_X (Figure 11f), show smoother elevation transitions, which might under-represent the actual terrain roughness. Also, the contour lines vary based on the spatial resolution of each DEM, as observed in Figure 10a–f. This variation influences the hillshade, which in turn affects the rockfall trajectories modeled by the HY-STONE software. Consequently, these differences in terrain representation explain the diversity of the rockfall simulations generated by the software, concluding that the path of the rockfall trajectory in the Platiana site is an issue of the topography.

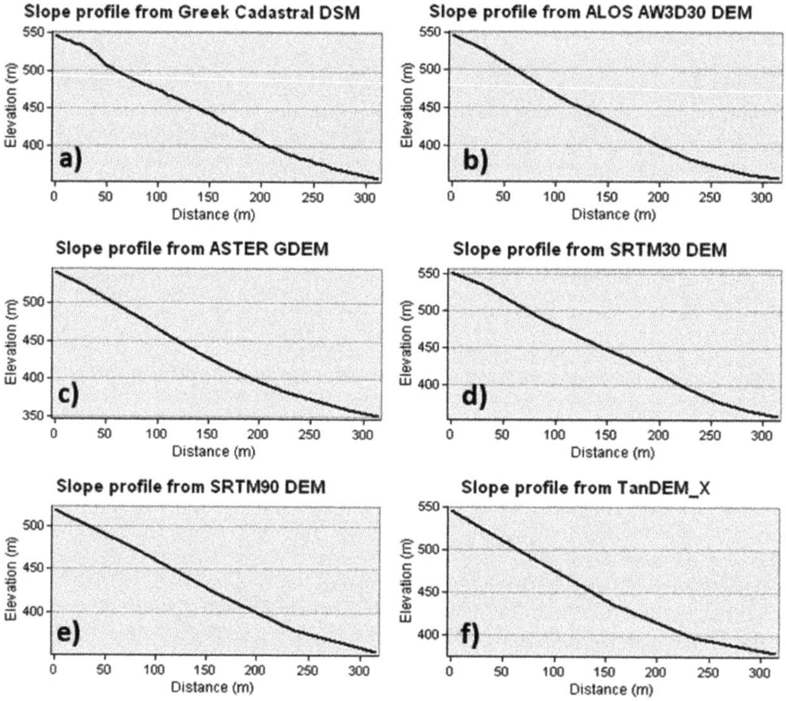

Figure 11. Slope profiles from (**a**) Greek Cadastral DEM, (**b**) ALOS AW3D30 DEM, (**c**) ASTER GDEM, (**d**) SRTM30 DEM, (**e**) SRTM90 DEM, and (**f**) TanDEM_X.

4.2. Results by Using Different Restitution Coefficients and Rolling Friction Coefficient

Considering the overestimation of the maximum runout distance projected by the simulations of both case studies with the low-resolution DEMs, a calibration of the normal (E_N) and tangential (E_T) coefficients of restitution, as well as the rolling friction coefficient (A_T) is required. In this section, the values of these coefficients are manually adjusted for each DEM scenario to obtain simulations that more closely resemble actual events. Specifically, to reduce the simulated runout distances, progressively lower restitution coefficients and higher friction coefficients are tested until the observation is matched. These adjustments compensate for different morphological roughness represented in each DEM.

4.2.1. Myloi Rockfall

After the coefficient adjustment procedure, it was determined that, for the Myloi case, only the Greek Cadastral DEM yielded reliable results, as the other DEMs (ALOS AW3D30 DEM, ASTER GDEM, SRTM30 DEM, SRTM90 DEM, and TanDEM_X) proved unsuitable for small-scale events like the Myloi rockfall, leading to unrealistic outputs. Regarding the UAV DEM, no new trials were conducted as the results with the initial coefficient values already align with the in situ observations. The forthcoming results showcase the transit frequency and maximum translational kinetic energy from the Greek Cadastral DEM, obtained with the revised coefficients shown in Table 3.

Table 3. Optimal coefficients for the Myloi rockfall using the Greek Cadastral DEM.

Description	E_N	E_T	A_T
Houses	0.20	0.20	1.00
Road-asphalt	0.35	0.55	0.50
Debris well vegetated	0.20	0.45	0.80
Debris without vegetation	0.25	0.50	0.70

The results from the Greek Cadastral DEM reveal that the transit frequency values were high, nearing 1000, in the rockfall source area and along the main rockfall path (Figure 12a). In terms of the maximum translational kinetic energy, the HY-STONE calculations showed energy levels ranging from 0 to 500 kJ. The rockfall source area and the endpoint recorded minimal energies, whereas the main body exhibited significantly higher energies, up to 500 kJ (Figure 12b). This simulation using the Greek Cadastral DEM accurately reflects the actual rockfall event.

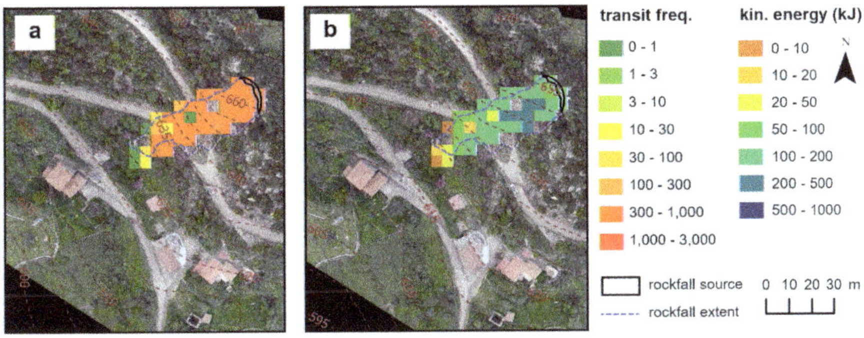

Figure 12. HY-STONE results by using the Greek Cadastral DEM with the optimal coefficients shown in Table 3 for the site of Myloi: (**a**) the transit frequency and (**b**) the maximum translation kinetic energy.

Figures 6b and 12a from the HY-STONE simulations illustrate the effect of coefficient customization on rockfall modeling accuracy for the Myloi site by using the Greek Cadastral DEM at a 5 m resolution. In Figure 12a, where the values of the coefficients were fine-adjusted to match the in situ observations, the depicted rockfall trajectories accurately terminate at the exact rock endpoints that were observed on-site, demonstrating a precise localization of rockfall impact. In contrast, Figure 6b uses standard empirically derived coefficients, resulting in trajectories that extend beyond the actual observed endpoints, suggesting an overestimation of the impact area. This comparison highlights the importance of adjusting the simulation parameters to specific site conditions to ensure the accuracy and reliability of rockfall risk assessments. Furthermore, Figures 7b and 12b illustrate the HY-STONE results for the maximum translation kinetic energy, also at the Myloi site, using the Greek Cadastral DEM at a 5 m resolution. As can be seen in Figure 12b, utilizing the adjusted coefficients for a closer match to the in situ observations shows a precise distribution of kinetic energy with high values concentrated near the rockfall source, thus closely reflecting the actual rockfall paths. In contrast, Figure 7b employs empirically derived coefficients, resulting in a more dispersed and extended kinetic energy distribution across the area.

4.2.2. Platiana Rockfall

For the Platiana rockfall, the coefficients E_N, E_T, and A_T have been adjusted for each DEM scenario, except for the Greek Cadastral DEM, which was already calibrated in the previous section against the in situ observations. Presented next are the outcomes for the transit frequency and maximum translational kinetic energy from the simulations using the ALOS AW3D30 DEM, ASTER GDEM, SRTM30 DEM, SRTM90 DEM, and TanDEM_X using the optimal restitution and rolling friction coefficients for each scenario.

Table 4 and Figure 13 showcase the optimal coefficients and simulation results in terms of the transit frequency and kinetic energies for the ALOS AW3D30 DEM.

Table 4. Optimal coefficients for the Platiana rockfall using the ALOS AW3D30 DEM.

Description	E_N	E_T	A_T
Houses	0.20	0.20	1.00
Road-asphalt	0.45	0.60	0.40
Debris well vegetated	0.25	0.45	0.64
Debris without vegetation	0.30	0.55	0.50
Stone wall	0.20	0.20	1.00

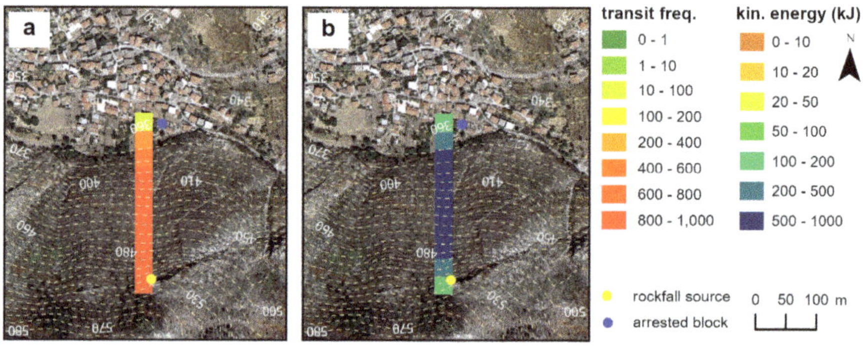

Figure 13. HY-STONE results by using the ALOS AW3D30 DEM with optimal coefficients shown in Table 4 for the site of Platiana: (**a**) the transit frequency and (**b**) the maximum translation kinetic energy.

According to Figure 13a, the ALOS AW3D30 DEM exhibited a single corridor of trajectories that pass on the left side of the real rock endpoint and show only low transit frequencies close to the endpoint. The computation of maximum translational kinetic energy displayed a range of values, beginning with low energies (50–100 kJ), escalating to very high energies (500–1000 kJ), and finally ending again with low energies (50–100 kJ) before reaching the endpoint (Figure 13b).

The comparison between Figures 13a and 9b highlights the differences in the transit frequency before and after the parameter adjustment. While Figure 13a depicts a very concentrated narrow path, suggestive of a well-defined rockfall trajectory, Figure 9b shows a more dispersed and broad frequency distribution that extends beyond the actual observed endpoints, implying either different simulation settings or more complex terrain interactions. Similarly, Figures 10b and 13b contrast in their portrayal of kinetic energy; Figure 13b presents a linear decrease in energy from the source along a controlled path, whereas Figure 10b depicts a diffuse energy pattern with values further from the rock endpoint.

Table 5 and Figure 14 showcase the optimal coefficients and simulation results in terms of the transit frequency and kinetic energies for the ALOS AW3D30 DEM.

Regarding the transit frequency maps, the ASTER GDEM documented (Figure 14a) high to low transit frequencies during the rock path crossing the rock stop point.

Table 5. Optimal coefficients for the Platiana rockfall using the ASTER GDEM.

Description	E_N	E_T	A_T
Houses	0.20	0.20	1.00
Road-asphalt	0.45	0.58	0.38
Debris well vegetated	0.22	0.42	0.69
Debris without vegetation	0.27	0.52	0.48
Stone wall	0.20	0.20	1.00

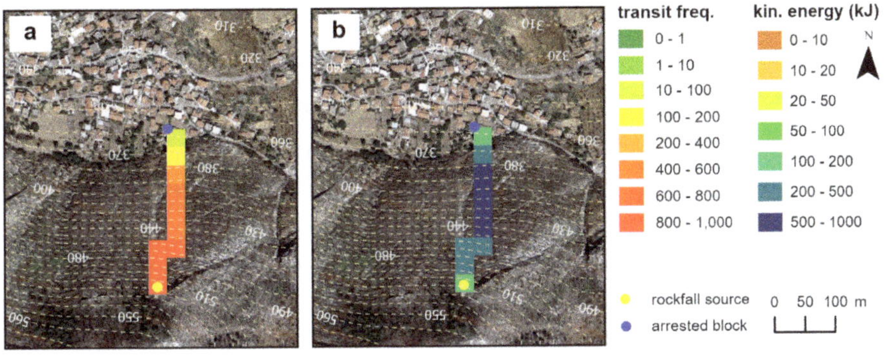

Figure 14. HY-STONE results by using the ASTER GDEM with the optimal coefficients shown in Table 5 for the site of Platiana: (**a**) the transit frequency and (**b**) the maximum translation kinetic energy.

Concerning the kinetic energy calculations, the ASTER GDEM develops similarly to the ALOS AW3D30 DEM simulation, starting with low values, followed by an increase until the final part of the trajectories where the blocks stopped.

Figure 14a presents a single concentrated transit frequency along a narrow path, reflecting a precise and realistic depiction of rockfall trajectory. In contrast, Figure 9c results in a broader and less defined distribution of transit frequencies, indicating multiple potential trajectories and a potential overestimation of the impact area. According to the kinetic energies, Figure 10c shows a detailed gradient of kinetic energy with multiple zones ranging from low- to high-energy values, reflecting a nuanced understanding of energy dissipation along the trajectory. This gradient indicates a more refined simulation where kinetic energy decreases progressively from the source, offering a detailed perspective on energy distribution during rockfall. On the other hand, Figure 14b exhibits a more simplified energy distribution, primarily concentrating high kinetic energy near the rockfall source with a sharp drop to lower values, suggesting a less complex interaction with the terrain.

Table 6 and Figure 15 showcase the optimal coefficients and simulation results in terms of the transit frequency and kinetic energies for the SRTM30 DEM.

Table 6. Optimal coefficients for the Platiana rockfall using the SRTM30 DEM.

Description	E_N	E_T	A_T
Houses	0.20	0.20	1.00
Road-asphalt	0.45	0.60	0.40
Debris well vegetated	0.25	0.45	0.64
Debris without vegetation	0.30	0.55	0.50
Stone wall	0.20	0.20	1.00

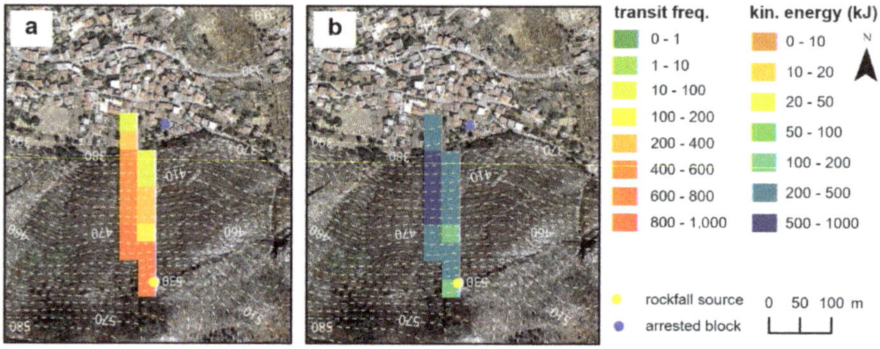

Figure 15. HY-STONE results by using the SRTM30 DEM with the optimal coefficients shown in Table 6 for the site of Platiana: (**a**) the transit frequency and (**b**) the maximum translation kinetic energy.

The SRTM30 DEM displayed medium transit frequencies over an extended section of the rock path, high frequencies in the source area, and no recorded values at the rock endpoint (Figure 15c). The kinetic energy results delineated two rock paths: the left, with energy levels up to 500–1000 kJ, and the right, with lower values around 200 kJ (Figure 15d).

Figure 9d, Figure 10d, and Figure 15a,b present the rockfall simulation results for the transit frequency and kinetic energies, correspondingly, using the SRTM30 DEM. Figure 15a,b uses the adjusted coefficients, while Figures 9d and 10d use coefficients on the highest resolution DEM. Figure 15a shows a more diverse color spectrum, indicating a dynamic interaction with a varied terrain that ends at the left side of the rock endpoint, while Figure 9d presents a more uniform and concentrated frequency, suggesting simpler terrain interactions or model settings that end after the rock endpoint. Figure 10d displays a linear distribution of kinetic energy with clear high-energy zones concentrated near the rockfall source, suggesting a direct rockfall trajectory. In contrast, Figure 15b shows a more uniform distribution, with fewer fluctuations in energy levels along the trajectory, indicating a potential oversimplification of the terrain's impact on energy dissipation.

Table 7 and Figure 16 showcase the optimal coefficients and simulation results in terms of the transit frequency and kinetic energies for the SRTM90 DEM.

Table 7. Optimal coefficients for the Platiana rockfall using the SRTM90 DEM.

Description	E_N	E_T	A_T
Houses	0.20	0.20	1.00
Road-asphalt	0.42	0.55	0.45
Debris well vegetated	0.22	0.40	0.70
Debris without vegetation	0.28	0.50	0.55
Stone wall	0.20	0.20	1.00

The SRTM90 DEM shows a constant decrease in the transit frequencies along the slope (Figure 16a), resembling the results of the ASTER GDEM and ALOS AW3D30 DEM. Regarding the kinetic energy (Figure 16b), the SRTM90 DEM exhibited initial lower energies of 200–500 kJ at the source area, increasing to 500–1000 kJ along the slope, without any decrease in energy before the rock stop point (Figure 16b).

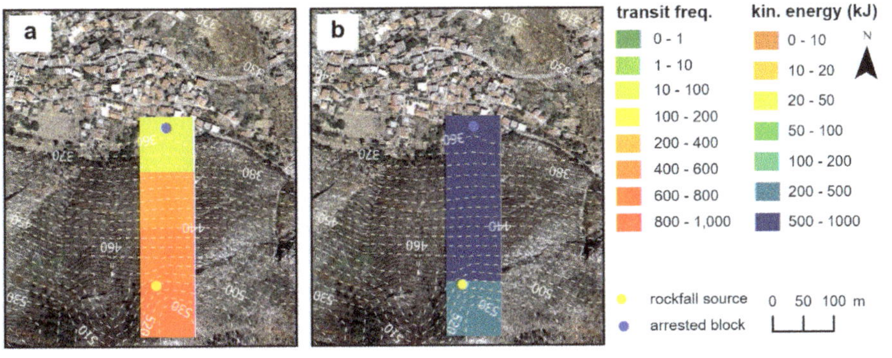

Figure 16. HY-STONE results by using the SRTM90 DEM with the optimal coefficients shown in Table 7 for the site of Platiana: (**a**) the transit frequency and (**b**) the maximum translation kinetic energy.

The transit frequency results shown in Figures 9e and 16a demonstrate a vertical layout with varying degrees of distribution; Figure 9e shows a poor gradation of the frequency zones along the slope continuing after the rock endpoint, whereas Figure 16a presents a more streamlined frequency layout ending at the rock endpoint. Similarly, the kinetic energy results in Figures 10e and 16b display linear energy distributions along the rockfall path. Figure 10e shows a complex pattern of low- and high-energy zones near the rockfall source and at the endpoint. On the other hand, Figure 16b portrays a simpler and more uniform distribution of kinetic energy, highlighting the role of the model's parameters in shaping the energy dissipation outcomes.

Table 8 and Figure 17 showcase the optimal coefficients and simulation results in terms of the transit frequency and kinetic energies for the TanDEM_X.

Table 8. Optimal coefficients for the Platiana rockfall using the TanDEM_X.

Description	E_N	E_T	A_T
Houses	0.20	0.20	1.00
Road-asphalt	0.35	0.52	0.48
Debris well vegetated	0.20	0.38	0.72
Debris without vegetation	0.25	0.47	0.58
Stone wall	0.20	0.20	1.00

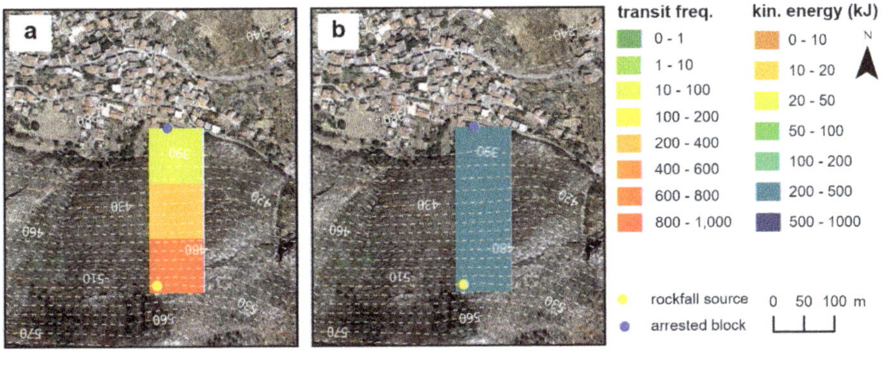

Figure 17. HY-STONE results by using the TanDEM_X with the optimal coefficients shown in Table 8 for the site of Platiana: (**a**) the transit frequency and (**b**) the maximum translation kinetic energy.

According to the TanDEM_X rockfall simulation map, the transit frequency for the Platiana rockfall is high in the source area and low at the endpoint of the rockfall (Figure 17a). The maximum translational kinetic energies show little variation along the slope, ranging in the 200–500 kJ class (Figure 17b).

The comparison between Figures 17a and 9f concluded that the simulation with the parameters optimized on the TanDEM_X presents a simplified and very linear representation of the transit frequency ending at the rock endpoint, which strictly follows a vertical pattern, indicating a highly targeted and specific rockfall path. Meanwhile, Figure 9f displays a more complex distribution with a smaller spectrum of transit frequencies, suggesting more variability in rockfall behavior. Last but not least, the compared Figures 17b and 10f sum up that both figures show a constant vertical distribution of kinetic energy values, but Figure 17b depicts a less spread-out energy distribution with a more realistic ending point.

To explore how the spatial resolution of the DEMs influences the normal (E_N) and tangential (E_T) restitution coefficients and the rolling friction coefficient (A_T), a plot reporting the differences in the coefficients for the most representative land cover class (debris without vegetation) is shown for the Platiana case (Figure 18). This plot clearly shows that the optimal normal (E_N) and tangential (E_T) coefficients decrease with increasing spatial resolution, whereas the rolling friction coefficient (A_T) increases with a higher spatial resolution.

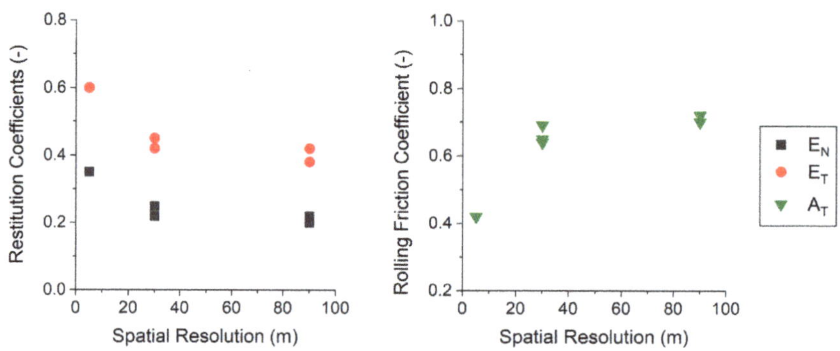

Figure 18. The effect of spatial resolution in the normal (E_N) and tangential (E_T) restitutions and the rolling friction (A_T) coefficients. For 30 m and 90 m, three and two DEMS are available, respectively.

The graph shows a roughly logarithmic relationship between the spatial resolution of DEMs and the optimal rockfall simulation coefficients. As the pixel size increases, indicative of coarser spatial resolution, the rolling friction coefficient (A_T) exhibits an increasing trend. This suggests that larger pixel sizes, which represent less detailed terrain data, demand high values of rolling friction coefficient to halt rocks more rapidly on the slope.

In contrast, the restitution coefficients, E_N and E_T, show a decreasing trend with increasing pixel size. This pattern indicates that, in simulations, rocks lose more energy upon impact with coarser-resolution DEMs, resulting in shorter bounces and less travel distance, which is characteristic of more inelastic collisions.

The correlation depicted here highlights the significance of selecting DEMs with appropriate pixel sizes for rockfall modeling to ensure that the physical behavior of rockfalls is realistically replicated in the simulation outcomes. The variability of the coefficients for a certain resolution is due to slight differences in the morphological description of the different DEMs. Future research will expand the dataset to validate these findings further. However, the observed trend provides valuable insights into the behavior of each coefficient in relation to each spatial resolution.

5. Discussion

The use of HY-STONE in analyzing rockfall simulations across different DEMs has highlighted the pivotal role of spatial resolution in the accuracy of rockfall modeling outcomes. The detailed analyses from the rockfall events at Myloi and Platiana, utilizing DEMs with a broad range of resolutions, confirm that higher-resolution DEMs, such as the UAV DEM and Greek Cadastral DEM, produce more accurate simulations of rockfall trajectories, aligning with our field observations. This finding is in line with previous studies suggesting that finer resolutions enhance the simulation's precision due to their ability to capture critical topographical details [34,36,41]. Remote sensing techniques, as used in this research, offer significant advantages by minimizing the data collection time and enabling access to otherwise inaccessible areas, thus enhancing the efficiency of rockfall hazard assessments [11,29,42]. This study further highlights the variations in performance among high-resolution DEMs, highlighting the importance of careful DEM selection based on the specific characteristics of a rockfall area and the methodologies behind data collection and processing [36]

This study extends the previous findings by not only confirming the influence of spatial resolution on simulation accuracy but also exploring how HY-STONE's capabilities can be optimized through strategic DEM selection. Large differences in the results can be observed by using the same parameter sets with different DEMs, with a longer runout simulated for coarser DEMs and a shorter runout with more resolute DEMs [36]. This effect can be attributed to the different ruggedness of DEMs as a function of the cell size [30]. For instance, the discrepancies in simulation outcomes among various high-resolution DEMs prompt a recommendation for future research to develop more sophisticated algorithms within HY-STONE. While the influence of DEM resolution is clear, DEMs created using distinct algorithms require unique parameter sets, even when sharing the same cell size [36]. These algorithms would compensate for the limitations of lower-resolution DEMs, especially in economically constrained settings where high-resolution DEMs are not feasible [32].

Previous studies using HY-STONE software for rockfall simulations also emphasize the critical role of the selected DEM and the calibration of its three coefficients concluding to similar results. It has been noted that an improved DEM accuracy correlates directly with higher precision in the HY-STONE simulation results, particularly with a target grid resolution of one meter by one meter or less in the volume scenarios from local-scale rockfalls [37,38,43,44] and a grid resolution of five meters by five meters for larger-scale rockfalls.

In this paper, it has been highlighted that the proper calibration of restitution and friction coefficients can compensate for the effects of different morphological roughness, represented in each DEM, by reaching satisfactory results, even with coarser-resolution DEMs. Different degrees of accuracy compared to the high spatial resolution simulations are achieved in terms of the transit frequencies and kinetic energies. Furthermore, for small-scale events, low-resolution DEMs are unsuitable for reliable predictions, such as the case of the Myloi rockfall. Another limitation of DEMs with coarser resolutions is how the rockfall source area is located inside large pixels, which could lead to potential misfits of the initial trajectory, leading to very large errors in the final outcome. This limitation cannot be overcome by parameter optimization; however, it is a potential improvement in the development of the HY-STONE software.

Additionally, the role of vegetation cover in influencing rockfall trajectories has been reaffirmed [33,45,46]. Studies have shown that vegetation can significantly impact the path and energy of falling rocks, emphasizing the need for HY-STONE to incorporate the vegetation parameters effectively [33]. Future studies should also investigate the integration of multiple DEMs to refine the predictive capabilities and adapt to the constraints of resource-limited settings. Ultimately, this discussion emphasizes the necessity of using high-resolution DEMs for detailed and accurate rockfall modeling, particularly when employing HY-STONE software. The continuous refinement of DEMs and the integration of new data sources, such as UAV photogrammetry, are crucial for advancing rockfall simulation techniques and improving the reliability of rockfall hazard assessments.

This research provides a substantial contribution to the geoscience field, particularly in rockfall hazard management and mitigation. By evaluating the impact of a DEM's spatial resolution on 3D rockfall simulations within a GIS environment, this study offers key insights into the optimization of rockfall prediction models. It shows how different DEM resolutions affect the accuracy of rockfall trajectory simulations, which is crucial for effective hazard assessments and planning in vulnerable areas. This advancement allows for better-informed decision-making regarding land use and civil protection measures in areas prone to rockfalls. Moreover, the study's methodological approach, utilizing the HY-STONE software for detailed multiscale analyses, sets a new benchmark in the field. It paves the way for further research on the integration of various DEMs and modeling techniques to refine rockfall hazard predictions. Ultimately, this research not only progresses our scientific understanding but also contributes to improving public safety and infrastructure protection in rockfall-prone regions.

6. Conclusions

This research exemplifies the application of HY-STONE in modeling rockfalls of varying scales. The software performs effectively for both small-scale events, as shown in the Myloi settlement, and larger-scale occurrences, such as the Platiana rockfall, by appropriately selecting and adjusting DEMs and simulation parameters. The variability in results between different DEMs emphasizes the importance of selecting the correct spatial resolution for a given rockfall scenario. The spatial resolution directly influences the ability of the HY-STONE software to produce simulations that closely resemble the physical behavior observed in real-world events.

The results highlight the critical role of spatial resolution in rockfall modeling, with finer-resolution DEMs providing the superior detail necessary for precise modeling outcomes. The analysis indicates that high-resolution DEMs not only provide a detailed terrain representation crucial for identifying rockfall sources but also enhance the accuracy of kinetic energy distribution in the simulations. In this study, the comparative analysis of a multitude of freely available DEMs within HY-STONE reveals the significance of choosing the right model for a given scenario, with higher resolutions preferred for detailed local-scale analyses. This study also highlights the importance of empirical knowledge and calibration against field data in setting the restitution and rolling friction coefficients, which significantly influence the behavior of simulated rockfalls.

More specifically, the results show that:

1. By maintaining constant values for the restitution and rolling friction coefficients (determined empirically) for the Myloi rockfall, the UAV DEM with a spatial resolution of 0.05 m led to the most reliable simulation via HY-STONE, as the transit frequency stopped precisely where the rocks did, thus matching the in situ observations.
2. In the case of the Platiana rockfall, keeping the aforementioned empirically defined coefficients constant, the Greek Cadastral DEM with a 5 m spatial resolution yielded results that most closely resembled the actual rockfall.
3. By altering the E_N, E_T, and A_T parameters, for the Myloi rockfall, only the Greek Cadastral DEM provided a reliable representation. DEMs with larger spatial resolutions did not lead to an accurate rockfall simulation due to the small scale of the event and the DEMs' pixel size being too large to capture detailed results.
4. In contrast, for the Platiana rockfall, which is a larger-scale event, changing the E_N, E_T, and A_T parameters—specifically, decreasing E_N and E_T and increasing A_T as the spatial resolution increased—resulted in simulations through HY-STONE that increasingly resembled the actual rockfall event.
5. In all instances, the topography depicted by the DEMs with a spatial resolution exceeding 30 m showed a minimal variation in the HY-STONE rockfall simulations.

To sum up, the most accurate results are those obtained by keeping the restitution and rolling friction coefficients constant, as they have been established based on empirical experience. Regarding the spatial resolution of the DEMs, the finer the pixel size, the closer to reality the simulation will be. This suggests that high-resolution DEMs should be prioritized when available and are applicable to rockfall studies.

Author Contributions: Conceptualization, K.G.N. and M.P.K.; methodology, M.P.K. and P.F.; software, P.F., A.P. and M.P.K.; validation, P.F. and A.P.; investigation, K.G.N. and M.P.K.; data curation, K.G.N. and M.P.K.; writing—original draft preparation, M.P.K., P.F. and A.P. writing—review and editing, K.G.N. and P.F.; supervision, K.G.N. All authors have read and agreed to the published version of the manuscript.

Funding: This research received no external funding.

Data Availability Statement: The data are part of the 1st author PhD thesis, therefore there are restrictions to share.

Conflicts of Interest: The authors declare no conflict of interest.

References

1. Fell, R.; Corominas, J.; Bonnard, C.; Cascini, L.; Leroi, E.; Savage, W.Z. Guidelines for landslide susceptibility, hazard and risk zoning for land use planning. *Eng. Geol.* **2008**, *102*, 85–98. [CrossRef]
2. Farvacque, M.; Eckert, N.; Candia, G.; Bourrier, F.; Corona, C.; Toe, D. Holistic rockfall risk assessment in high mountain areas affected by seismic activity: Application to the Uspallata valley, Central Andes, Chile. *Risk Anal.* **2024**, *44*, 1021–1045. [CrossRef]
3. Van Veen, M.; Hutchinson, D.J.; Bonneau, D.A.; Sala, Z.; Ondercin, M.; Lato, M. Combining temporal 3-D remote sensing data with spatial rockfall simulations for improved understanding of hazardous slopes within rail corridors. *Nat. Hazards Earth Syst. Sci.* **2018**, *18*, 2295–2308. [CrossRef]
4. Yomralıoğlu, T. *Geographic Information Systems: Basic Concepts and Applications*; Akademi Publishing House: İstanbul, Turkyie, 2009; p. 480. ISBN 975-97369-0-X.
5. Çellek, S. Effect of the Slope Angle and Its Classification on Landslide. *Nat. Hazards Earth Syst. Sci.* 2020, *in press*. [CrossRef]
6. Keijsers, J.G.S.; Schoorl, J.M.; Chang, K.-T.; Chiang, S.-H.; Claessens, L.; Veldkamp, A. Calibration and resolution effects on model performance for predicting shallow landslide locations in Taiwan. *Geomorphology* **2011**, *133*, 168–177. [CrossRef]
7. Pradhan, B.; Sameen, M.I. Effects of the Spatial Resolution of Digital Elevation Models and Their Products on Landslide Susceptibility Mapping. In *Laser Scanning Applications in Landslide Assessment*; Springer: Berlin/Heidelberg, Germany, 2017; pp. 133–150.
8. Acosta, E.; Agliardi, F.; Crosta, G.B.; Rıos Aragues, S. Regional Rockfall Hazard Assessment in the Benasque Valley (Central Pyrenees) Using a 3D Numerical Approach. In Proceedings of the 4th EGS Plinius Conference—Mediterranean Storms, Mallorca, Spain, 2–4 October 2002; pp. 555–563.
9. Baillifard, F.; Jaboyedoff, M.; Sartori, M. Rockfall hazard mapping along a mountainous road in Switzerland using a GIS-based parameter rating approach. *Nat. Hazards Earth Syst. Sci.* **2003**, *3*, 435–442. [CrossRef]
10. Marquínez, J.; Menéndez Duarte, R.; Farias, P.; Jiménez Sánchez, M. Predictive GIS-Based Model of Rockfall Activity in Mountain Cliffs. *Nat. Hazards* **2003**, *30*, 341–360. [CrossRef]
11. Jaboyedoff, M.; Baillifard, F.; Philippossian, F.; Rouiller, J.-D. Assessing fracture occurrence using "weighted fracturing density": A step towards estimating rock instability hazard. *Nat. Hazards Earth Syst. Sci.* **2004**, *4*, 83–93. [CrossRef]
12. Derron, M.-H.; Jaboyedoff, M.; Blikra, L.H. Preliminary assessment of rockslide and rockfall hazards using a DEM (Oppstadhornet, Norway). *Nat. Hazards Earth Syst. Sci.* **2005**, *5*, 285–292. [CrossRef]
13. Lan, H.; Derek Martin, C.; Lim, C.H. RockFall analyst: A GIS extension for three-dimensional and spatially distributed rockfall hazard modeling. *Comput. Geosci.* **2007**, *33*, 262–279. [CrossRef]
14. Loye, A.; Jaboyedoff, M.; Pedrazzini, A. Identification of potential rockfall source areas at a regional scale using a DEM-based geomorphometric analysis. *Nat. Hazards Earth Syst. Sci.* **2009**, *9*, 1643–1653. [CrossRef]
15. Ravanel, L.; Allignol, F.; Deline, P.; Gruber, S.; Ravello, M. Rock falls in the Mont Blanc Massif in 2007 and 2008. *Landslides* **2010**, *7*, 493–501. [CrossRef]
16. Zieher, T.; Formanek, T.; Bremer, M.; Meissl, G.; Rutzinger, M. Digital Terrain Model Resolution and its Influence on Estimating the Extent of Rockfall Areas. *Trans. GIS* **2012**, *16*, 691–699. [CrossRef]
17. Palma, B.; Parise, M.; Reichenbach, P.; Guzzetti, F. Rockfall hazard assessment along a road in the Sorrento Peninsula, Campania, southern Italy. *Nat. Hazards* **2011**, *61*, 187–201. [CrossRef]
18. Corona, C.; Trappmann, D.; Stoffel, M. Parameterization of rockfall source areas and magnitudes with ecological recorders: When disturbances in trees serve the calibration and validation of simulation runs. *Geomorphology* **2013**, *202*, 33–42. [CrossRef]

19. Nikolakopoulos, K.; Depountis, N.; Vagenas, N.; Kavoura, K.; Vlaxaki, E.; Kelasidis, G.; Sabatakakis, N. Rockfall Risk Evaluation Using Geotechnical Survey, Remote Sensing Data, and GIS: A Case Study from Western Greece. In Proceedings of the Third International Conference on Remote Sensing and Geoinformation of the Environment (RSCy2015), Paphos, Cyprus, 19 June 2015.
20. Volkwein, A.; Schellenberg, K.; Labiouse, V.; Agliardi, F.; Berger, F.; Bourrier, F.; Dorren, L.K.A.; Gerber, W.; Jaboyedoff, M. Rockfall characterisation and structural protection—A review. *Nat. Hazards Earth Syst. Sci.* **2011**, *11*, 2617–2651. [CrossRef]
21. Stevens, W.D. RocFall, a Tool for Probabilistic Analysis, Design of Remedial Measures and Prediction of Rockfalls. Ph.D. Thesis, University of Toronto, Toronto, ON, Canada, 1998.
22. Rocscience. RocFall software—For risk analysis of falling rocks on steep slopes. In *Rocscience User's Guide*; Rocscience: Toronto, ON, Canada, 2002.
23. Hind, H. Comparison of 3D and 2D Rockfall Models: Considering Terrain Model Quality Effect on RespectiveModel Performances. Master's Thesis, UiT the Arctic University of Norway, Tromsø, Norway, 2018.
24. Kakavas, M.; Nikolakopoulos, K. Rock-fall simulation and validation with in situ data: The case of Moira settlement in Western Greece. In *Earth Resources and Environmental Remote Sensing/GIS Applications XII*; SPIE: Bellingham, Washington, USA, 2021; Volume 11863.
25. Abellán, A.; Vilaplana, J.M.; Martínez, J. Application of a long-range Terrestrial Laser Scanner to a detailed rockfall study at Vall de Núria (Eastern Pyrenees, Spain). *Eng. Geol.* **2006**, *88*, 136–148. [CrossRef]
26. Žabota, B.; Repe, B.; Kobal, M. Influence of digital elevation model resolution on rockfall modelling. *Geomorphology* **2018**, *328*, 183–195. [CrossRef]
27. Guzzetti, F.; Crosta, G.B.; Detti, R.; Agliardi, F. STONE: A computer program for the three-dimensional simulation of rockfalls. *Comput. Geosci.* **2002**, *28*, 1081–1095. [CrossRef]
28. Crosta, G.B.; Agliardi, F. A methodology for physically based rockfall hazard assessment. *Nat. Hazards Earth Syst. Sci.* **2003**, *3*, 407–422. [CrossRef]
29. Kakavas, M.; Nikolakopoulos, K. Digital Elevation Models of Rockfalls and Landslides: A Review and Meta-Analysis. *Geosciences* **2021**, *11*, 256. [CrossRef]
30. Crosta, G.B.; Agliardi, F. Parametric evaluation of 3D dispersion of rockfall trajectories. *Nat. Hazards Earth Syst. Sci.* **2004**, *4*, 583–598. [CrossRef]
31. Bühler, Y.; Christen, M.; Glover, J.; Bartelt, P. Significance of Digital Elevation Model Resolution for Numerical Rockfall Simulations. In Proceedings of the 3rd International Symposium Rock Slope Stability C2ROP, Lyon, France, 15–17 November 2016; pp. 15–17.
32. Frattini, P.; Crosta, G.; Carrara, A.; Agliardi, F. Assessment of rockfall susceptibility by integrating statistical and physically-based approaches. *Geomorphology* **2008**, *94*, 419–437. [CrossRef]
33. Agliardi, F.; Crosta, G.B.; Frattini, P. Integrating rockfall risk assessment and countermeasure design by 3D modelling techniques. *Nat. Hazards Earth Syst. Sci.* **2009**, *9*, 1059–1073. [CrossRef]
34. Agliardi, F.; Crosta, G.B. High resolution three-dimensional numerical modelling of rockfalls. *Int. J. Rock Mech. Min. Sci.* **2003**, *40*, 455–471. [CrossRef]
35. Themistocleous, K.; Danezis, C. Monitoring cultural heritage sites affected by geo-hazards using in situ and SAR data: The Choirokoitia case study. In *Remote Sensing for Archaeology and Cultural Landscapes: Best Practices and Perspectives Across Europe and the Middle East*; Springer: Berlin/Heidelberg, Germany, 2020; pp. 285–308.
36. Frattini, P.; Crosta, G.B.; Agliardi, F.; Imposimato, S. Challenging Calibration in 3D Rockfall Modelling. In *Landslide Science and Practice*; Springer: Berlin/Heidelberg, Germany, 2013; pp. 169–175.
37. Kusak, M.; Valagussa, A.; Frattini, P. Key issues in 3d rockfall modeling, natural hazard and risk assessment for rockfall protection in hřensko (Czechia). *Acta Geodyn. Et Geomater.* **2019**, *16*, 393–408. [CrossRef]
38. Pilz, J.; Agliardi, F.; Crosta, G.B.; Zavodni, Z.M. Three–dimensional rock fall simulation in the mining environment using Hy_Stone. In Proceedings of the Slope Stability 2011: International Symposium on Rock Slope Stability in Open Pit Mining and Civil Engineering, Vancouver, BC, Canada, 18–21 September 2011.
39. Depountis, N.; Lainas, S.; Pyrgakis, D.; Sabatakakis, N.; Koukis, G. Engineering geological and geotechnical investigation of landslide events in wildfire affected areas of Ilia Prefecture, Western Greece. *Bull. Geol. Soc.* **2010**, *43*, 1138–1148. [CrossRef]
40. Vagenas, N. Rockfall Simulation by Analytical Methods. Correlation of Rockfall Energy Dissipation Parameters and Rock Mass Characteristics. Ph.D. Thesis, University of Patras, Greece, Department of Geology, Patras, Greece, 2020.
41. Kakavas, M.P.; Nikolakopoulos, K.G.; Kyriou, A.; Koukouvelas, I. The Influence of the DEM Spatial Resolution in Rockfall Simulation and Validation with In Situ Data. *Geosciences* **2023**, *13*, 57. [CrossRef]
42. Jaboyedoff, M.; Labiouse, V. Preliminary estimation of rockfall runout zones. *Nat. Hazards Earth Syst. Sci.* **2011**, *11*, 819–828. [CrossRef]
43. Lanfranconi, C.; Frattini, P.; Sala, G.; Dattola, G.; Bertolo, D.; Sun, J.; Crosta, G.B. Accounting for the effect of forest and fragmentation in probabilistic rockfall hazard. *Nat. Hazards Earth Syst. Sci.* **2023**, *23*, 2349–2363. [CrossRef]
44. Margottini, C.; Spizzichino, D.; Crosta, G.B.; Frattini, P.; Mazzanti, P.; Scarascia Mugnozza, G.; Beninati, L. Rock fall instabilities and safety of visitors in the historic rock cut monastery of Vardzia (Georgia). In *Volcanic Rocks and Soils*; CRC Press: Boca Raton, FL, USA; Balkema: Cape Town, South Africa; Taylor and Francis Group: Abingdon, UK, 2016; pp. 371–378.

45. Crosta, G.B.; Agliardi, F.; Frattini, P. Modelling rockfall impact onstructures. In *Geophysical Research Abstracts*; EGU05-A-08555; H3.01-1WE4P-0110; Copernicus GmbH: Göttingen, Germany, 2005.
46. Crosta, G.B.; Frattini, P.; Imposimato, S.; Agliardi, F. Modeling vegetation and fragmentation effects on rockfalls. *Geophys. Res.* **2006**, *8*, 07694.

geosciences

MDPI

Article

Vegetation Monitoring of Palm Trees in an Oasis Environment (Boudenib, Morocco) Using Automatic Processing of Medium-Resolution Remotely Sensed Data

Kaoutar Badioui [1,*], Ann Van Griensven [1,2] and Boud Verbeiren [1]

[1] Department of Water and Climate, VUB Vrije Universiteit Brussel, Vrije Universiteit Brussels, Pleinlaan 2, B-1050 Brussels, Belgium; ann.van.griensven@vub.be (A.V.G.); boud.verbeiren@vub.be (B.V.)
[2] Water Science and Engineering Department, IHE Delft Institute for Water Education, 2611 AX Delft, The Netherlands
* Correspondence: kaoutar.badioui@vub.be

Abstract: Oases are part of the natural wealth and heritage of Morocco and contribute to the social, economic, and touristic environment. Morocco has lost more than 2/3 of its oases during the past century due to water scarcity, succession of drought periods, climate change and over-exploitation of groundwater resources. Palm trees are strongly dependent on irrigation and availability of surface water as soon as the water table depth falls below the root zone of 9 m. Improving management and monitoring of oasis ecosystems is strongly encouraged by UNESCO Biosphere Reserve and RAMSAR guidelines. The Boudenib and Tafilalet oases are among the biggest palm groves located in the south-eastern part of Morocco. These oases belong to catchments of the rivers Guir and Ziz, respectively. This paper uses remotely sensed data from PROBA-V for monitoring vegetation in oases, and linking vegetation characteristics to water availability, water management and quality and quantity of date crops. The Normalized Differential Vegetation Index (NDVI) derived from optical images provides a good estimation of changes in vegetation cover over time. Images of various spatial resolutions (100 m, 300 m and 1 km) obtained with the frequently revisiting Belgian satellite PROBA-V and available since 2014, can be successfully used for deriving time series of vegetation dynamics. TREX—Tool for Raster data Exploration—is a Python-GDAL processing tool of PROBA-V NDVI images for analyzing vegetation dynamics, developed at the Vrije Universiteit Brussel and available online. TREX has various applications, but the main functionality is to provide an automatic processing of PROBA-V satellite images into time series of NDVI and LAI, used in vegetation monitoring of user-defined points of interest. This study presents the results of application of TREX in the arid ecosystems of the Boudenib oasis for the period 2014–2018. The resulting NDVI and LAI time series are also compared to time series of groundwater depth and date crops quantity and quality. Low LAI is observed when water depth is low, and the palm trees lose their greenery. Low LAI is also correlated to low quantity and quality of dates in October 2015 and October 2017. PROBA-V images can therefore be used for monitoring the health of palm trees in oasis environments. However, considering the fact that the PROBA-V satellite mission has ended, this approach could instead be applied to Sentinel-3 data using the same analysis. These results have important implications for water management in the area and can help decision-makers to make better decisions about prevention of water scarcity in the region.

Keywords: Oasis; Palms trees; Guir watershed; TREX; remote sensing; NDVI; LAI; PROBA-V; MODIS; LANDSAT; VITO; ESA

https://doi.org/10.3390/geosciences15030104

1. Introduction

The Boudenib Oasis is the largest palm grove in the southeastern region of Morocco, located within the Guir watershed, close to the Moroccan–Algerian border. The region, characterized by an arid climate, faces significant water resource shortages and relies on irrigation from dams [1]. This oasis has recently been the subject of agricultural and hydrological studies due to growing concerns about water scarcity and stress [2]. These challenges, including recurrent droughts and poverty, threaten one of the country's most vital ecosystems, which is also listed under the Ramsar Convention [3].

In recent years, the Moroccan government has placed increased emphasis on the conservation of its oases, which are recognized as key cultural and ecological heritage sites. UNESCO and Ramsar guidelines have also encouraged efforts to improve the sustainability of these ecosystems (https://www.ramsar.org/news/morocco-names-20-varied-new-ramsar-sites, accessed on 15 January 2005). The Sahara region, where the oasis is located, is home to an estimated five million date palms, some of which can live up to 150 years. These palms are crucial to the local economy, providing the primary source of income for the population. Additionally, the palm trees offer essential shade for other crops and fruit trees that grow beneath their broad fronds (https://maroc.costasur.com/en/arbres-de-paume.html, accessed on 21 January 2025).

Morocco not only meets domestic demand for dates but also exports over 90,000 tons annually (https://maroc.costasur.com/en/arbres-de-paume.html, accessed on 21 January 2025). The deep, binding roots of the date palms play a crucial role in preventing soil erosion caused by desert winds, contributing to the stabilization of the ecosystem. Historically, the oases were created by Arab warriors through the khettara network, further enhancing the region's landscape. Date palms thrive in a delicate ecosystem with high evapotranspiration, making groundwater an essential resource for their growth. However, the oasis faces increasing degradation driven by periodic droughts, rising temperatures due to climate change, and the overexploitation of groundwater resources, which has led to the drying of wells in the region [4].

Oases in arid regions such as Boudenib are vital ecosystems that sustain biodiversity, provide water resources, and support agricultural practices, particularly date palms. These oases, with their fragile water availability, are increasingly vulnerable to degradation due to a variety of factors including water scarcity, climate change, and overexploitation of groundwater resources. Monitoring these ecosystems is crucial for their sustainable management, particularly given the significant role they play in the local economy and the cultural heritage of Morocco. The decline in the health of these oases, driven by periodic droughts, soil salinity, and poor land management practices, highlights the importance of continuous monitoring to ensure the sustainability of these ecosystems.

The use of remote sensing (RS) for vegetation monitoring has advanced significantly, particularly in spectral indices, high-resolution, and hyperspectral data time series analysis. Studies in Morocco's Tafilalet Oasis have utilized Sentinel-1 and Sentinel-2 data to investigate irrigation patterns, vegetation moisture, and the impacts of climate change on water availability [5,6]. Sentinel-2 imagery, in particular, has shown a consistent decline in the Leaf Area Index (LAI) from July 2015 to May 2018, indicating deterioration in the structure of palm trees. This trend suggests that the oasis is facing degradation due to factors such as water scarcity and land mismanagement. Additionally, soil salinity in the region has led to water pollution [2]. These studies have provided valuable insights into water management and the effects of water scarcity on vegetation health.

However, similar remote sensing applications are lacking for the Boudenib Oasis, which remains under-monitored using geospatial technologies. While the Tafilalet Oasis has been extensively studied using conventional satellites such as MODIS, LANDSAT, and Sentinel, the spatial and temporal resolution of these satellites varies. Sentinel-2 offers higher resolution (10–60 m) but has a revisit time of 5 days, while Sentinel-3 provides broader spectral coverage but with a less frequent revisit. Landsat (30 m, 16-day revisit) and MODIS (250–1000 m, daily revisit) offer either limited spatial detail or temporal frequency.

In contrast, Boudenib Oasis has not been monitored using remote sensing technologies. This research follows a similar approach to previous remote sensing studies, but with the distinction of using alternative input data applied to the Boudenib Oasis, which has been under-researched in terms of geospatial monitoring. The primary contribution of this study is to fill this gap, enriching previous research in the region and providing decision-makers with valuable data to help preserve this important heritage.

Despite the growing recognition of the importance of oases like Boudenib, there is a significant lack of in-depth research focusing on the environmental and hydrological conditions of this specific oasis. While studies have been conducted in other Moroccan oases, such as Tafilalet, which have used remote sensing technologies to monitor vegetation health and water resources, Boudenib has remained underexplored in this regard. Furthermore, existing studies on Boudenib have not employed remote sensing validation to assess the extent of environmental degradation or its link to groundwater depletion. This research gap limits the ability to accurately assess the current state of Boudenib and to develop strategies for its preservation and sustainable management.

Over the past decade, there has been increasing interest among scientists in using Earth observation satellite data to monitor land surface parameters [7]. Remote sensing (RS) is particularly valuable in regions where field data are scarce or unavailable, such as the Boudenib Oasis. Remote sensing involves detecting and monitoring an area's characteristics from a distance, typically using satellites or aircraft. In countries with limited access to real-time data, RS provides a vital alternative for gathering information across various domains. It offers substantial insights for mapping and assessing environmental challenges, especially in areas facing water scarcity and desertification, which are among the most significant threats in Morocco, particularly in the oases of the southeastern part of the country [8].

While RS data have been widely utilized for vegetation monitoring, particularly in the Middle East to assess water quality and the effects of salinity on palm tree growth [9], fewer studies have focused on applying spatial tools and satellite imagery to monitor spatiotemporal dynamics in arid regions like Boudenib. Notable studies in this context include long-term vegetation change detection using Landsat data from 1974 to 1999 and the monitoring of land cover in the Zagora Oasis from 1984 to 2015 [8].

The potential of remote sensing in vegetation monitoring and water resource management has been widely acknowledged. Research in crop water parameters using RS techniques has been growing since the 1970s [10], with applications related to water consumption and irrigation water supply for crop growth. Additionally, advances in geoinformatics software have provided users with sophisticated tools to handle large datasets, further enhancing the ability of remote sensing to monitor the extent of irrigated vegetated areas [11].

Previous studies on vegetation monitoring in Morocco's oases have primarily relied on Landsat imagery and GIS-based software for image processing [12]. To overcome the lack of field measurements in the region, this study will use remote sensing to monitor key vegetation parameters, specifically utilizing the PROBA-V Belgian satellite. Launched in 2013, PROBA-V provides daily, medium-resolution imagery with spatial resolutions of 100 m, 300 m, and 1 km, and is focused on land use and crop monitoring (https://earth.esa.int/eogateway/missions/proba-v, accessed on 21 January 2025). The TREX algorithm, developed at VUB, will be employed to automatically process these satellite images into time series of NDVI (Normalized Difference Vegetation Index) and LAI (Leaf Area Index) [13]. This tool has not yet been applied to monitor palm trees, making this study a novel application of PROBA-V data in the context of Boudenib Oasis.

While Sentinel-2 and Sentinel-3 satellites offer higher spatial resolution and more spectral bands, PROBA-V's combination of medium resolution, high temporal frequency, wide swath width, and optimized spectral bands make it an ideal choice for vegetation monitoring in oasis environments. Its near-daily revisit time allows for frequent land observations, making it particularly well-suited for large-scale, long-term trend analysis. PROBA-V's optimized spectral bands—Blue, Red, NIR, and SWIR—are specifically designed for deriving vegetation indices like NDVI and LAI, which makes it a better fit for this research compared to the more complex datasets of Sentinel products that require additional image processing.

The choice of PROBA-V is driven by its mission objectives, which align with the goals of this research: land use and crop monitoring. The 100 m resolution strikes a balance between spatial detail and coverage, allowing us to observe vegetation changes across Boudenib Oasis. Given the region's low landscape contrast, green pixels correspond to palm tree areas, while brown represents desert areas. This provides a clear indication of crop yield, which in turn reflects the quality of dates produced in Boudenib Oasis [8]. This research expands the use of remotely sensed data in arid regions by applying an automated method to process NDVI and LAI maps from PROBA-V. The results are compared with groundwater depth data, using piezometric level time series provided by the area's watershed agency (ABHZGR). The outcome will aid in monitoring crop yield, offering insights into water availability for date productivity, especially during the harvest season.

The analysis focuses on three key parameters: NDVI, LAI, and water availability. Remote sensing data from PROBA-V enables continuous monitoring of the oasis environment, and, notably, few studies have applied PROBA-V data to Moroccan oases. Thus, this research represents a significant contribution, particularly in areas where field measurements are difficult to obtain.

This research will contribute to the existing literature by extending the use of remote sensing in Morocco, particularly in regions like the Boudenib Oasis that have not been previously studied using PROBA-V data. The findings will enhance the understanding of vegetation–water interactions in arid environments, specifically in oasis ecosystems. Moreover, the results will support decision-making processes for sustainable water management and provide a scientific basis for the protection and conservation of this crucial oasis. By linking vegetation health with groundwater data, this study aims to offer practical solutions for improving irrigation practices and safeguarding water resources in the Boudenib Oasis, thereby contributing to the preservation of this important cultural and environmental heritage.

This study uses remote sensing techniques to monitor vegetation dynamics in Boudenib Oasis, specifically employing PROBA-V satellite data to derive vegetation indices like NDVI and LAI. The goal is to assess the health of date palms in relation to groundwater levels. The hypothesis is that combining remote sensing data with groundwater depth measurements will reveal vegetation stress patterns due to water scarcity and environmental pressures. This research aims to improve understanding of water-vegetation interactions in arid environments, offering solutions for better irrigation and water management in the oasis. Additionally, it seeks to provide data for decision-makers to support the protection and conservation of the oasis, vital for the local economy and environmental heritage.

2. Materials and Methods

2.1. Case Study: Boudenib Oasis and Guir Catchment

The palm grove of Boudenib oasis belongs to the Guir watershed (Figure 1), located in the south-east of Morocco with an area of 5 km^2, surrounded by the high Atlas Mountains.

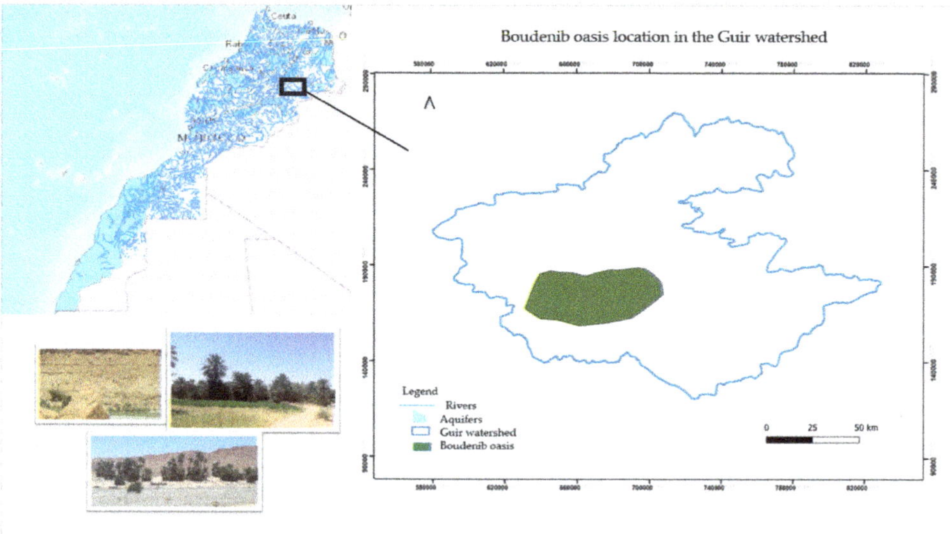

Figure 1. The geography of Boudenib oasis and the Guir watershed. Coordinate unit (m).

The palm grove of Boudenib is situated near the town of Boudenib in the Guir Valley (Figure 1). This oasis is renowned for its diverse flora, structured into distinct layers that serve vital ecological and agricultural roles. The ecosystem is primarily composed of date palms, which thrive in the arid climate, along with other vegetation sustained by traditional oasis irrigation systems.

The date palms of Boudenib provide much-needed shade and produce high-quality dates, which form the cornerstone of the region's economy and the primary livelihood for its population. These palms are still supported by the traditional khettara irrigation methods, an ancient system that channels groundwater to sustain the oasis.

In addition to its agricultural importance, the oasis serves as a habitat for various bird species and wildlife, enhancing its ecological richness. The palm leaves provide shelter from the intense heat, contributing to a dynamic landscape that transitions from lush green to desert brown.

Boudenib itself is a small city of approximately 10,000 inhabitants, known for its agricultural activities and desert tourism. The oasis relies heavily on its water resources for irrigation and domestic use. However, the palm groves have been significantly affected by water scarcity, losing their greenery due to the impacts of recent climate changes in Morocco, including severe droughts and occasional floods.

The Boudenib oasis is characterized by flat to gently undulating terrain, where seasonal streams (oueds) and traditional irrigation systems influence the landscape. The elevation of the area is approximately 1200 m above sea level, which contributes to its semi-arid climate with significant diurnal temperature variations. The soils in the Boudenib oasis are primarily alluvial, deposited by periodic flooding of the Guir River. These soils are generally sandy to silty with moderate clay content, and are low in organic matter. The soil in Boudenib suffers from a significant salinity issue exacerbated by over-irrigation and water evaporation, which leaves salt deposits on the surface.

This area is known for an arid climate characterized by low and irregular rainfall, coupled with significant temperature variations. The mean annual rainfall is 157 mm and mean annual temperature is 25 °C [14].

Boudenib oasis has two main aquifers, a shallow aquifer (Quaternary) and a deep aquifer (Cretaceous) (Figure 2). Groundwater is mainly supplied by temporary surface water infiltration or flooding. These aquifers are therefore a product of irrigation and infiltration in arid areas, but originate from a humid area. The flow of the aquifers varies irregularly from upstream to downstream with an irregular piezometric level [15].

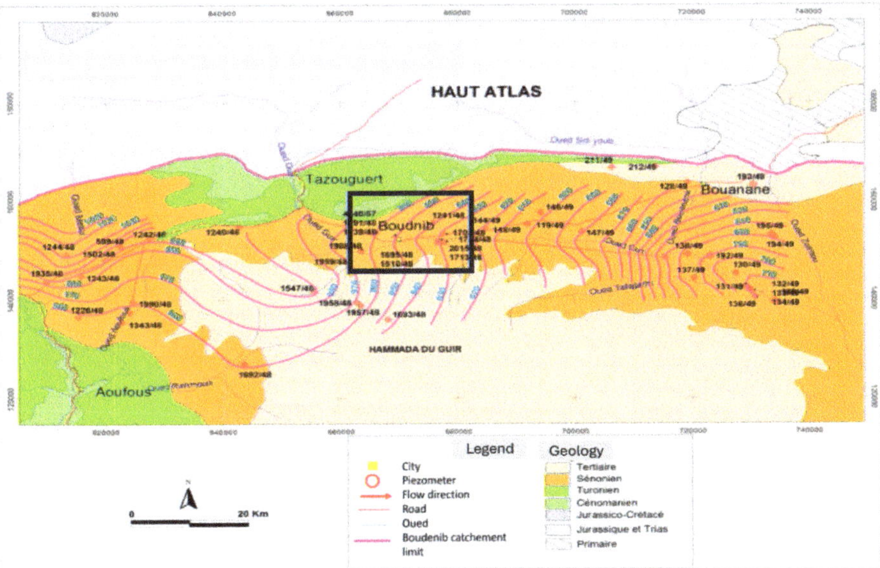

Figure 2. Study area: geological cross section (Boudenib) [15]. Coordinate unit (m).

The hydrographic network of Boudenib oasis is provided by the water catchments of the rivers Guir and Bouanane which flow on into Algerian territory from their origin in the high Atlas Mountains [14].

The Guir catchment occupies an area of 19,000 km^2 in which there are several hydrometric and climatological stations (around 20 stations) for the assessment of surface water resources. The basin has historical rainfall potential of 400 mm/year, and 5% of its area is cultivated [1].

2.2. Field Visit and Data Collection

This study aims to evaluate groundwater dynamics by conducting field measurements and analyzing water quality.

The field program was designed to obtain precise groundwater depth measurements, collect hydrological data from the Catchment Agency ABH-ZGR of Errachidia, identify dry wells, and perform chemical analyses of groundwater in the Guir watershed. Water sampling was systematically conducted for each well, with two samples per well collected in specialized bottles to ensure proper preservation for laboratory analysis. The results of this study will contribute to a better understanding of groundwater availability and quality, providing valuable insights for sustainable water management in the region.

2.2.1. Water-Depth Measurements in Boudenib Oasis and Water Quality

The Boudenib oasis (Figure 1) is characterized by the presence of groundwater resources containing different types of geological sediments and formation (clay, sand and sandy rock).

The piezometric measurements (Figure 3) conducted on these aquifers provide insights into the temporal evolution of groundwater levels in both shallow and deep aquifers. These measurements also facilitate the identification of dry wells and contribute to monitoring both the quantity and quality of groundwater resources, which are essential for local water supply. The black dots indicate the wells where water depth measurements were recorded during the field survey. Groundwater levels were manually measured using an electric probe meter, which emits a signal upon contact with the water surface, ensuring accurate depth assessment.

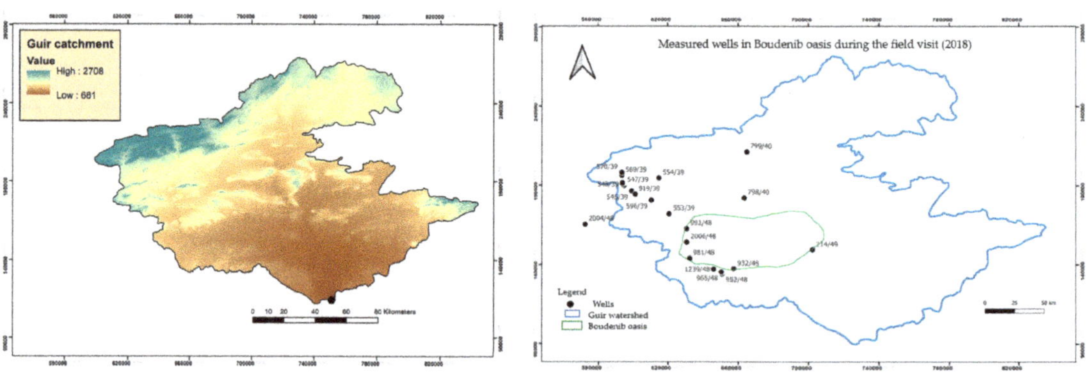

Figure 3. Wells where the piezometric measurements were taken (Boudenib oasis and Guir watershed). Digital elevation model (DEM) of the Guir catchment (https://earthexplorer.usgs.gov/, accessed on 21 January 2025). Ordinate unite (km).

Table 1 provides an overview of the water quality in the study area, based on laboratory chemical analyses of two samples collected from each well. The analysis was limited to the wells listed in the table, as the remaining wells shown in the map (Figure 3) were either dry due to the seasonal aridity during the summer or inaccessible for sampling. These constraints highlight the challenges associated with data collection in the region, particularly under seasonal and logistical limitations.

Physicochemical analyses were conducted on water samples collected from wells to assess the water quality in the region. The results indicate that the water quality is generally good and suitable for various uses, including drinking, irrigation, and domestic purposes. However, the groundwater in the region tends to have elevated levels of salinity, while maintaining a neutral to alkaline pH, which is typically favorable for agricultural and domestic use.

It was concluded that the water resources in the area are generally safe for irrigation after basic filtration or treatment. However, the high salinity levels can negatively affect soil fertility and crop yield, particularly for salt-sensitive plants. Therefore, careful management is needed to mitigate these impacts on agriculture.

Table 1. Table of water quality results June/2018.

Well ID.	PL-06/2018 (m)	PL-05/2018 (m)	Water Depth (m)	Water Quality	Salinity mg/L	pH	T°C
1029/48	17.61	17.54	18.28	Drinking water	2	7	23
952/48	6.11	5.71	7.05	Drinking water	4.5	7.5	22
965/48	** dry	** dry		**			**
969/48	5.41	5.17	6.92	Drinking water	3.9	8	21
932/48	5.3	5.15	5.95	Polluted water	3.7	7.5	22
981/48	9.6	9.27	10	Drinking water	5	7	22
992/48	11.83	11.67	13.35	Drinking water	3.4	7.4	23
553/39	12.34	12.28	13.92	Drinking water	4.8	7.5	22
581/39	8.97	9.28	16.86	Drinking water	3	8.5	21
596/39	18.09	18.25	18.3	Drinking water	5	8	21
554/34	13.29	13.52	17.06	Polluted water	3	7	20
569/39	11.6	11.59	15.66	Drinking water	4	7.5	18
798/40	17.26	16.91	20.26	Drinking water	2.9	8	19
214/49	** dry	** inaccessible	12.26	Drinking water			**

Table legend: **: dry well or inaccessible well. Drinking water: Water with good quality (matching Moroccan water norms for drinking). **PL:** Piezometric level. **T°:** Temperature.

2.2.2. Aquifers Delineation

Data obtained from ABHZGR-Errachidia identified two aquifers within the Guir catchment area (Figure 4). These aquifers primarily belong to the Cretaceous and Quaternary geological periods, providing valuable groundwater resources for the region; however, Boudenib oasis is primarily located over the Quaternary aquifer which is relatively recent in geological terms and formed by the deposition of alluvial materials like sand and clay. This aquifer is the dominant source of groundwater in the Boudenib oasis.

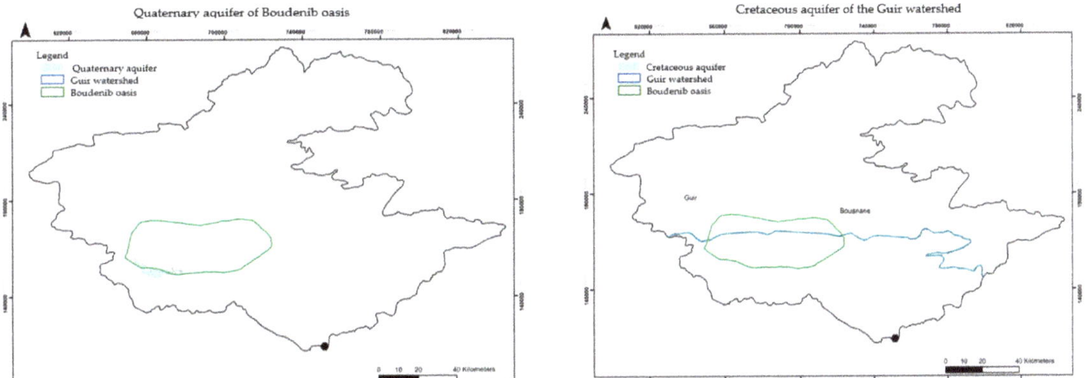

Figure 4. Aquifers in the Guir basin.

Figure 5 illustrates the groundwater measurements for May and June 2018 conducted during the field visit, revealing a slight variation in the water level of the Boudenib oasis, ranging from 2 to 8 cm in June. This fluctuation can be attributed to the high temperatures during this period, which often exceed 40 °C, leading to rapid water evaporation and, in some cases, dry wells.

larger-scale vegetation patterns, aligning with its primary mission objectives of land use and crop vegetation monitoring. This resolution is well-suited for monitoring the state of vegetation in the Boudenib oasis. Furthermore, the 100 m resolution provides a good balance between spatial detail and coverage. It is sufficient to monitor palm trees due to the low land-use contrast in the area, where green typically represents palm tree zones and brown indicates desert.

LAI can be used to predict primary productivity and crop growth and it has an important influence on exchanges of energy, water vapor and carbon dioxide between plants and the atmosphere. This index can be easily derived from NDVI or other equations (Figure 6) [21].

Various applications of NDVI and LAI require image processing and application of mathematical functions on satellite image bands. This may look easy in terms of single images, but when we have a monthly variability of NDVI we might easily have issues with manually processing big datasets. It is also a time-consuming operation.

Pixels of PROBA-V images after re-projecting from original degree units into metric system are reshaped into rectangles, thus enabling their use for multi-spatial analysis, through the improved utilization of GDAL software 2.2.2 and programming in Python (2.7/3.X). Based on initial user inputs, this tool can automatically check radiometric quality, select cloudless images, re-project, resample, clip and adjust NDVI products of 100 m PROBA-V images. After specifying the reference raster of the shapefile, this tool can produce the following:

- NDVI raster maps in .tif format for each input image;
- Monthly NDVI raster maps in .tif or .asc format;
- LAI raster maps in .tif or .asc format for each input image;
- Monthly LAI raster maps in .tif or .asc format;
- Interpolated raster LAI maps in .asc format;
- Color representation of LAI maps in .jpg format;
- LAI time series extracted for specific points in .csv format;
- Color representation of LAI time series in .jpg format [13];
- Color representation NDVI maps and time series.

The easiest way to set up the tool is to install Anaconda for Python 2.7.14, which provides the Spyder Python environment, managing the installation of Python packages and handling environmental variables for Windows. This allows us to do the following:

- Check radiometric quality with PROBA-V status map (SM) and exclude from processing images with too many invalid pixels;
- Select cloudless images with automatic resampling, reprojection and clipping to the exact extent of the study area;
- Convert images from digital values to physical values;
- Re-project maps based on the user input;
- Resample and filter maps to a given grid based on user input;
- Clip and separate out regions based on user input;
- Deliver NDVI maps in .asc or .tif format;
- Compute LAI maps from NDVI maps;
- Simulate values for missing pixels if enough information is available;
- Deliver 3 types of LAI maps in .asc, or .tif: LAI maps based on 5 days synthesized NDVI maps and monthly maps of average LAI with or without simulating missing data;
- Extract time series of NDVI or LAI for selected points and save in .csv format (readable by MS Excel);
- Prepare and save in .PNG format visualization of LAI maps.

The radiometric quality of NDVI maps with the PROBA-V SM allows quick filtering out of those maps which contain too many invalid pixels (clouds). GDAL in TREX consists of translating digital numbers (DN) in PROBA-V to physical values (PV) [11].

$$PV = ((DN - offset))/scale \tag{1}$$

PV—a value of NDVI;
DN—digital number ranging from 0 to 250 assigned to each pixel;
offset and scale—take constant values of 20 and 250, respectively.

TREX always processes all maps to match the exact properties of the input reference (Figure 7) such as raster extent, masked values, projections, or spatial resolution. Using the 'Reference Map' as input, all selected images are clipped to the area of interest, potentially reducing the subsequent processing time and power needed significantly. The leaf area index (LAI) is computed from NDVI using the following Equation (2) [19].

$$\sqrt{(NDVI \times \frac{1+NDVI}{1-NDVI})} \tag{2}$$

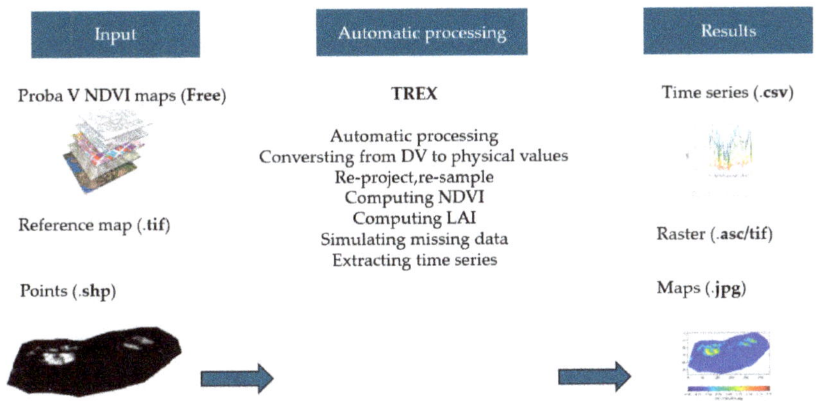

Figure 7. TREX processing chain (https://github.com/VUB-HYDR, accessed on 21 January 2025).

LAI Equation (2) is applied to NDVI maps assisted by TREX which produce LAI values above 0. Monthly maps of LAI are generated based on the aggregation of five-day synthesized maps whenever data is available. When TREX completes processing of NDVI and LAI maps, the tool generates time series of the same parameters (Figure 7).

There are three types of input data:

1. Reference map: Contour map or shapefile of Boudenib oasis, in .tif format. The extension, resolution, projection and masked areas within this file will be used for processing raw PROBA-V images, which means that all properties of the output map will match with this reference map. Size and resolution of the reference map are only limited by the processing capacity of the local machine;
2. Reference point: in a SHP format. Each point should have an attribute specified named 'ID' which could be any number or name. The geographic projection system of the reference map and the reference point must be similar;
3. Free PROBA-V NDVI images obtained from the website (https://www.vito-eodata.be, accessed on 21 January 2025), with a resolution of 100 m products specified as 'S5-TOC—5-daily global composites, Top-Of-Canopy'. Each NDVI image downloaded from the website should be associated to a status map (SM).

The following scripts were applied to these input data:

TREX_dowload.py: this script is used for downloading satellite images from an FTP server provided by VITO, i.e, PROBA-V NDVI images with the status map (MP); **TREX_processing.py**: this script calls GDAL functions for processing raster data, checks radiometric quality of images, discards clouded images, converts to physical values and produces maps in .tif or .asc format suitable for ARCGIS and QGIS; **TREX_visualisation.py**: this script is focused on producing secondary output time series or color representation of processed maps.

The TREX scripts have been structured, incorporating cloud and atmospheric condition handling. The scripts automatically filter out maps with dominant cloud cover and apply cloud removal techniques for cases where cloud presence is minimal, ensuring cleaner and more reliable vegetation indices.

2.4. TREX Inputs and Application in Boudenib Oasis

Three types of input data for TREX were complemented with meteorological, hydrological, and agricultural data to link vegetation characteristics to water availability, water management and crop health:

1. PROBA-V satellite images of NDVI (and corresponding SM), five-day synthesis, TOC correction for the period April 2014–October 2018 (52 images for the year 2014 and 72 for each year 2015, 2016, and 2017), with a spatial resolution of 100 m;
2. A reference map of Boudenib oasis, with a spatial resolution of 30 m, and a size of 2×10 km;
3. A point map with several GPS points of palm trees and desert;
4. The time series of meteorological data (temperature and precipitation) and hydrogeological data obtained from La Diréction génerale de la Météorologie de Rabat, and ABH Errachidia;
5. Agricultural data about dates productivity in Morocco obtained from Ministére d'Agriculture du Maroc.

The first three data sources were used as geospatial references to extract NDVI maps from the PROBA-V satellite dataset and reshape them automatically using the oasis contour map. Only these input data were required for executing the TREX script and producing NDVI/LAI maps or deriving time series of NDVI/LAI for selected points. Meteorological data (4) and agricultural data (5) were only used for analyzing the outcomes and data validation.

These three types of input data are required to execute TREX scripts for producing NDVI/LAI maps and time series for selected points for vegetation monitoring, the evolution of the vegetation in an oasis environment, spatial and temporal variability of some physical parameters, and the prediction of some natural hazards such us drought periods and erosion. The combination of TREX results and hydrological modelling helps in prediction of natural hazards and future climate change scenarios in the area.

2.5. Statistical Analysis and Correlation Methodoloy

The Spearman correlation coefficient is a non-parametric statistical method used to measure the relation and the direction of the relationship between two variables. This is unlike the Pearson methodology which assumes a linear correlation and normal distribution of data; therefore, Spearman is ideal for nonlinear or monotonic relationships. This analysis evaluates whether there is an increase or decrease in one variable and whether it consistently corresponds to an increase or decrease in the other variable. The coefficient varies between a range of -1 (which corresponds to negative correlation) and 1 (which corresponds to positive correlation). The value 0 indicates that there is no correlation [22].

This method is used in different fields like ecology and hydrology, where data may not meet the assumptions of parametric tests.

The statistical analysis was conducted using the R programming language after preprocessing the datasets for both groundwater levels and LAI time series in order to obtain averaged PL measurements and averaged LAI values for the correlation analysis. The data preparation process involved the following steps: data cleaning and transformation, data analysis, statistical correlation analysis, and the correlation matrix using the Spearman coefficient [23].

Data cleaning and transformation: A critical step in statistical analysis, this helps to ensure that the data are accurate and formatted correctly, improves data quality and prepares the dataset for analysis. This step involves many actions, such as handling missing data by replacing the missing values with average, mean, median or mode, removing rows/columns with excessive missing values, removing duplicates to avoid skewed results, correcting errors such as out-of-range values, and finally standardizing the data by ensuring the consistency of units, formats, and naming conventions. Data transformation includes scaling, encoding categorical variables, and data aggregation to combine multiple data points into summary metrics. In this study, averaged values were used for both variables.

Data analysis: This refers to the process of examining and interpreting the data to uncover meaningful patterns, trends and relationships or insights. It involves applying statistical methods and techniques to transform raw data into actionable information that can support decision-making and validate hypotheses.

Statistical correlation: This describes the strength and direction of the relationship between two variables and quantifies how changes in one variable are associated with the other. The correlation can have several interpretations: Positive Correlation: as one variable increases, the other also increases; Negative Correlation: as one variable increases, the other decreases; and finally, No Correlation: there is no discernible relationship between the two variables [24]. To decide on the on the sign of the correlation we have to create a correlation matrix using the Spearman method.

Analyzing the relationship between groundwater levels and the Leaf Area Index (LAI) is crucial for understanding the effects of fluctuating water availability on vegetation dynamics. Groundwater levels, measured through piezometry, indicate the availability of water resources, while LAI quantifies leaf area density, serving as a vital metric for plant growth and ecosystem health. Identifying correlations between these two time series provides valuable insights into the ecological and hydrological interactions in the region.

Spearman's correlation coefficient is a statistical measure of the strength of a monotonic relationship between paired data. Its interpretation is similar to that of Pearson's method, i.e., the closer the correlation is, the stronger the monotonic relationship. Correlation is an effect size, so we can verbally describe the strength of the correlation using the following guide for absolute values, as follows:

- 0.00–0.19 "very weak"
- 0.20–0.39 "weak"
- 0.40–0.59 "moderate"
- 0.60–0.79 "strong"
- 0.80–1.0 "very strong"

Note that, unlike Pearson's correlation, there is no requirement for normality or monotonic data; the relationship between data when using Pearson's method needs to be linear [25].

The Spearman correlation coefficient ranges between -1 and 1. A value close to 1 indicates a strong positive monotonic correlation, while a value close to -1 indicates a strong negative monotonic correlation (Figure 8). A value close to 0 indicates an absence of monotonic relationship. This matrix summarizes the correlation between the wells located in the same area of LAI surfaces. This method enables the analysis of correlations across a wide range of time series data, providing a general overview of their relationships. Alternatively, a more focused approach can be applied by considering only a specific number of wells for each corresponding surface. In this case, a yearly timestep was chosen to study this correlation.

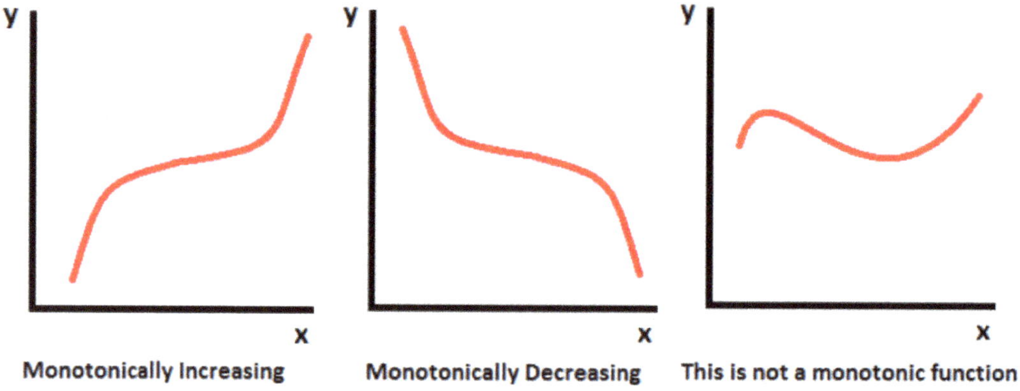

Figure 8. Monotonic data representation [22].

2.6. Limitation of the Study

The data used in this study to characterize crop growth of palm trees in the Boudenib oasis were derived from PROBA-V satellite products with a 100 m resolution. In this study, we did not include a dedicated section for in situ data validation due to the lack of direct field measurements for NDVI and LAI. Instead, our validation approach relied on statistical analyses to establish the relationship between groundwater levels (GW) and LAI trends. Since groundwater depth and date yield are officially measured and recorded by government agencies, we used these datasets to assess the consistency of our TREX-derived NDVI and LAI maps. By applying statistical correlation analyses, we confirmed that both groundwater levels and LAI exhibited similar temporal patterns, supporting the reliability of our remotely sensed vegetation indices. While field validation remains an important future step, our current approach provides a robust preliminary assessment of the data's accuracy in capturing vegetation dynamics in the Boudenib Oasis. Furthermore, PROBA-V provided valuable and reliable dataset inputs for our analysis; however, its relatively coarse spatial resolution imposed limitations on accurately capturing fine-scale vegetation variations and dynamics, particularly in heterogeneous arid environments.

The spatial resolution of remote sensing data plays a critical role in the accuracy and reliability of the results, particularly when characterizing crop growth and land cover. In this study, 100 m resolution provides a balance between spatial detail and the ability to cover large areas. However, the influence of this resolution on the study's findings warrants further analysis. Higher spatial resolution data could offer more precise measurements of vegetation characteristics, potentially revealing finer-scale patterns of crop growth. On the other hand, coarser resolution data may lead to generalized or averaged results, potentially overlooking small-scale variations in vegetation. A comprehensive analysis of how spatial resolution impacts the interpretation of NDVI and LAI trends in relation to groundwater levels will be crucial for understanding the limitations of the current approach. Discussing these factors will provide greater context for the study's conclusions and highlight areas for improvement in future remote sensing-based research.

Additionally, it is necessary to expand the discussion on potential sources of error in remote sensing analysis and their impact on result interpretation. Factors such as atmospheric disturbances, sensor calibration, and cloud contamination can influence NDVI and LAI calculations. To mitigate these issues, TREX was specifically chosen over other available NDVI/LAI processing methods due to its automated cloud detection and removal capabilities, as well as its ability to process time-series data efficiently. TREX's integration with Python and GDAL ensures reliable data handling and minimizes errors caused by atmospheric interference, making it a suitable choice for long-term vegetation monitoring in arid environments like Boudenib Oasis.

3. Results and Discussion

3.1. Spatial and Temporal Variation of NDVI and LAI

The results obtained using the TREX scripts provide a detailed assessment of vegetation dynamics over the study area, with a monthly temporal resolution spanning from 2014 to 2017 and a spatial resolution of 100 m. This specific time frame was selected based on the availability of satellite and observational data, ensuring a consistent and reliable dataset for analyzing vegetation trends. Figures 9 and 10 present the spatial and temporal variations of two key vegetation indices: the Normalized Difference Vegetation Index and the Leaf Area Index (LAI). NDVI is widely used as an indicator of vegetation health, biomass density, and photosynthetic activity, whereas LAI quantifies the total leaf surface area per unit ground area, providing insights into canopy structure and productivity [26]. By analyzing these indices, we aim to assess seasonal fluctuations, interannual variability, and their relationship with water availability.

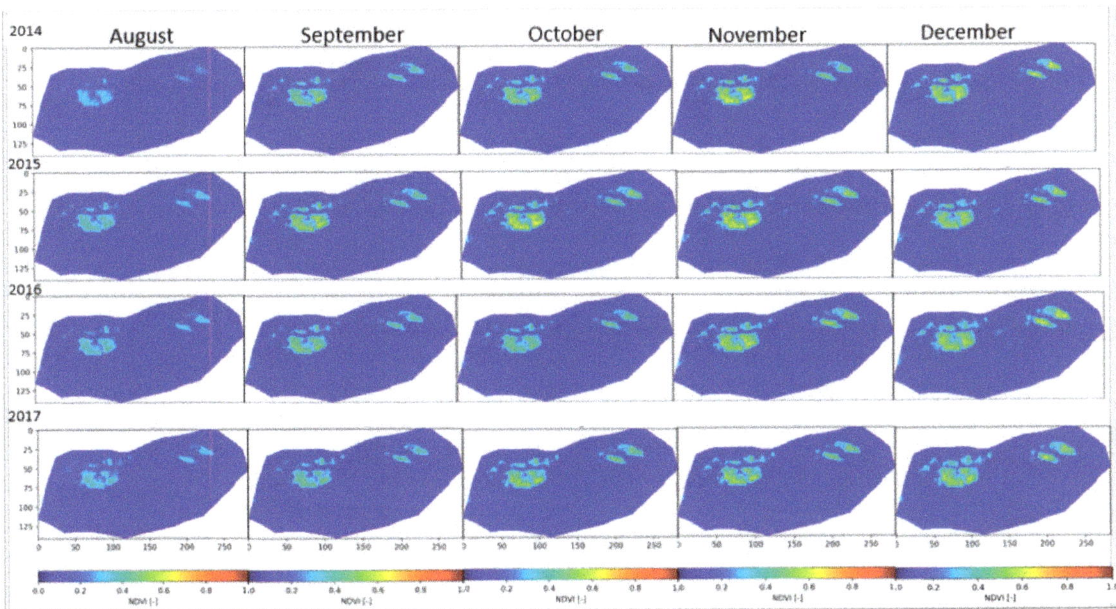

Figure 9. NDVI monthly aggregation variability (August 2014–December 2017).

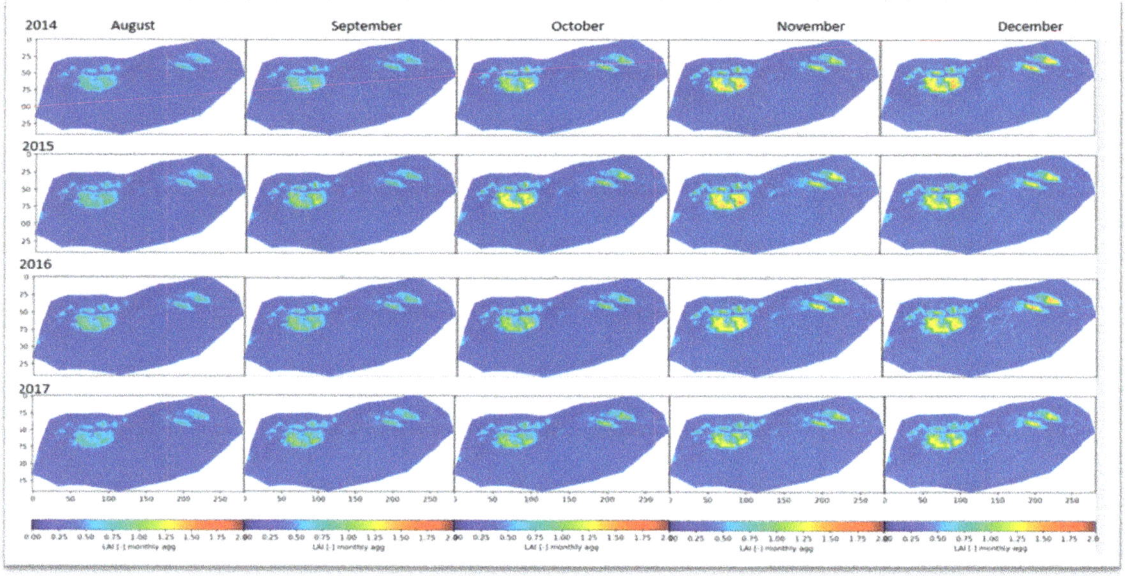

Figure 10. LAI monthly aggregation variability (August 2014–December 2017).

NDVI analysis:

Figure 9 highlights significant spatial and temporal variability in NDVI values within the palm tree areas, as reflected by changes in color brightness. According to the color scale bar, regions with blue indicate low NDVI values, suggesting sparse or stressed vegetation, while areas shaded in yellow to red correspond to high NDVI values (~1), representing dense and healthy vegetation cover.

Although the overall annual NDVI pattern remains relatively stable, notable monthly variations can be observed. Higher NDVI values (yellow to red) are most prominent during October and November, coinciding with the peak growth phase of palm trees. In Morocco, this period aligns with the date palm harvest season, reflecting enhanced photosynthetic activity and optimal canopy development. The increase in NDVI during these months suggests favorable environmental conditions, including adequate moisture availability and optimal temperatures, which support vigorous vegetation growth.

Conversely, lower NDVI values, particularly in drier months, indicate moisture-stressed vegetation, which may be linked to seasonal drought conditions, water scarcity, or physiological stress in palm trees. These fluctuations underscore the sensitivity of NDVI as a key indicator of vegetation health, providing critical insights into the ecological and climatic factors affecting palm tree development. The ability to monitor NDVI variability over time is essential for assessing crop health, optimizing irrigation strategies, and predicting potential yield fluctuations in date palm plantations.

LAI analysis:

LAI varies with canopy architecture, which is influenced by geographic conditions, environmental factors, and agricultural practices. As illustrated in Figure 10, LAI exhibits seasonal fluctuations, with the lowest values recorded from July to September. This decline can be attributed to increased evapotranspiration, heat stress, and potential leaf senescence during the peak summer months. However, as the harvest season approaches in October and November, LAI progressively increases with a value of 1.25, indicating enhanced leaf development and photosynthetic activity. This seasonal variability reflects the growth cycle of date palms, where optimal canopy expansion before harvest plays a crucial role in fruit development.

While higher LAI generally contributes to improved date yield by maximizing photosynthetic capacity, excessive canopy density can negatively impact fruit quality and quantity. In late summer, when LAI is high, dense foliage may restrict light penetration to lower canopy layers, leading to suboptimal photosynthesis and uneven fruit ripening. This phenomenon is particularly pronounced in September, where shading effects may reduce the accumulation of sugars and essential metabolites in the fruit, ultimately affecting market quality.

Furthermore, LAI serves as a key indicator of leaf health and physiological efficiency, as it directly influences carbon assimilation, photosynthetic performance, and oxygen exchange. Given its strong correlation with crop productivity, LAI variability is a critical metric for assessing date palm yield in terms of both quality and quantity. Effective canopy management, such as strategic pruning, may help optimize LAI to balance light distribution and fruit development, ensuring both high yield and superior fruit quality.

LAI profile using the coordinates of the extraction points (2014–2017):

The variability in the Leaf Area Index (LAI) profile across the extraction points (Figure 11) provides valuable insights into the evolution of this key eco-physiological parameter. Palm trees, which predominantly grow along streams and irrigated areas, rely on water sources supplied by geological faults and structural features originating from the High Atlas Mountains. Consequently, points 1 through 6 exhibit significant dynamic variability in LAI, reflecting the favorable conditions for palm tree development in these regions. In contrast, points 7, 8, and 9, located in a desert area, show LAI values approaching ~0, indicating minimal vegetation due to arid conditions. These findings underscore the critical role of LAI in influencing palm tree yield, both in terms of quality and quantity, and highlight its sensitivity to water availability and environmental conditions.

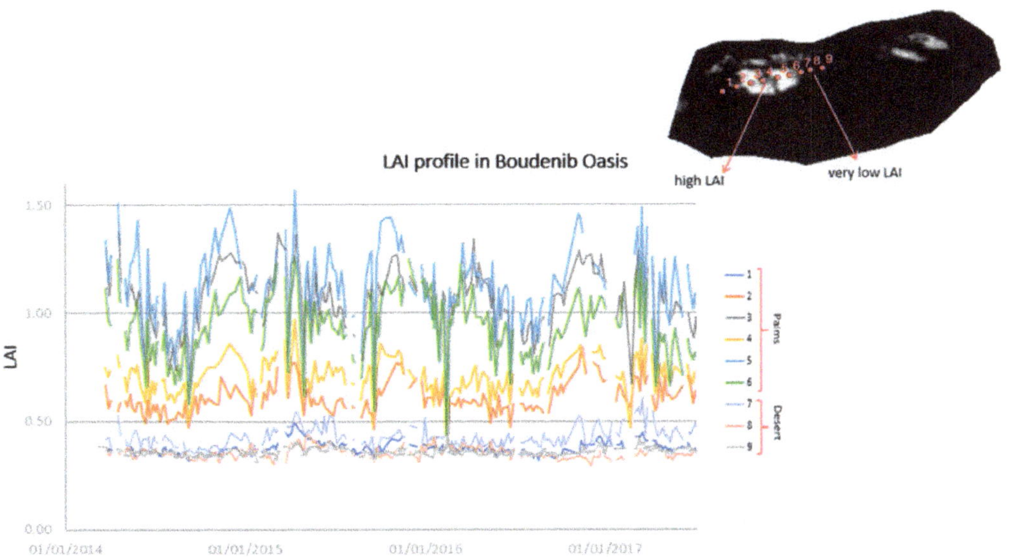

Figure 11. LAI time series of the extraction points (2014–2017).

3.2. *Statistical Analysis Proving the Correlation Between LAI and the Piezometry of Boudenib*

Analyzing the relationship between groundwater levels and leaf area index (LAI) is essential to understanding the impact of changing water levels on vegetation monitoring.

Piezometry reflects groundwater availability, while LAI measures leaf area density, which is a key indicator of plant growth. Understanding how these two time series are correlated can provide valuable insights into ecological and hydrological dynamics and understanding.

This statistical analysis was done following the steps below:

Data preparation: The databases for both water levels and LAI time series were cleaned and preprocessed. This step (Tables 2 and 3) involved transforming all non-numeric values into numerical formats and addressing missing data by replacing empty fields with mean or median values. This approach ensures the dataset remains meaningful and usable for analysis while preserving the integrity of the original data.

Table 2. Example of final dataset after computing and preparing the original database (Piezometry) for the period (2014–2017).

Region	Data	Variable	Effective Value	Valid Values		Non-Existent Values		Average (m)		Median (m)		Standard Deviation	
				Before	After	Before	After	Before	after	Before	After	Before	After
Boudenib	Piezo level	1107/48	37	36	37	1	0	30.05	30.13	28.83	29.28	4.28	4.32
Boudenib	Piezo level	1201/48	37	29	37	8	0	14.10	14.39	13.78	14.27	1.73	1.81
Boudenib	Piezo level	1203/48	37	29	37	8	0	27.89	28.12	28.44	29.51	5.46	6.13
Boudenib	Piezo level	1204/48	37	33	37	4	0	26.53	26.64	26.55	26.55	2.40	2.45
Boudenib	Piezo level	1206/48	37	29	37	8	0	14.74	15.03	14.38	14.92	1.85	1.93
Boudenib	Piezo level	1207/48	37	29	37	8	0	27.46	28.93	25.65	29.44	6.23	6.19

Table 3. Example of final dataset after computing and preparing the original database (LAI) for the period (2014–2017).

S	SURFACE	X	Min	1st_Qrt	Median	Mean	3rd_Qrt	Max
S1	SURFACE	LAI	0.34	0.36	0.38	0.40	0.42	0.46
S2	SURFACE	LAI	0.52	0.57	0.61	0.62	0.65	0.78
S3	SURFACE	LAI	0.79	0.96	1.07	1.06	1.16	1.27
S4	SURFACE	LAI	0.61	0.67	0.72	0.72	0.75	0.83
S5	SURFACE	LAI	0.79	1.06	1.11	1.13	1.21	1.42
S6	SURFACE	LAI	0.64	0.85	0.94	0.94	1.04	1.17
S7	SURFACE	LAI	0.36	0.41	0.43	0.43	0.45	0.54
S8	SURFACE	LAI	0.32	0.34	0.35	0.35	0.36	0.41
S9	SURFACE	LAI	0.34	0.35	0.35	0.36	0.37	0.39

Data analysis: These graphs show the different evolution patterns obtained after preparing the piezometry data of the wells in the Boudenib oasis (the Y axis represents the well ID). Interpolation was employed to estimate missing values based on neighboring data points, ensuring a continuous graph that preserves the overall trend and direction of data evolution. This step was crucial for analyzing the patterns of LAI and PL, providing valuable insights to support the statistical analysis.

Figure 12 illustrates the evolution of groundwater levels (piezometric levels) in the Boudenib oasis from 2014 to 2017, revealing key trends and patterns. Most graphs display gradual declines in water levels over time, indicating the depletion of groundwater resources due to recurring environmental and climatic factors, such as rising temperatures and reduced precipitation. However, some plots exhibit stable or slightly increasing trends, which can be explained by seasonal changes linked to rainfall or irrigation cycles, where groundwater recharge occurs during wet seasons. Notably, the piezometric levels show an increase in 2015 for the majority of the wells, likely reflecting temporary recharge events or improved water management practices. Some time series include flat sections, which may indicate periods of no data collection, interpolation, or static conditions where groundwater levels remained unchanged. These findings highlight the complex interplay between climatic variability, seasonal recharge, and anthropogenic pressures on groundwater resources in arid regions. The results underscore the need for sustainable water management strategies to mitigate groundwater depletion and ensure the long-term resilience of the Boudenib oasis. In terms of the statistical significance of trends, we can visually conclude (Figure 12) that some series show persistent increases or decreases over time while others exhibit short-term fluctuations without a consistent trend, and finally, the presence of sharp dips or peaks suggests potential structural breaks or nonlinear relationships.

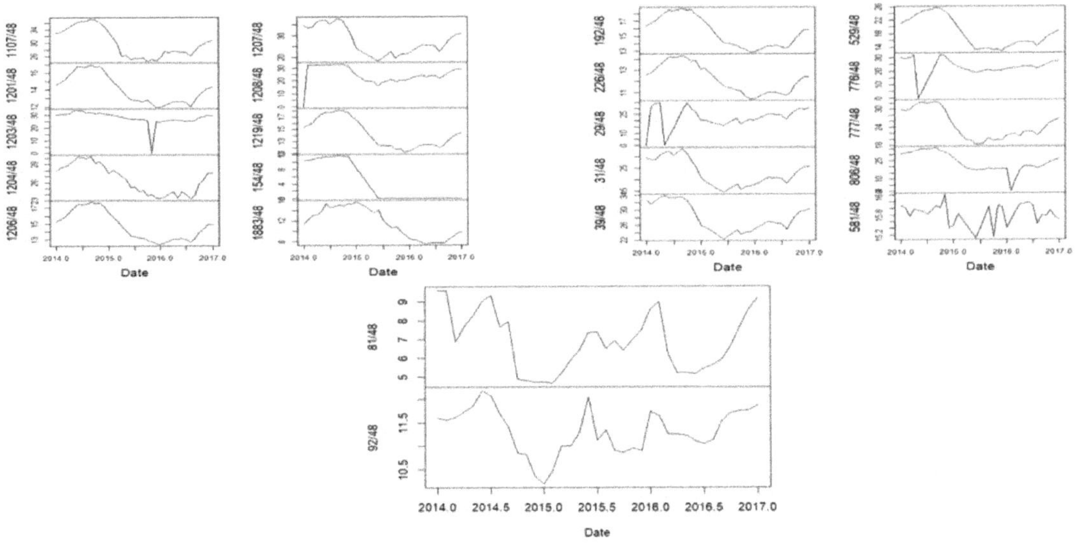

Figure 12. Yearly piezometric level evolution for the period 2014 to 2017 in the wells of Boudenib.

The Leaf Area Index (LAI) exhibited temporal variability across all selected surfaces from 2014 to 2017 (Figure 13). An initial decline in LAI was observed during the first half of 2014, followed by a gradual increase in the second half, reaching its peak in 2015. This pattern demonstrated a consistent cyclical trend approximately every six months, suggesting seasonal influences on vegetation cover. Surfaces S1 to S6 displayed higher LAI values, with noticeable seasonal fluctuations, while surfaces S7, S8, and S9 exhibited consistently lower LAI values due to their extraction points being located in desert areas characterized by sparse vegetation and low LAI. The observed patterns emphasize the impact of both seasonal variations and surface-specific characteristics on LAI dynamics over time. Visually, surfaces S1–S4 show clear seasonal trends, and surfaces S7–S9 display relatively flat patterns, suggesting no significant upward or downward trends, though their variability might still be significant depending on the test used.

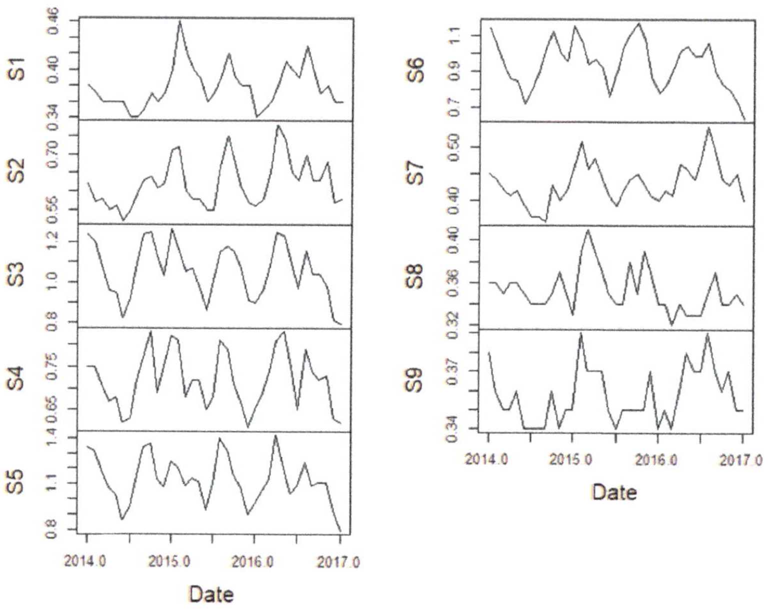

Figure 13. Yearly LAI evolution for the period 2014 to 2017.

Statistical correlation analysis: From the above, we observed that the variations of the piezometric level and LAI depend largely on the factor of time. It emerged that the behavior of these two indicators is strongly linked to the time factor, which determines the period of heavy precipitation and the season of low precipitation in the region of Boudenib. An autocorrelation analysis is used in this study to confirm this aspect.

The question to be answered with this analysis is the following: Are these two indicators linked, and can they impact each other?

The methodology of cross-correlation was chosen to answer the question of the putative link between the piezometric level and LAI. This method consists of calculating the cross-correlation for each value in the piezometric level (X) using a value of the leaf area index (Y). For the correlation analysis, only the wells associated with the LAI surfaces were considered, which means wells in the same location; furthermore, the correlation was executed for the period 2014–2019, with a yearly timestep.

The first step of any correlation analysis of two quantitative parameters with a heterogeneous and diverse time series of data is checking the linearity of the relationship of the data before calculating the correlation. This step ensures that the analysis is relevant and allows the appropriate statistical tools to be chosen to capture the real relationship between the time series.

This study shows that the relationship between the LAI and piezometric level time series is non-linear, as illustrated in Figure 14, below. The data utilized in this model consist of yearly averaged values for both Leaf Area Index (LAI) and Piezometric Level (PL) for the period 2014–2017. This approach effectively reduces the influence of short-term fluctuations and enables a focus on the long-term trends and underlying relationships between these variables.. All results exhibited a similar pattern, indicating the absence of linearity.

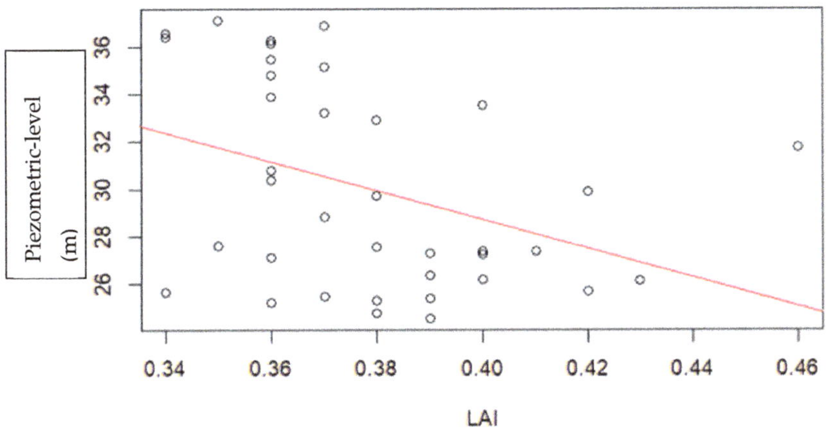

Figure 14. Linear regression model for time series of yearly averaged LAI and piezometric level for the period (2014–2017).

This computation was performed using the Pearson correlation coefficient, and each test generated a *p* value which determined the statistical significance of the observed correlations.

The distribution of colors in Figure 15 looks approximately equal; purple areas indicate a negative monotonic correlation between the piezometric level of certain wells and LAI, which means that when LAI increases the PL decreases; however, orange indicates a positive monotonic correlation which means when the LAI decreases, the PL also decreases, thus confirming the hypothesis of this research.

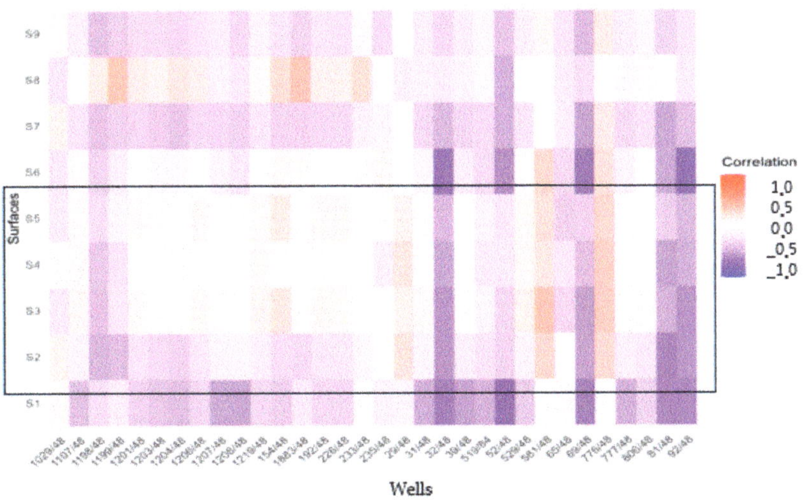

Figure 15. Yearly correlation matrix for piezometry and LAI for surfaces S2,3,4 and 5 (Spearman's method).

The areas with the highest level of LAI say a lot about the palm yield, and contain more data than other surfaces that may be desert or semi-desert; therefore, the surface samples of LAI used for the correlation analysis are S3, S4, and S5, corresponding to the palm trees area, based on the extraction points used previously. For these surfaces, the values of the correlation correspond to a strong positive monotonic correlation; therefore, when the LAI decreases, the PL also decreases, which confirms the hypothesis of an existing link between both parameters. The resulting LAI time series will be compared to time series of groundwater depth, and this comparison will confirm that low LAI is correlated to low PL.

3.3. Relations Between Vegetation Characteristics, Water Availability and Crop Health

LAI vs. Water depth data: The water depth of wells in the Boudenib oasis, as measured during the sampling campaign, does not exceed ±40 m, indicating that the water is primarily extracted from the shallow Quaternary aquifer, which contains fresh, potable water. However, excessive pumping in the region has led to a decline in water availability, adversely affecting irrigation for palm trees. To analyze this trend, monthly averaged time series of piezometric levels (PL) for wells in the Boudenib oasis were obtained from the Watershed Agency (ABHZ) for the period 2014–2017 and compared to monthly averaged time series of LAI for the same period.

The analysis reveals a strong correlation between the Leaf Area Index (LAI) and groundwater levels, as illustrated in Figure 16. A decrease in groundwater levels corresponds to a reduction in LAI values, highlighting the dependence of palm tree growth on shallow groundwater availability, particularly in arid regions. The results indicate that optimal palm tree growth in the Boudenib oasis occurs when groundwater depths are maintained between 8 and 9 m. This range corresponds to the piezometric levels associated with the highest LAI values, underscoring the critical role of groundwater in sustaining palm tree health and productivity in the oasis.

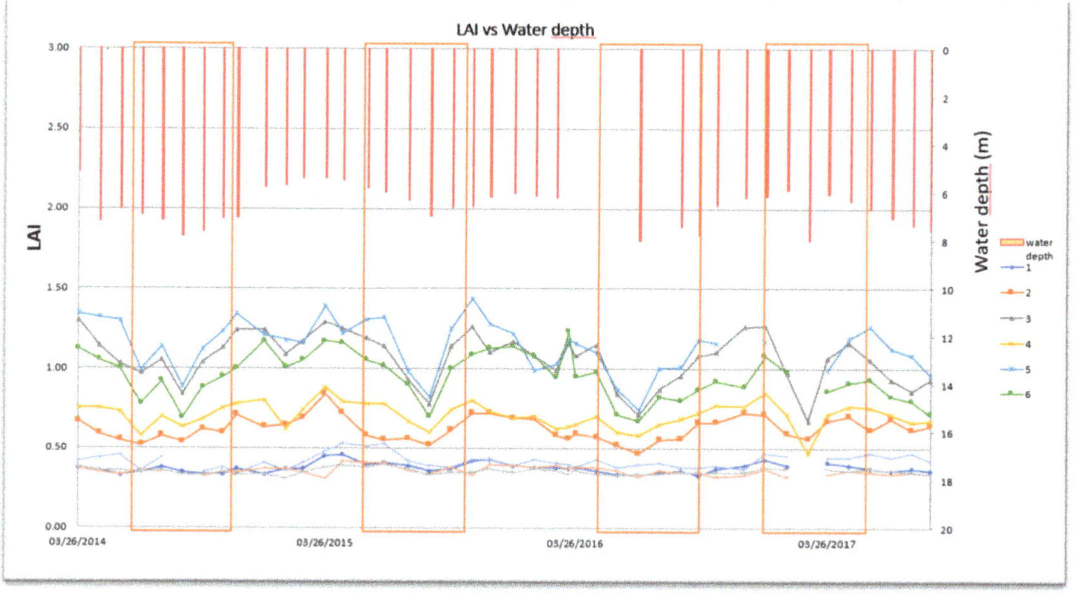

Figure 16. Monthly averaged LAI vs. monthly averaged PL data (2014–2017).

Palm trees are relatively moderate consumers of water compared to other crops, but they rely significantly on groundwater for root development and growth. However, when the piezometric level declines due to factors such as excessive groundwater extraction, prolonged droughts, or climate change, palm trees are forced to extend their root systems deeper to access the receding water table. Research indicates that palm roots can grow to a maximum depth of approximately 9 m [27] beyond which they cannot effectively access groundwater.

The deep rooting system of palm trees enables them to maintain sufficient water supply to their canopies, even under arid conditions. However, when the groundwater table falls below the maximum rooting depth, the vegetation is unable to sustain adequate water uptake, leading to reduced canopy health and productivity. This highlights the critical relationship between groundwater availability and the physiological resilience of palm trees, particularly in water-scarce environments.

LAI vs Dates yield (October 2015/October 2017): In Morocco, the date harvest season typically occurs in October [28]. To assess the relationship between vegetation health and date production, Leaf Area Index (LAI) time series values were compared with date yield data for 2015 and 2017, as these years were identified as significant for date production in the country. In 2015, date production reached a high yield, whereas in 2017, production declined by 12% compared to 2015 (Figure 17). LAI experienced a decline in that same year. This reduction in yield was accompanied by a decline in LAI and quality, which was attributed to several factors including insufficient precipitation, reduced nutrient availability from palm leaves, and the overexploitation of groundwater resources due to excessive pumping—a critical issue in the region.

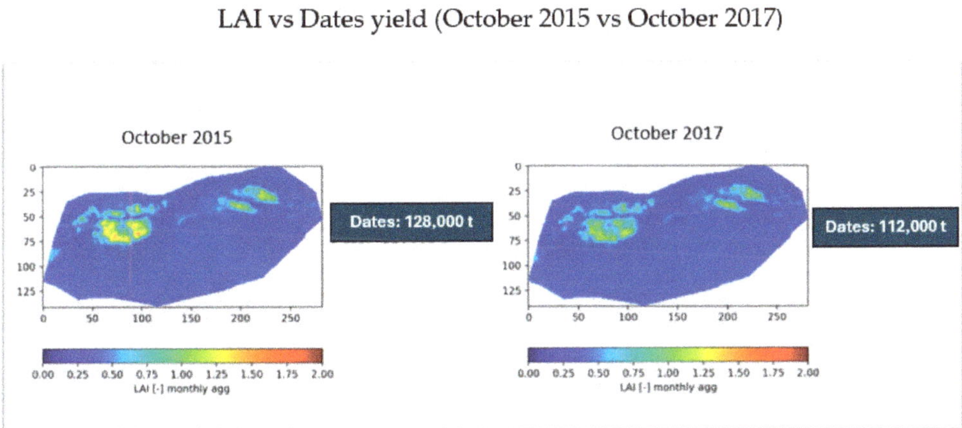

Figure 17. Dates productivity (2015 and 2017).

In response to the unsustainable extraction of groundwater, the Moroccan government implemented regulatory measures to limit water pumping, including restrictions on the volume of water that can be extracted per individual [14]. These measures aim to mitigate the adverse effects of groundwater depletion on agriculture and ecosystems.

The analysis demonstrates that LAI values derived from TREX align with observed trends in groundwater levels and date yields in the study area. These findings highlight the utility of LAI as an indicator of vegetation health and its sensitivity to groundwater availability. However, further validation of these results through additional in situ measurements would strengthen the reliability of the findings and provide a more comprehensive understanding of the interactions between groundwater dynamics, vegetation health, and agricultural productivity.

3.4. General Interpretation and Recommendations

Numerous studies have demonstrated a strong positive correlation between vegetation indices, such as NDVI and LAI, and water availability in arid environments. Annual precipitation is a primary driver of vegetation growth and crop yield, while temperature variations, though less influential, also affect vegetation health and distribution. Human activities, including crop selection, sowing dates, and irrigation practices, further modify vegetation indices. Climate change exacerbates these challenges by increasing aridity and altering precipitation patterns, leading to long-term changes in vegetation cover. This analysis is thoroughly addressed in our ongoing future work, which aims to investigate the impact of climate change on the Boudenib oasis.

Our study of the Boudenib oasis revealed a significant correlation between LAI and groundwater levels, consistent with findings from other arid regions. For example, research in the Namoi catchment, Australia, highlighted the influence of groundwater levels on riparian vegetation, while a study in Ejina, China [29–31], demonstrated that areas with shallow groundwater depths, exhibited higher NDVI values. These findings underscore the critical role of groundwater in sustaining vegetation in arid ecosystems especially in the Moroccan oasis as cited before in previous studies [32].

The observed changes in NDVI and LAI in the Boudenib oasis are likely driven by a combination of natural climatic variability and anthropogenic factors. This analysis is further addressed in our ongoing research, which investigates the impact of climate change on the Boudenib oasis. Integrating our results with existing studies enhances the understanding of the intricate relationship between groundwater resources and vegetation dynamics, emphasizing the importance of sustainable groundwater management for maintaining vegetation cover in fragile arid ecosystems. This comprehensive approach provides a clearer understanding of the factors influencing vegetation changes and informs effective conservation strategies to ensure ecosystem resilience in the face of environmental change.

Based on these results, we recommend implementing sustainable groundwater management practices to prevent resource depletion by agricultural management policies and regulation of groundwater extraction to ensure long-term water availability. This should include promoting water-efficient irrigation techniques such as adopting drip irrigation to optimize water usage and reduce wastage, encouraging the cultivation of drought-tolerant crops, and establishing remote sensing-based monitoring systems for real-time vegetation and water resource assessment. Additionally, reforestation and land rehabilitation initiatives, investment in climate impact research, stakeholder training on sustainable practices, and collaboration with research institutions are essential for enhancing ecosystem resilience in arid regions like the Boudenib oasis.

4. Conclusions

This study demonstrates the potential of integrating remotely sensed data with GIS-based numerical models to monitor and analyze vegetation dynamics in oasis environments, particularly in areas where field access is limited or equipment for instant data collection is unavailable. By leveraging medium-resolution satellite imagery, such as PROBA-V and Sentinel-3, we employed the TREX method to derive key physical parameters, including the Leaf Area Index (LAI) and Normalized Difference Vegetation Index (NDVI), for vegetation monitoring. The integration of these remotely sensed datasets with field observations provides a robust framework for validating the reliability of this approach for future projections and analyses.

Our findings reveal a clear correlation between groundwater depth and vegetation health, as indicated by LAI. Specifically, low LAI values (0.5–0.8) were observed during periods of reduced groundwater availability with a groundwater depth ranging between 3 and 5 m, leading to decreased vegetation density in palm trees and a decline in both the quantity and quality of date production, as noted in October 2015 and October 2017. Conversely, high LAI values (1.25–1.5) were recorded when groundwater levels were elevated, with the piezometric level (PL) reaching 8 m, indicating a strong dependence of palm tree vitality on groundwater availability. These results underscore the utility of medium-resolution satellite imagery for monitoring palm tree health in oasis ecosystems, with significant implications for sustainable water resource management. By linking vegetation monitoring outcomes to groundwater depth, this study provides actionable insights for optimizing water use in arid regions.

However, the relationship between NDVI-derived vegetation indices and underlying groundwater resources remains complex. While automated time-series analysis of remotely sensed data has elucidated correlations between groundwater patterns and vegetation canopy dynamics, further research is needed to refine these findings.

Future research should focus on enhancing vegetation monitoring accuracy by integrating in situ NDVI and LAI measurements with high-resolution hyperspectral imagery. Incorporating land use change analysis will provide a more comprehensive understanding of NDVI and LAI variations, enabling a more precise assessment of vegetation dynamics in response to environmental and anthropogenic factors. Additionally, the transition to Sentinel-3 data should be explored to evaluate its potential for improving long-term vegetation monitoring at a broader spatial scale. The use of higher-resolution hyperspectral datasets is also recommended to refine both spatial and temporal analyses, allowing for more detailed assessments of vegetation health and productivity. These advancements will contribute to more accurate and effective ecosystem monitoring, supporting sustainable land and water resource management strategies. From an ecological perspective, the dual reliance of palm trees on surface water for irrigation and groundwater for root development highlights the need for a balanced approach to water resource management. Protecting groundwater reserves is critical for sustaining root growth, while surface water resources are essential for irrigation and crop nutrition. Achieving this balance is vital for the long-term health of oasis ecosystems.

Finally, from a socio-economic standpoint, the preservation and development of agriculture—central to the livelihoods of oasis communities—must be prioritized. Continuous vegetation monitoring, coupled with data-driven policies such as regulating water extraction rates, can safeguard these ecosystems for future generations. This study not only advances the application of remote sensing techniques in arid environments but also provides a foundation for informed decision-making to promote sustainable development in oasis regions.

Author Contributions: Conceptualization, K.B.; Methodology, K.B. and B.V.; Software, K.B.; Validation, B.V. and A.V.G.; Formal analysis, K.B.; Investigation, K.B.; Resources, K.B.; Data curation, K.B. and B.V.; Writing—original draft preparation, K.B.; Writing—review and editing, K.B. and V.B; Visualization, K.B. All authors have read and agreed to the published version of the manuscript.

Funding: This work was funded by the VLIR-UOS program under the WASIS project VLIR344.

Data Availability Statement: An ethical approval was not required for this research.

Acknowledgments: To Steven Eisenreich, and Boud Verbeiren, for their constant advice that helped in making and editing this manuscript. To Suliga Joanna, James Chawanda and the other colleagues of the HYDR department for their support and help in the methodology and their encouragement in getting me integrated into the department. To the team of ABH-GZR Errachidia, for their time in the field program and providing the input data we needed for this study.

Conflicts of Interest: The authors declare that the research was conducted in the absence of any commercial or financial relationships that could be construed as a potential conflicts of interest.

References

1. Combe, M.; Simonot, M. Ressources en eaux du Maroc: La haute moulouya, le sillon d'Itzer-Enjil et le massif de Bou-Mia-Aouli. *Notes Mémoires Serv. Géologique Maroc* **1971**, *231*, 193–201.

2. Eddahby, L.; Popov, M.A.; Stankevich, S.A.; Kozlova, A.A.; Svideniuk, M.O.; Mezzane, D.; Ibouh, H. Assessing vegetation structural changes in Oasis agro-ecosystems using sentinel-2 image time series: Case study for Drâa-Tafilalet region Morocco. *Int. Arch. Photogramm. Remote Sens. Spat. Inf. Sci.-ISPRS Arch.* **2019**, *42*, 69–73. [CrossRef]
3. Bouaamlat, I.; Larabi, A.; Faouzi, M. Assessment of Water Resources in the Tafilalet Oasis System by a Mathematical Model (South East of Morocco). *Hydrogeol. J.* **2012**, *24*, 1479–1496. [CrossRef]
4. Oufkir, M.A. Changements socio-économiques et recompositions territoriales dans le bassin du Guir (Sud-Est marocain). *GéoDév* **2020**, *8*, 1–10.
5. Rafik, A.; Ibouh, H.; El Alaoui El Fels, A.; Eddahby, L.; Mezzane, D.; Bousfoul, M.; Amazirh, A.; Ouhamdouch, S.; Bahir, M.; Gourfi, A.; et al. Soil Salinity Detection and Mapping in an Environment under Water Stress between 1984 and 2018 (Case of the Largest Oasis in Africa-Morocco). *Remote Sens.* **2022**, *14*, 1606. [CrossRef]
6. Rayne, L.; Brandolini, F.; Makovics, J.L.; Hayes-Rich, E.; Levy, J.; Irvine, H.; Assi, L.; Bokbot, Y. Detecting desertification in the ancient oases of southern Morocco. *Sci. Rep.* **2023**, *13*, 19424. [CrossRef]
7. Bastiaanssen, W.G.M. *Remote Sensing in Water Resources Management: The State of the Art*; International Water Management Institute: Colombo, Sri Lanka, 1998. Available online: https://api.semanticscholar.org/CorpusID:127205500 (accessed on 8 February 2021).
8. Lamqadem, A.A.; Saber, H.; Rahimi, A. Spatiotemporal Changes of Vegetation in the Middle Draa Valley Oasis: A Study Case of M'hamid El Ghizlane Oasis (Morocco). *Eur. Sci. J.* **2017**, *13*, 115–132. [CrossRef]
9. Rashid, N.S.A.; Askari, M.; Tanaka, T.; Simunek, J.; van Genuchten, M.T. Inverse estimation of soil hydraulic properties under oil palm trees. *Geoderma* **2015**, *241*, 306–312. [CrossRef]
10. Hiler, E.; Clark, R. stress Day Index to Characterize Effects of Water Stress on Crop Yields. *Trans. ASAE* **1971**, *14*, 757–761. Available online: https://api.semanticscholar.org/CorpusID:110532365 (accessed on 3 March 2019).
11. Aldakheel, Y.Y. Assessing NDVI Spatial Pattern as Related to Irrigation and Soil Salinity Management in Al-Hassa Oasis, Saudi Arabia. *J. Indian Soc. Remote Sens.* **2011**, *39*, 171–180. [CrossRef]
12. Meskour, A.; Ahattab, J.; Aachib, M.; Hasnaoui, M.D. Assessing the Impact of Drought and Upstream Dam Construction on Agriculture in Arid and Semi-Arid Regions: A Case Study of the Middle Draa Valley, Morocco. *Environ. Monit. Assess.* **2025**, *197*, 236. [CrossRef] [PubMed]
13. Suliga, J.; Bhattacharjee, J.; Chormański, J.; van Griensven, A.; Verbeiren, B. Automatic Proba-V Processor: TREX—Tool for Raster Data Exploration. *Remote Sens.* **2019**, *11*, 2538. [CrossRef]
14. Ministère de L'équipement et de L'eau du MAROC. Available online: https://www.equipement.gov.ma/eau/Pages/Accueil.aspx# (accessed on 2 June 2023).
15. Tlemcani, J.; Hilali, M.; Mahboub, A.; Moudden, H.; Outaleb, A.; Boumeshoul, S.; Outaleb, A.; Boumeshoul, S. Hydrogeology and groundwater modeling of the Boudenib senonian aquifer (Southeastern Morocco). *Environ. Water Sci. Public Health Territ. Intell. J.* **2019**, *3*, 221–230. Available online: https://revues.imist.ma/?journal=ewash-ti/ (accessed on 4 April 2022).
16. Eckert, S.; Hüsler, F.; Liniger, H.; Hodel, E. Trend analysis of MODIS NDVI time series for detecting land degradation and regeneration in Mongolia. *J. Arid Environ.* **2015**, *113*, 16–28. [CrossRef]
17. Moumane, A.; Al Karkouri, J.; Benmansour, A.; El Ghazali, F.E.; Fico, J.; Karmaoui, A.; Batchi, M. Monitoring long-term land use, land cover change, and desertification in the Ternata oasis, Middle Draa Valley, Morocco. *Remote Sens. Appl. Soc. Environ.* **2022**, *26*, 100745. [CrossRef]
18. Fayech, D.; Tarhouni, J. Climate variability and its effect on normalized difference vegetation index (NDVI) using remote sensing in semi-arid area. *Model. Earth Syst. Environ.* **2021**, *7*, 1667–1682. [CrossRef]
19. Wolters, E.; Dierckx, W.; Iordache, M.-D.; Swinnen, E. *PROBA-V Products User Manual v3.01*; Image: Rochester, NY, USA, 2018; pp. 1–110. [CrossRef]
20. Alhammadi, M.S.; Kurup, S.S. Impact of Salinity Stress on Date Palm (*Phoenix dactylifera* L.)—A Review. *Crop Prod. Technol.* **2012**, *9*, 169–173. [CrossRef]
21. Parker, G. Tamm review: Leaf Area Index (LAI) is both a determinant and a consequence of important processes in vegetation canopies. *For. Ecol. Manag.* **2020**, *477*, 118496. [CrossRef]
22. Available online: https://statstutor.ac.uk/ (accessed on 12 January 2023).
23. Senthilnathan, S. Usefulness of Correlation Analysis. *SSRN Electron. J.* **2019**, *122*, 200. [CrossRef]
24. Available online: https://www.scribbr.com/statistics/correlation-coefficient/ (accessed on 5 August 2023).
25. Ahmadi, H.; Fatemizadeh, E.; Motie-Nasrabadi, A. A Comparative Study of Correlation Methods in Functional Connectivity Analysis Using fMRI Data of Alzheimer's Patients. *J. Biomed. Phys. Eng.* **2023**, *13*, 125–134. [CrossRef] [PubMed] [PubMed Central]
26. Weiss, M.; Baret, F.; Smith, G.J.; Jonckheere, I.; Coppin, P. Review of methods for in situ leaf area index (LAI) determination Part II. Estimation of LAI, errors and sampling. *Agric. For. Meteorol.* **2004**, *121*, 37–53. [CrossRef]

27. Moradkhani, H. Hydrologic Remote Sensing and Land Surface Data Assimilation. *Sensors* **2008**, *8*, 2986–3004. [CrossRef] [PubMed]
28. Mazri, M.A.; Meziani, R.; El Fadile, J.; Ezzinbi, A.E. Optimization of medium composition for in vitro shoot proliferation and growth of date palm cv. Mejhoul. *3 Biotech* **2016**, *6*, 111. [CrossRef]
29. Acworth, R.I.; Timms, W.A.; Kelly, B.F.J.; Mcgeeney, D.E.; Ralph, T.J.; Larkin, Z.T.; Rau, G.C. Late Cenozoic paleovalley fill sequence from the Southern Liverpool Plains, New South Wales—implications for groundwater resource evaluation. *Aust. J. Earth Sci.* **2015**, *62*, 657–680.
30. Xi, H.; Feng, Q.; Liu, W.; Si, J.; Chang, Z.; Su, Y. The research of groundwater flow model in Ejina Basin, Northwestern China. *Environ. Earth Sci.* **2010**, *60*, 953–963. [CrossRef]
31. Jin, X.M.; Schaepman, M.E.; Clevers, J.G.P.W.; Su, Z.B.; Hu, G.C. Groundwater Depth and Vegetation in the Ejina Area, China. *Arid. Land Res. Manag.* **2011**, *25*, 194–199. [CrossRef]
32. Margat, J. Hydrogéologie de la plaine du Tafilalt (Maroc Pré-Saharien). *Houille Blanche* **1961**, *47* (Suppl. S2), 678–682. [CrossRef]

MDPI AG
Grosspeteranlage 5
4052 Basel
Switzerland
Tel.: +41 61 683 77 34

Geosciences Editorial Office
E-mail: geosciences@mdpi.com
www.mdpi.com/journal/geosciences

www.ingramcontent.com/pod-product-compliance
Lightning Source LLC
Chambersburg PA
CBHW061514210926
51850CB00008B/1912
9783725862887